11th International Conference on Civil Engineering Design and Construction (DCB'2020)

Held online due to COVID-19

IOP Conference Series: Materials Science and Engineering
Volume 951

Varna, Bulgaria
10 - 12 September 2020

ISBN: 978-1-7138-3199-0
ISSN: 1757-8981

Printed from e-media with permission by:

Curran Associates, Inc.
57 Morehouse Lane
Red Hook, NY 12571

Some format issues inherent in the e-media version may also appear in this print version.

Institute of Physics
Dirac House, Temple Back
Bristol BS1 6BE UK

Phone: 44 1 17 929 7481
Fax: 44 1 17 920 0979

techtracking@iop.org

Additional copies of this publication are available from:

Curran Associates, Inc.
57 Morehouse Lane
Red Hook, NY 12571 USA
Phone: 845-758-0400
Fax: 845-758-2633
Email: curran@proceedings.com
Web: www.proceedings.com

TABLE OF CONTENTS

Preface

The International scientific conference on "Civil Engineering Design and Construction" (DCB) is held every two years, starting in 2000 as a national conference of the Union of Bulgarian Construction Engineers and the Bulgarian Scientific and Technical Union of Civil Engineering. This conference is of special importance for Bulgaria, where scientists, researchers and experts present their achievements and share their experience or innovative ideas in various fields of design, construction and architecture. Since 2006, the DCB conference has been held as an international scientific forum for in-depth discussions and fruitful meetings.

The organiser of the XI Conference on "Civil Engineering Design and Construction" (DCB'2020) was the Scientific and Technical Union of Civil Engineering in Bulgaria (NTSSB), strongly supported by the Chamber of Engineers in the Investment Design in Bulgaria, Bulgarian Construction Chamber, Federation of the Bulgarian Scientific and Technical Unions, the Varna Free University and "Evrika" Foundation. Special attention was paid to sustainable construction, digitalisation in the construction industry and building information modelling (BIM). During the conference, we celebrated the 220th anniversary of the birth of the Usta (Master) Kolyu Ficheto – famous Bulgarian National Revival architect and builder.

Due to the outbreak of COVID-19, this year's conference, which was supposed to be held in Varna Free University "Chernorizets Hrabar", Bulgaria, was organised as a virtual conference, using of Google G-suit platform. It was a fully synchronised online conference, that had live Q&A, participants jointly listened to presentations, and attended other live events associated with the conference. The dates of the virtual conference remained unchanged to the initially proposed dates, which were on September 10-12, 2020.

The conference program was organised in 9 sessions, in which 4 plenary and 68 specialized reports were presented. Apart from principal Bulgarian scientific organisers and the University of Architecture, Civil Engineering and Geodesy, University of Structural Engineering and Architecture "Lyuben Karavelov", Varna Free University, New Bulgarian University and Bulgarian Academy of Science BAS-, scientists from the UK, Greece, Poland, Serbia, Russia and Nigeria, took part in the scientific forum. In addition to thematic discussions on the main topics, a special session for young scientists was organised, supported by the "Evrika" Foundation. Leading companies working in the building and construction sectors were actively presented.

The accepted papers are published in the IOP Conference Series - Materials Science and Engineering, Great Britain and separate volume of Conference Proceedings and copyrighted with an ISSN.

Editors DCB'2020:
Prof. Yatchko Ivanov, DSc., Eng.
Member of BAS
Assoc. Prof. Ventseslav Stoyanov, PhD, Eng.
Assoc. Prof. Lachezar Hrischev, PhD, Eng

IOP Conf. Series: Materials Science and Engineering **951** (2020) 011002 doi:10.1088/1757-899X/951/1/011002

Conference Committees

International Scientific Committee

Chair:
Acad. Yachko Ivanov – Bulgaria

Vice Chairs:
Prof. Vlastimir Radonjanin – Serbia
Prof. Rao Surampalli – USA

Members:
Prof. Azariy Lapidus – Russia
Prof. Andrzcj Garbacz – Poland
Prof. Anna Nedyalkova – Bulgaria
Prof. Asterios Liolios – Greece
Prof. Dubravka Bjegović – Croatia
Prof. Faruk Karadoğan – Turkey
Prof. Giliano Panza – Italy
Prof. Heinz Brandl – Austria
Prof. Ivan Vaniček – Czech Republic
Prof. Ján Ravinger – Slovakia
Prof. Jurai Kralik – Slovakia
Prof. Liberato Ferrara – Italy
Prof. Marinela Barbuta – Romania
Prof. Petar Hristov - Bulgaria
Prof. Radomir Folić – Serbia
Prof. Rüdiger Höffer – Germany
Prof. Veronika Shendova – North Macedonia
Assoc. Prof. Asli Gurgun – Turkey
Assoc. Prof. Ewa Blazik-Borowa – Poland
Assoc. Prof. Milena Kichekova – Bulgaria
Assoc. Prof. Robert Eadie – UK
Eng. Ivan Karaleev, PhD – Bulgaria
Eng. Nabi Ibadov, PhD – Poland
Eng. Blashko Dimitrov – North Macedonia
Eng. Iliyan Terziev – Bulgaria

IOP Conf. Series: Materials Science and Engineering **951** (2020) 011002 doi:10.1088/1757-899X/951/1/011002

National Scientific Committee

Chairs:
Prof. Dimitar Nazarski
Prof. Konstantin Kazakov
Assoc. Prof. Lachezar Hrischev

Members:
Prof. Bogomil Petrov
Prof. Borianka Zaharieva-Georgieva
Prof. Dimitar Nazarski
Prof. Doncho Partov
Prof. Fantina Rangelova
Prof. Ivelin Ivanov
Prof. Ivan Markov
Prof. Ivanka Paskaleva
Prof. Konstantin Kazakov
Prof. Krasimir Enimanev
Prof. Lyuben Totev
Prof. Marina Traykova
Prof. Rumen Kalchev
Prof. Stoyan Bratoev
Prof. Teodor Roshavelov
Assoc. Prof. Ana Yanakieva
Assoc. Prof. Anita Handruleva
Assoc. Prof. Chavdar Kolev
Assoc. Prof. Daria Mihaleva
Assoc. Prof. Emad Abdulahad
Assoc. Prof. Georgi Linkov
Assoc. Prof. Ivan Torodov
Assoc. Prof. Vencislav Stoyanov
Assoc. Prof. Ventseslav Stoyanov
Eng. Antoni Chipev
Eng. Georgi Kordov
Eng. Dimitar Nachev

Organizing Committee

Chairs:
Assoc. Prof. Lachezar Hrischev, PhD
Assoc. Prof. Ventseslav Stoyanov, PhD

Secretary:
Svetulka Ruseva

Members:
Assos. Prof. Aneta Georgieva, PhD
Assos. Prof. Galina Todorova, PhD
Assos. Prof. Gencho Panicharov, PhD
Assos. Prof. Ivan Pavlov, PhD
Assoc. Prof. Tsveta Zhekova, PhD
Georgi Ivanov, PhD
Yana Kancheva, PhD
Natasha Baklarova, PhD

IOP Conf. Series: Materials Science and Engineering **951** (2020) 011002 doi:10.1088/1757-899X/951/1/011002

Radostina Stefanova

IOP Conf. Series: Materials Science and Engineering **951** (2020) 011003 doi:10.1088/1757-899X/951/1/011003

Peer review declaration

All papers published in this volume of *IOP Conference Series: Materials Science and Engineering* have been peer reviewed through processes administered by the Editors. Reviews were conducted by expert referees to the professional and scientific standards expected of a proceedings journal published by IOP Publishing.

- **Type of peer review:** Single-blind
  The review process is single-blind, the authors did not know reviewers, but reviewers know the authors.

- **Conference submission management system:** Emails
  All the submission and communication to all the DCB'2020 authors through conference emails – ntssb_sf@abv.bg and dcb2020.bg@gmail.com

- **Number of submissions received:** 37

- **Number of submissions sent for review:** 34

- **Number of submissions accepted:** 32

- **Acceptance Rate (Number of Submissions Accepted / Number of Submissions Received X 100):** 86.4 %

- **Average number of reviews per paper:** 2

- **Total number of reviewers involved:** 35

- **Any additional info on review process:** Reviewers were asked to judge the manuscripts based on ten criteria with four levels of assessments (good, acceptable, not acceptable, and poor). When the reviewer confirmed that all assessments as "good", the manuscript was accepted for publication. English language editing and formatting were used in the review process.

- **Contact person for queries:**
  Assoc. Prof. Ventseslav Stoyanov, PhD; e-mail: vensy.stoyanov@gmail.com

Analysis of internal delamination in multilayered rod in torsion

V Rizov

Department of Technical Mechanics, University of Architecture, Civil Engineering and Geodesy, 1046 - Sofia, Bulgaria

E-mail: v_rizov_fhe@uacg.bg

Abstract. A multilayered inhomogeneous rod with an internal delamination crack loaded in torsion is analyzed. The crack has a circular cross-section. The delamiantion is located symmetrically with respect to the middle of the rod. The layers exhibit continuous material inhomogeneity in radial and length directions. The material properties are distributed symmetrically with respect to the middle of the rod. In order to obtain the torsion moments in the crack arms, the rod is treated as a structure with one degree of internal static indeterminacy. The delamination is studied in terms of the strain energy release rate by using the compliance method. Effects of material inhomogeneity, length and location of the crack on the delamination behaviour are investigated.

1. Introduction

The multilayered inhomogeneous structural members and components are made of adhesively bonded layers of different materials. The use of multilayered structures in various load-bearing applications in different areas of practical engineering has been attempted mainly to achieve a substantial weight-reduction. Since multilayered structures have relatively low interlaminar strength, delamination fracture is one of the most common failure modes of these structures when they are subjected to various loading conditions [1, 2, 3, 4].

The present paper is concerned with a delamination fracture analysis of a multilayered inhomogeneous rod of circular cross-section loaded in torsion. It should be noted that in contrast to previous papers which deal with a delamination crack located in the end of rod [5, 6], the present paper studies an internal delamination.

2. Theoretical model of a multilayered rod with internal delamination

A multilayered inhomogeneous rod of circular cross-section of radius, R_E, is shown in figure 1. The length of the rod is $2l$. The rod is loaded by two torsion moments, T, applied at its ends. The rod is made of an arbitrary number of concentric layers. A delamination crack of length, $2a$, is located arbitrary between layers as shown in figure 1. The delamination is located in rod portion, $l-a \le x \le l+a$. In portions, $0 \le x \le l-a$ and $l+a \le x \le 2l$, the rod is un-cracked. It should be noted that the internal delamination crack arm is treated as a rod of length, $2a$, and radius of the cross-section, R_I. The external delamination crack arm is treated as a ring-shaped rod of radiuses of the internal and external surfaces, R_I and R_E, and length, $2a$.

Each layer of the rod exhibits continuous material inhomogeneity in both radial and length directions. The distribution of the shear modulus, G_i, of the i-th layer in radial direction is written as

$$G_i = G_{Bi} + \frac{G_{Di} - G_{Bi}}{\left(R_{i+1} - R_i\right)^{m_i}} \left(R - R_i\right)^{m_i}, \tag{1}$$

where G_{Bi} and G_{Di} are the values of the shear modulus in the internal and external surfaces of the layer, R_i and R_{i+1} are the radiuses of the internal and external surface of the layer, m_i is a material property that controls the material inhomogeneity in radial direction.

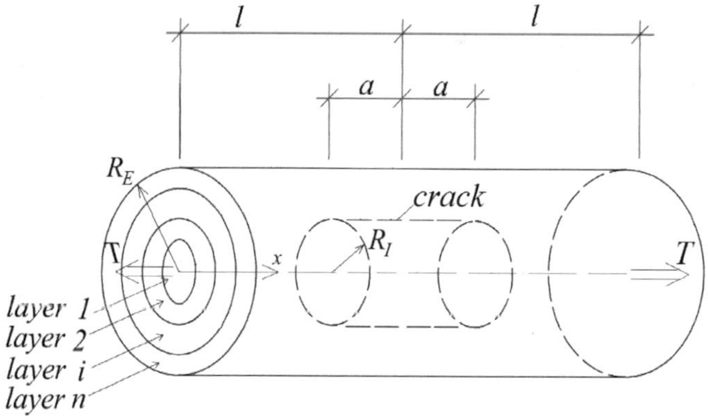

Figure 1. Geometry and loading of a multilayered inhomogeneous rod with an internal delamination crack.

The distribution of G_{Bi} and G_{Di} along the rod length is expressed as

$$G_{Bi} = G_{NB_i} + \frac{G_{SB_i} - G_{NBi}}{l^{q_i}} x^{p_i}, \; G_{Di} = G_{ND_i} + \frac{G_{SD_i} - G_{NDi}}{l^{q_i}} x^{p_i}, \; 0 \le x \le l, \tag{2}$$

$$G_{Bi} = G_{NB_i} + \frac{G_{SB_i} - G_{NBi}}{l^{q_i}} (2l - x)^{q_i}, \; G_{Di} = G_{ND_i} + \frac{G_{SD_i} - G_{NDi}}{l^{q_i}} (2l - x)^{q_i}, \; l \le x \le 2l, \tag{3}$$

where G_{NBi} and G_{NDi} are the values of G_{Bi} and G_{Di} in the two ends of the rod, G_{SBi} and G_{SDi} are the values of G_{Bi} and G_{Di} in the mid-span, $x = l$, p_i and q_i are material properties which control the material inhomogeneity along the rod length. Formulae (2) and (3) indicate that G_{Bi} and G_{Di} are distributed symmetrically with respect to the middle of the rod.

The delamination is analyzed in terms of the strain energy release rate, G. In order to derive the strain energy release rate, first, the torsion moments in the internal and external crack arms are obtained (the torsion moments are needed to calculate the shear strains which are used to determine the strain energy). The rod is treated as a structure with one degree of internal static indeterminacy. Thus, the torsion moment in the external crack arm, T_E, is obtained by the theorem of Menabrea

$$\frac{\partial U}{\partial T_E} = 0, \tag{4}$$

where U is the strain energy. Due to the symmetry, only half of the rod, $l \le x \le 2l$, is analyzed.

Thus, the strain energy is written as

$$U = U_I + U_E + U_U, \tag{5}$$

where U_I , U_E and U_U are the strain energies in the internal and external crack arms, and the un-cracked portion of the rod. The strain energy in the internal crack arm is expressed as

$$U_I = \sum_{i=1}^{i=n_1} \int_{l}^{l+a} \int_{R_i}^{R_{i+1}} u_{I0_i} 2\pi R dR dx, \tag{6}$$

where n_1 is the number of layers in the internal crack arm. The strain energy density in the *i*-th layer, u_{I0_i} , is obtained as

$$u_{I0_i} = \frac{1}{2} G_i \gamma_{II}^2 . \tag{7}$$

Since in the present paper rods of high length to diameter ratio are considered, the distribution of shear strain, γ_{II}, is treated by applying the Bernoulli's hypothesis. Thus, the distribution of γ_{II} is written as

$$\gamma_{II} = \frac{\gamma_I}{R_I} R, \tag{8}$$

where the shear strain at the periphery of the internal crack arm, γ_I, is obtained from the equation of equilibrium of the elementary forces in the cross-section of the internal crack arm

$$\sum_{i=1}^{i=n_1} \int_{R_i}^{R_{i+1}} G_i \gamma_{II} 2\pi R^2 dR = T_I , \tag{9}$$

where T_I is the torsion moment in the internal crack arm. It is obvious that (figure 1)

$$T_I = T - T_E . \tag{10}$$

After substituting of (1), (8) and (10) in (9), the equation of equilibrium is solved with respect to γ_I . In this way, γ_I is expressed as a function of T_E .

The strain energy in the external crack arm is written as

$$U_E = \sum_{i=1}^{i=n_2} \int_{l}^{l+a} \int_{R_i}^{R_{i+1}} u_{E0_i} 2\pi R dR dx, \tag{11}$$

where n_2 is the number of layers in the external crack arm. The strain energy density in the *i*-th layer of the external crack arm, u_{E0_i} , is obtained by replacing of γ_{II} with γ_{EE} in (7). The distribution of the shear strain, γ_{EE} , in the cross-section of the external crack arm is expressed by replacing of γ_I with γ_E in (8). The shear strain, γ_E , at the periphery of the external crack arm is obtained as a function of T_E from the following equation of equilibrium of the cross-section of the external crack arm:

$$\sum_{i=1}^{i=n_2} \int_{R_i}^{R_{i+1}} G_i \gamma_{EE} 2\pi R^2 dR = T_E . \tag{12}$$

After substituting of G_i and γ_{EE} in (12), the equation for equilibrium is solved with respect to γ_E . The strain energy in the un-cracked rod portion is written as

$$U_U = \sum_{i=1}^{i=n} \int_{l+a}^{2l} \int_{R_i}^{R_{i+1}} u_{U0_i} 2\pi R dR dx, \tag{13}$$

where n is the number of layers in the rod. The strain energy density in the i-th layer, u_{U0_i}, is obtained by formula (7). For this purpose, γ_{II} is replaced with γ_{UU}. The distribution of the shear strain, γ_{UU}, in the cross-section of the un-cracked rod portion is found by replacing of γ_I and R_I with γ_E and R_E in (8). The shear strain at the periphery of the rod, γ_E, is obtained as a function of T by equation of equilibrium (13). For this purpose, n_2, γ_{EE} and T_E are replaced, respectively, with n, γ_{UU} and T, and then the equation of equilibrium is solved with respect to γ_E.

After substituting of (5), (6), (7), (11) and (13) in (4), the equation is solved with respect to T_E. The strain energy release rate is obtained by applying the compliance method

$$G = 2\frac{T^2}{2l_{cf}}\frac{dC}{da}, \tag{14}$$

where the delamination crack front length, l_{cf}, is written as

$$l_{cf} = 2\pi R_I \; . \tag{15}$$

The compliance, C, is obtained as

$$C = \frac{\varphi}{T}. \tag{16}$$

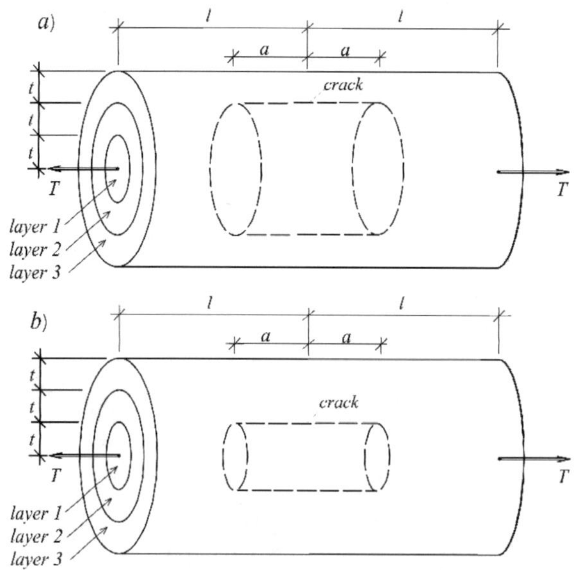

Figure 2. Two three-layered rods with a delamination crack located (a) between layers 2 and 3, and (b) between layers 1 and 2.

It should be noted that the right-hand side of (14) is doubled in view of the symmetry (figure 1). By using the integrals of Maxwell-Mohr, the angle of twist, φ, of the end section of the rod is found as

$$\varphi = \int_{l}^{l+a} \frac{\gamma_I}{R_I} dx + \int_{l+a}^{2l} \frac{\gamma_U}{R_E} dx. \tag{17}$$

By substituting of (15), (16) and (17) in (14), one derives the following expression for the strain energy release rate:

$$G = \frac{T}{2\pi R_I}\left(\frac{\gamma_I}{R_I} - \frac{\gamma_U}{R_E}\right). \quad (18)$$

In order to verify (18), the strain energy release rate is obtained also by differentiating the strain energy with respect to the delamination crack area

$$G = 2\frac{dU}{l_{cf}\,da}, \quad (19)$$

where da is an elementary increase of the delamination crack length.

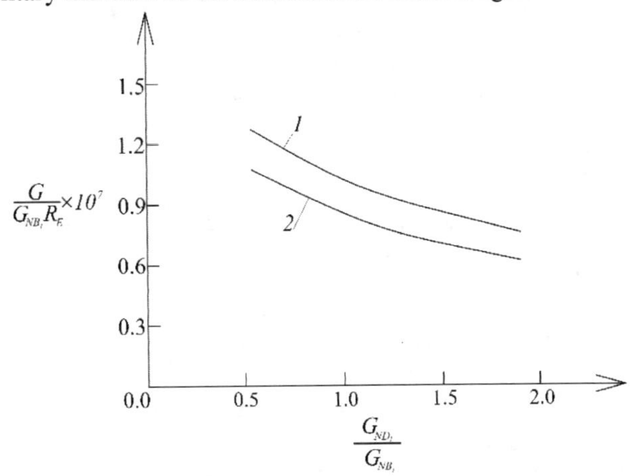

Figure 3. The strain energy release rate in non-dimensional form plotted against G_{ND_1}/G_{NB_1} ratio (curve 1 - for the three-layered rod configuration with a delamination between layers 1 and 2, curve 2 - for the three-layered rod configuration with a delamination between layers 2 and 3).

The right-hand side of (19) is doubled in view of the symmetry (figure 1). It should be noted that the strain energy release rate calculated by (19) is exact match of that obtained by (18). This fact is a verification of the strain energy release rate analysis developed in the present paper.

3. Parametric analysis

The solution to the strain energy release rate (18) is applied to evaluate the influence of material inhomogeneity, the delamination crack location and the delamination length on the delamination fracture behaviour. The strain energy release rate is presented in non-dimensional form by using the formula $G_N = G/(G_{NB1}R_E)$. Two three-layered rod configurations are analyzed (figure 2) in order to evaluate the effect of delamination crack location in radial direction on the delamination behaviour. A delamination crack is located between layers 2 and 3 in the rod shown in figure 2a. A rod with delamination between layers 1 and 2 is also investigated (figure 2b). The thickness of each layer is t . It is assumed that $l = 0.3$ m, $t = 0.001$, $m_i = p_i = q_i = 0.8$ and $T = 15$ kNm.

The material inhomogeneity in radial direction is characterized by G_{ND_1}/G_{NB_1} ratio. The strain energy release rate in non-dimensional form is plotted against G_{ND_1}/G_{NB_1} ratio in figure 3 for the two rod configurations shown in figure 2. One can observe in figure 3 that the strain energy release rate decreases with increasing of G_{ND_1}/G_{NB_1} ratio. The curves in figure 3 indicate also that the strain energy release rate is lower when the crack is located between layers 2 and 3.

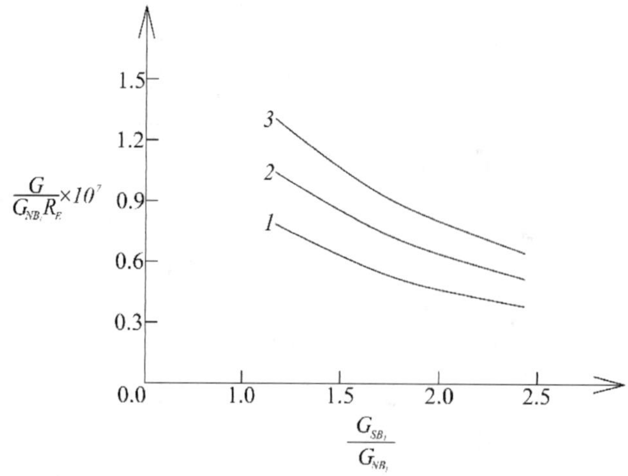

Figure 4. The strain energy release rate in non-dimensional form plotted against G_{SB_1}/G_{NB_1} ratio (curve 1 – at $a/l=0.2$, curve 2 - at $a/l=0.4$, curve 3 - at $a/l=0.6$).

The effect of material inhomogeneity in the length direction of the rod on the delamination behaviour is illustrated in figure 4 where the strain energy release rate in non-dimensional form is plotted against G_{SB_1}/G_{NB_1} ratio at three a/l ratios. The curves in figure 4 show that the strain energy release rate decreases with increasing of G_{SB_1}/G_{NB_1} ratio. The increase of a/l ratio leads to increase of the strain energy release rate (figure 4).

4. Conclusions

The delamination behaviour of a multilayered inhomogeneous rod with an internal delamination crack is analyzed. The crack is located symmetrically with respect to the middle of the rod. The rod is loaded in torsion. The layers of the rod exhibit continuous material inhomogeneity in both radial and length directions. The rod has a circular cross-section. In order to obtain the torsion moments in the delamination crack arms, the rod is treated as a structure with one degree of internal static indeterminacy by applying the theorem of Menabrea. The strain energy release rate is derived by using the compliance method. The strain energy release rate is obtained also by differentiating the strain energy for verification. Parametric analyses are carried-out in order to investigate influences of the material inhomogeneity, the location and the length of the crack on the strain energy release rate. Two three-layered rods with different location of the delamination crack are considered to assess the influence of the crack location in radial direction. It is found that the strain energy release rate is lower when the delamination is located near the periphery of the rod. The analysis reveals that the strain energy release rate increases whit increasing of the delamination crack length. The strain energy release rate decreases also with increasing of G_{ND_1}/G_{NB_1} and G_{SB_1}/G_{NB_1} ratios.

References

[1] Wang J and Zhang C 2007 Composites science and technology **67** 3323-3330
[2] Klingbeil NW and Beuth JI 1997 *Eng. Fract. Mech.* **56** 113-126
[3] Yang W and Shih CF 1994 Int. J. Solids Structures **31** 985-1002
[4] Ogasawara T, Yokozeki T and Aoki T 2008 Compos. Sci. Technol. **68** 760-767
[5] Rizov V 2020 IOP Conf. Ser.: Mater. Sci. Eng. **739** 012003
[6] Rizov V 2018 Multidiscipline Modelling in Materials and Structures **15** 156-169

Inhomogeneous beam with two internal vertical lengthwise cracks: a fracture study

V Rizov

Department of Technical Mechanics, University of Architecture, Civil Engineering and Geodesy, 1046 - Sofia, Bulgaria

E-mail: v_rizov_fhe@uacg.bg

Abstract. Fracture of an inhomogeneous cantilever beam whit two internal parallel lengthwise cracks is analyzed. Cracks have different lengths and are located arbitrary along the beam width. The beam exhibits continuous material inhomogeneity in both width and length directions. The fracture is studied in terms of the strain energy release rate. For this purpose, the beam is treated as a structure with two degrees of internal static indeterminacy. The solutions to the strain energy release rate, derived in the present paper, are verified by the using the J-integral approach. A parametric investigation of the lengthwise fracture behaviour of the beam is performed.

1. Introduction

Inhomogeneous materials are extensively used in many load-bearing structural applications where high performances are required. The properties of inhomogeneous materials vary continuously (smoothly) in the solid [1, 2, 3]. A certain kind of inhomogeneous materials such as functionally graded materials can be built-up layer by layer [4] which is a premise for appearance of lengthwise cracks. It is therefore very important to analyze various aspects of lengthwise fracture behaviour of inhomogeneous structural members and components.

The goal of the present paper is to analyze lengthwise fracture of an inhomogeneous beam with two internal parallel cracks in contrast to previous papers [5, 6] which are focussed on inhomogeneous beams with one lengthwise crack located in the end of the beam.

2. Solution to the strain energy release rate for the two internal cracks

The present paper studies the lengthwise fracture in the inhomogeneous cantilever beam with two parallel lengthwise cracks (figure 1). The lengths of crack 1 and crack 2 are a_1 and a_2, respectively. The left-hand crack arm has length, a_1, and width, b_1. The central and right-hand crack arms have length, a_2. The widths of the central and right-hand crack arm are b_2 and b_3, respectively. The beam is clamped in section, B. The length of the beam is l. The cross-section of the beam is a rectangle of width, b, and height, h. The loading of the beam consists of one bending moment, M, applied at the free end. The beam exhibits continuous material inhomogeneity in width and length directions. The distribution of the modulus of elasticity, E, along the beam width is written as

$$E = E_0 e^{f \frac{y_1 + \frac{b}{2}}{b}}, \qquad (1)$$

where E_0 is the value of the modulus of elasticity at the left-hand lateral surface of the beam, f is a material property that controls the material gradient along the beam width, y_1 is the horizontal centroidal axis (figure 1).

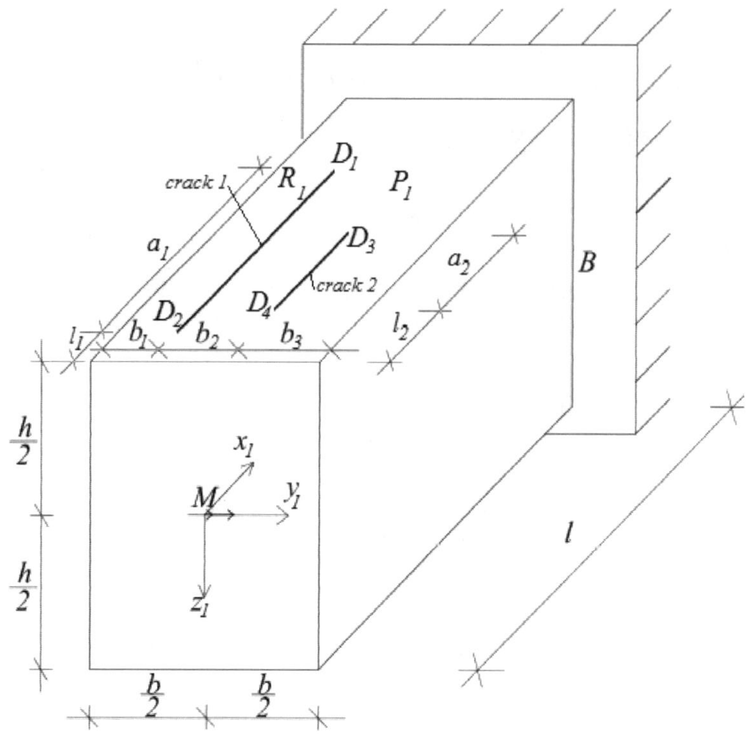

Figure 1. Inhomogeneous cantilever beam with two parallel lengthwise cracks.

The distribution of E_0 along the beam length is expressed as

$$E_0 = E_{L0} e^{g \frac{x_1}{l}}, \tag{2}$$

where E_{L0} is the value of E_0 at the free end of the beam.

The fracture behaviour of the beam is analyzed in terms of the strain energy release rate, G. For this purpose, first, a small increase, Δa_1, of the length of crack 1 in crack tip, D_1, is assumed. By using linear-elastic fracture mechanics, the strain energy release rate is expressed as (figure 1)

$$G = \frac{1}{h\Delta a_1}\left(\Delta a_1 \int_{-\frac{b_1}{2}}^{\frac{b_1}{2}} \int_{-\frac{h}{2}}^{\frac{h}{2}} u_{0R_1} dy_2 dz_2 + \Delta a_1 \int_{-\frac{b_2+b_3}{2}}^{\frac{b_2+b_3}{2}} \int_{-\frac{h}{2}}^{\frac{h}{2}} u_{0P_1} dy_3 dz_3 - \Delta a_1 \int_{-\frac{b}{2}}^{\frac{b}{2}} \int_{-\frac{h}{2}}^{\frac{h}{2}} u_{0Q_1} dy_4 dz_4 \right), \tag{3}$$

where u_{0R_1} and u_{0P_1} are the strain energy densities in the cross-sections, R_1 and P_1, of the two crack arms behind the crack tip, u_{0Q_1} is the strain energy density in the beam cross-section ahead of the crack tip, y_2 and z_2, and y_3 and z_3 are the centroidal axes of the cross-sections, R_1 and P_1, of the crack arms, y_4 and z_4 are the centroidal axes of the beam cross-section ahead of the crack tip, D_1.

The strain energy density, u_{0R_1}, is written as

$$u_{0R_1} = \frac{1}{2} E \varepsilon_{R_1}^2 , \tag{4}$$

where the distribution of the strain, ε_{R_1}, is expressed as

$$\varepsilon_{R_1} = \kappa_{R_1} z_2 . \tag{5}$$

In (5), κ_{R_1} is the curvature of the crack arm in cross-section, R_1. It should be noted that the strain is distributed linearly in the cross-section since in the present paper beams of high length to height ratio are under consideration. Thus, validity of the Bernoulli's hypothesis for plane sections is assumed.

The following equation of equilibrium of the elementary forces in the crack arm cross-section, R_1, is used to obtain the curvature, κ_{R_1}:

$$\int_{-\frac{b_1}{2}}^{\frac{b_1}{2}} \int_{-\frac{h}{2}}^{\frac{h}{2}} \sigma_{R_1} d y_2 dz_2 = M_H , \tag{6}$$

where M_H is the bending moment in the left-hand crack arm. The stress, σ_{R_1}, is found by the Hooke's law

$$\sigma_{R_1} = E \varepsilon_{R_1} , \tag{7}$$

It is obvious that in order to obtain κ_{R_1} from (6), one needs M_H. The bending moments in the crack arms are derived by treating the beam as a structure with two degrees of internal static indeterminacy by using the theorem of Menabrea

$$\frac{\partial U}{\partial M_H} = 0, \; \frac{\partial U}{\partial M_T} = 0 , \tag{8}$$

where U is the strain energy in the beam, M_T is the bending moment in the central crack arm. The strain energy is written as

$$U = \int_{0}^{l_1} \int_{-\frac{b}{2}}^{\frac{b}{2}} \int_{-\frac{h}{2}}^{\frac{h}{2}} \frac{E\varepsilon_{\alpha 1}^2}{2} dx_1 dy_1 dz_1 + \int_{l_1}^{l_2} \int_{-\frac{b_2+b_3}{2}}^{\frac{b_2+b_3}{2}} \int_{-\frac{h}{2}}^{\frac{h}{2}} \frac{E\varepsilon_{\alpha 2}^2}{2} dx_1 dy_5 dz_5 + \int_{l_1}^{l_1+a} \int_{-\frac{b_1}{2}}^{\frac{b_1}{2}} \int_{-\frac{h}{2}}^{\frac{h}{2}} \frac{E\varepsilon_{\alpha 3}^2}{2} dx_1 dy_2 dz_2 +$$

$$+ \int_{l_2}^{l_2+a_2} \int_{-\frac{b_2}{2}}^{\frac{b_2}{2}} \int_{-\frac{h}{2}}^{\frac{h}{2}} \frac{E\varepsilon_{\alpha 4}^2}{2} dx_1 dy_6 dz_6 + + \int_{l_2}^{l_2+a_2} \int_{-\frac{b_3}{2}}^{\frac{b_3}{2}} \int_{-\frac{h}{2}}^{\frac{h}{2}} \frac{E\varepsilon_{\alpha 5}^2}{2} dx_1 dy_7 dz_7 + \int_{l_2+a_2}^{l_1+a_1} \int_{-\frac{b_2+b_3}{2}}^{\frac{b_2+b_3}{2}} \int_{-\frac{h}{2}}^{\frac{h}{2}} \frac{E\varepsilon_{\alpha 6}^2}{2} dx_1 dy_8 dz_8 +$$

$$+ \int_{l_1+a_1}^{l} \int_{-\frac{b}{2}}^{\frac{b}{2}} \int_{-\frac{h}{2}}^{\frac{h}{2}} \frac{E\varepsilon_{\alpha 7}^2}{2} dx_1 dy_4 dz_4 , \tag{9}$$

where $\varepsilon_{\alpha 1}$ is the strain in the un-cracked part of the beam, $0 \le x_1 \le l_1$, $\varepsilon_{\alpha 2}$ is the strain in the beam part, $l_1 \le x_1 \le l_2$, $-(b_2 + b_3)/2 \le y_5 \le (b_2 + b_3)/2$, $\varepsilon_{\alpha 3}$ is the strain in the left-hand crack arm, $\varepsilon_{\alpha 4}$ is the strain in the central crack arm, $\varepsilon_{\alpha 5}$ is the strain in the right-hand crack arm, $\varepsilon_{\alpha 6}$ is the strain in the beam part, $l_2 + a_2 \le x_1 \le l_1 + a_1$, $-(b_2 + b_3)/2 \le y_8 \le (b_2 + b_3)/2$, $\varepsilon_{\alpha 7}$ is the strain in the un-cracked part of the beam, $l + a_1 \le x_1 \le l$, y_i and z_i where $i = 1,...,8$ are the centroidal axes of the cross-sections of the corresponding portions of the beam.

The distribution of strain, $\varepsilon_{\alpha 1}$, in the cross-section of the un-cracked portion of the beam, $0 \le x_1 \le l_1$, is written as

$$\varepsilon_{\alpha 1} = \kappa_{\alpha 1} z_1, \tag{10}$$

where the curvature, $\kappa_{\alpha 1}$, is obtained by using the equation of equilibrium of the cross-section

$$\int_{-\frac{b}{2}}^{\frac{b}{2}} \int_{-\frac{h}{2}}^{\frac{h}{2}} E\varepsilon_{\alpha 1} dy_1 dz_1 = M, \tag{11}$$

After substituting of (1) and (10) in (11), the equation of equilibrium is solved with respect to $\kappa_{\alpha 1}$. In this way, $\kappa_{\alpha 1}$ and $\varepsilon_{\alpha 1}$ through (10) are expressed as functions of the bending moment. Analogically, the curvatures and strains in each part of the beam and in the three crack arms are obtained as functions of the corresponding bending moments. The bending moment, M_C, in the right-hand crack arm is expressed as function of M_H and M_T by using the equation for equilibrium of moments

$$M_C + M_H + M_T = M. \tag{12}$$

By substituting of strains, $\varepsilon_{\alpha i}$, in (9), the strain energy in the beam is obtained as a function of M_H and M_T. After substituting of the strain energy in (8), the two equations are solved with respect to M_H and M_T.

After substituting of (5), (7) and M_H in (6), the equation of equilibrium is solved with respect to κ_{R_1}. Then the strain energy density, u_{0R_1}, is obtained by substituting of κ_{R_1} and (5) in (4). The strain energies, u_{0P_1} and u_{0Q_1}, are found by considering the equilibrium of the elementary forces in the cross-sections of the right-hand crack arm behind the crack tip and the beam cross-section ahead of the crack tip, D_1. The strain energy release rate is derived by substituting of u_{0R_1}, u_{0P_1} and u_{0Q_1} in (3).

The fracture is analyzed also assuming a small increase, Δa_1^*, of the length of crack 1 at crack tip, D_2. The strain energy release rate is obtained as

$$G = \frac{1}{h\Delta a_1^*}\left(\Delta a_1^* \int_{-\frac{b_1}{2}}^{\frac{b_1}{2}} \int_{-\frac{h}{2}}^{\frac{h}{2}} u_{0R_2} dy_2 dz_2 + \Delta a_1^* \int_{-\frac{b_2+b_3}{2}}^{\frac{b_2+b_3}{2}} \int_{-\frac{h}{2}}^{\frac{h}{2}} u_{0P_2} dy_5 dz_5 - \Delta a_1^* \int_{-\frac{b}{2}}^{\frac{b}{2}} \int_{-\frac{h}{2}}^{\frac{h}{2}} u_{0Q_2} dy_1 dz_1 \right), \tag{13}$$

where the strain energy densities, u_{0R_2}, u_{0P_2} and u_{0Q_2}, are derived by considering the equilibrium of the cross-sections of the crack arms behind the crack tip and the beam cross-section ahead of the crack tip, D_2.

A fracture analysis is carried-out also assuming a small increase, Δa_2, of the length of crack 2 at crack tip, D_3. The following expression for the strain energy release rate is derived:

$$G = \frac{1}{h\Delta a_2}\left(\Delta a_2 \int_{-\frac{b_2}{2}}^{\frac{b_2}{2}} \int_{-\frac{h}{2}}^{\frac{h}{2}} u_{0R_3} dy_6 dz_6 + \Delta a_2 \int_{-\frac{b_3}{2}}^{\frac{b_3}{2}} \int_{-\frac{h}{2}}^{\frac{h}{2}} u_{0P_3} dy_7 dz_7 - \Delta a_2 \int_{-\frac{b_2+b_3}{2}}^{\frac{b_2+b_3}{2}} \int_{-\frac{h}{2}}^{\frac{h}{2}} u_{0Q_3} dy_8 dz_8 \right). \tag{14}$$

The equations for equilibrium of the cross-sections behind and ahead of the crack tip, D_3, are used to obtain the strain energy densities, u_{0R_3}, u_{0P_3} and u_{0Q_3}.

IOP Conf. Series: Materials Science and Engineering **951** (2020) 012002 doi:10.1088/1757-899X/951/1/012002

Finally, the fracture is analyzed assuming a small increase, Δa_2^*, of the length of crack 2 at crack tip, D_4. The strain energy release rate is written as

$$G = \frac{1}{h\Delta a_2^*}\left(\Delta a_2^* \int\limits_{-\frac{b_2}{2}}^{\frac{b_2}{2}} \int\limits_{-\frac{h}{2}}^{\frac{h}{2}} u_{0R_4} dy_6 dz_6 + \Delta a_2^* \int\limits_{-\frac{b_3}{2}}^{\frac{b_3}{2}} \int\limits_{-\frac{h}{2}}^{\frac{h}{2}} u_{0P_4} dy_7 dz_7 - \Delta a_2^* \int\limits_{-\frac{b_2+b_3}{2}}^{\frac{b_2+b_3}{2}} \int\limits_{-\frac{h}{2}}^{\frac{h}{2}} u_{0Q_4} dy_5 dz_5 \right), \quad (15)$$

where the strain energy densities are obtained from equations for equilibrium of the cross-sections behind and ahead of the crack tip.

In order to verify solutions (3), (13), (14) and (15), the fracture is studied also by applying the *J*-integral approach. Solutions to the *J*-integral are derived for crack tips, D_1, D_2, D_3 and D_4.

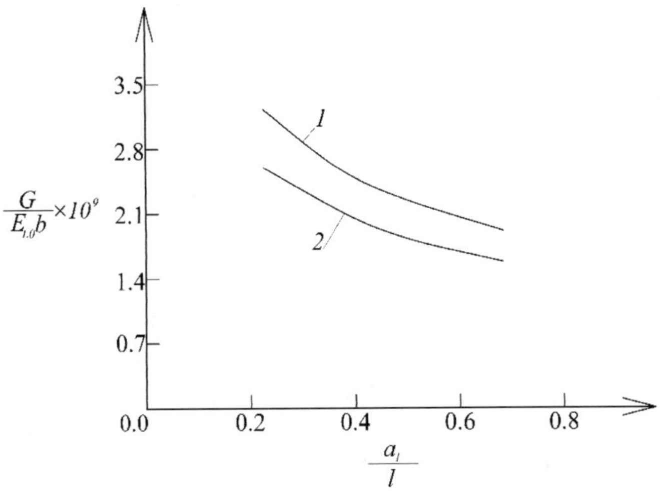

Figure 2. The strain energy release rate in non-dimensional form plotted against a_1/l ratio (curve 1 – at $b_1/b = 0.2$, curve 2 – at $b_1/b = 0.4$).

The fact that the *J*-integral values are exact matches of the strain energy release rates is a verification of the fracture analysis developed in the present paper.

3. Numerical results

Numerical results are obtained by using the solutions to the strain energy release rate in order to investigate the influence of locations and lengths of the two cracks and material inhomogeneity on the lengthwise fracture. The strain energy release rate is presented in non-dimensional form by using the formula $G_N = G/(E_{L0}b)$. It is assumed that $l = 0.2$ m, $b = 0.008$, $h = 0.006$, $l_1 = 0.03$ m, $l_2 = 0.04$ m and $T = 15$ kNm.

The influence of length and location of crack 1 on the lengthwise fracture behaviour is illustrated in figure 2 where the strain energy release rate in non-dimensional form is plotted against a_1/l ratio at two b_1/b ratios (the strain energy release rate is obtained by (3)). The curves in figure 2 indicate that the strain energy release rate decreases with increasing of a_1/l ratio (this behaviour is due to the fact that the modulus of elasticity of the beam cross-section in which the tip of crack 1 is located increases with increasing of a_1/l ratio). One can observe in figure 2 also that the strain energy release rate at $b_1/b = 0.2$ is higher than that at $b_1/b = 0.4$.

The effect of material inhomogeneity on lengthwise fracture is investigated too. For this purpose, the strain energy release rate in non-dimensional form is plotted against f in figure 3 (the strain energy release rate is found by (13) and (15)).

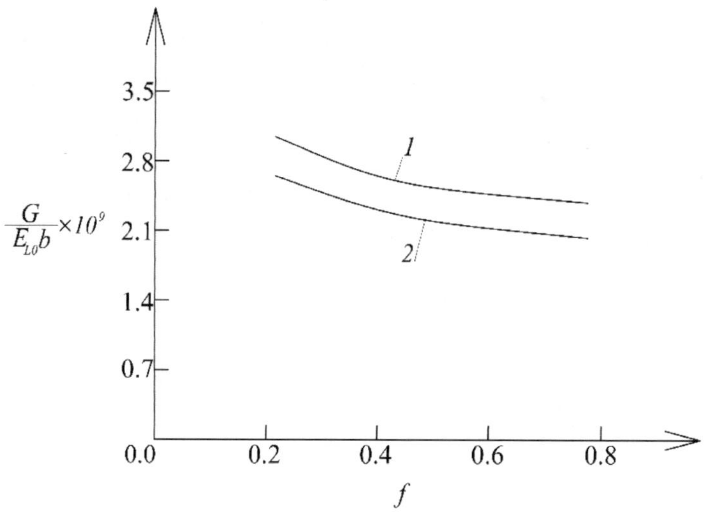

Figure 3. The strain energy release rate in non-dimensional form plotted against f (curve 1 – at increase of crack 1 at crack tip D_2, curve 2 – at increase of crack 2 at crack tip D_4).

It can be observed in figure 3 that the strain energy release rate decreases with increasing of f . Besides, the strain energy release rate found by (13) is higher than that found by (15).

4. Conclusions

A lengthwise fracture analysis of an inhomogeneous cantilever beam with two internal parallel lengthwise cracks of different lengths is carried-out. The cracks are located arbitrary along the beam width. Thus, the widths of the cross-sections of the three crack arms are different. The beam exhibits continuous material inhomogeneity in width and length directions. The bending moments in the crack arms are derived by treating the beam as a structure with two degrees of internal static indeterminacy. Four solutions to the strain energy release rate are obtained by assuming increases of the cracks in the four crack tips. The solutions are verified by the *J*-integral approach. The influence of the lengths and locations of cracks on the fracture behaviour is evaluated. It is found that the strain energy release rate decreases with increasing of a_1 / l ratio. The increase of b_1 / b ratio leads to decrease of the strain energy release rate. The analysis reveals also that the strain energy release rate decreases with increasing of the material property, f .

References

[1] Hirai T and Chen L 1999 *Material Science Forum* **308-311** 509-514
[2] Uslu Uysal M 2016 *Steel and Composite Structures, An International Journal* **21** 849-862
[3] Tokovyy Y and Ma C-C 2019 *Journal of Mechanics* **18** 1-14
[4] Mahamood RM and Akinlabi ET 2017 *Functionally Graded Materials* (Springer)
[5] Rizov V 2020 *IOP Conf. Ser.: Mater. Sci. Eng.* **739** 012004
[6] Rizov V 2019 *IOP Conf. Ser.: Mater. Sci. Eng.* **473** 012002

Review of open source software for modelling and analysis of structures

T Tsvetkova* and N Kerenchev

University of Architecture, Civil Engineering and Geodesy (UACEG), 1 Hristo Smirnenski Blvd., 1046 Sofia, Bulgaria

* E-mail: tr.tsvetkova_fce@uacg.bg

Abstract. The open-source software attracts the engineering society with the accessibility, the chasing of processes, the flexibility and the possibility of making changes to the interface or in the program code processes. This paper presents the results of a comparative review of several groups of open-source software (OSS) for modelling and analysis of constructions. The advantages and challenges of working with OSS in comparison to the use of widespread products with commercial licenses are analyzed. Software products of two categories are considered: (a) geometrically /graphically modelling/ drawing of the structures - 2D, 3D CAD; (b) calculation and analysis of building structures under different load conditions and geometry including geotechnical problems. The software products selected for the purpose of the report are compared based on the performance of a simple engineering task from practice. The sufficiency of the results, the options for presenting the data and other additional characteristics of the products are commented.

1. Introduction

Despite the vast publications, related to BIM modeling in the field of structural engineering and architecture Free/Libre Open Source Software (FOSS), still represents an unexplored territory, which interests the engineering society. To investigate the potential inherent in software for the main structural engineering essential parts – modeling, analyses, and geotechnical problems, imprimis we have to describe the benefits in the OSS [1].

The OSS movement was pushed by the Symbolic's (AI LAB) resolution to use non-free timesharing systems instead of ITS and IMB's policy changes in 80's, which contains closing the software architecture thought object-code only (OCO) [2,3]. The customers weren't allowed to use the source code of programs, because of the restrictions, therefore they were not able to control or govern the program result. These drastic changes in distributing software empowered the IT community to developed free/open software. The well-known results of FOSS movement are GNU Emac free/libre text editor, developed by Richard Stallman, and GNU Operational System (OS). Stallman started GNU (GNU Not Unix) project, which objectives exemplify the idea of "copyleft" – the method for redistributing and modifying free software [4]. Late Stallman founded Free Software Foundation (FSF), a nonprofit organization, which published one of the most popular free license General Public License GPL, the legalization of copyleft, and documentations for redistribution, modifying and developing free software, but without adding additional cost or restrictions for the users. [5]

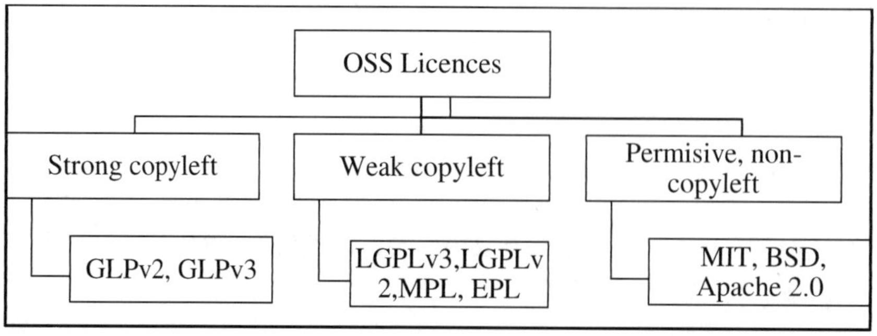

Figure 1. The OSS Licenses types, approved by OSI

FSF disseminated the substantial for the OSS movement the strong copyleft GPLv2, GPLv3 licenses, which been approved by Open Source Initiative (OSI) [5]. They are applicable for widespread open-source software and ensure the user's freedom, but coerces the customers to releases the final product of modified source code, under the same license type as the original software, also including the availability of source code. Lesser General Public License (LGPL), considered to be a weak copyleft license like Mozilla Public License (MPL), Eclipse Public License (EPL), permit distribution of application under the proprietary license agreement, in fulfilling certain requirements. The weak copyleft licenses necessitate the distribution of the binary code with corresponding copyleft-covered sections of source code. The permissive non-copyleft licenses Berkeley Software Distribution (BSD), Apache license, MIT License, approved by OSI, permit using the source code and redistribute it under the proprietary license.

The preference of the OSS application is not only under reliability, a result of accessibility to source code, but the addition of contribution of customers to the final product and maintenance of the source code in internet OS libraries or software development platforms. [6]

2. Geometrically and graphically modelling of structures

Computer-Aided Design (CAD) is prevalent in the field of structural engineering and architecture in the last few decades. The CAD software applications, part of OSS movement, for the purpose of the report, will be considered as 2D CAD and 3D CAD.

The AEC /Architecture, Engineering and Construction/ companies are providing CAD performance of their project, consequently, they have to consent with licensing policies of the closed software distributors, which dominates the market [7]. The usage of CAD software is also included in engineering education in AEC universities with student's versions, limited in provided functions and/or period of usage, commonly year or all-time of education. This established financial issues of graduated students and small AEC firms in the field with the acquiring CAD software. The OSS alternatives of CAD proprietary software provide to the customers equal functionalities without restrictions and additional costs.

2.1. 2D modelling

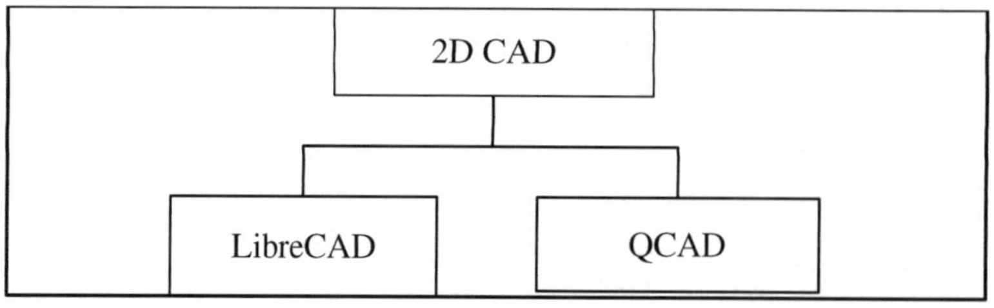

Figure 2. 2D CAD softwares, considered for the purposes to the report

The selection of relevant criteria for evaluation of 2D OSS from a scientific perspective is of particular importance. As consequence, the criteria are summarized in two different sections – "documentations" and "functions" in the evaluation list, in which the assessment is shown by a range (1) for poor to (5) for fulfilled. The section "documentation" includes the valuation of intelligibility of existing manuals and tutorials and the customers' support in forums, and also readily available GUI of the software. A variety of accessible geometric features, annotations and generated libraries are included like objectives in "functions" criteria. The researched date established more CAD software closed or limited, nevertheless the report considered to represent only OSS software and evaluated it. [8, 9]

2.1.1. LibreCAD

LibreCAD is an open-source 2D CAD application, released under GPLv2 for Windows, Linux, macOS and Unix operating systems. The latest version of the product LibreCad 2.2.0 series is based on Qt5 libraries. The native file format of the application is dxf, furthermore, LibreCAD supports pdf and image formats. The inability to export to dwg file format is one of the major limitations of LibreCAD, nevertheless, it supports partially dwg for import.

The LibreCAD 2.2.0 series are available in desktop and portable version, which doesn't need installation. The application is suitable for 2D representation of the project, as a result of a wide range of geometric features, annotation's tool and print options. Different libraries and blocks are generated in LibreCAD, which considerably simplifies the operating with complex projects. Online training, manuals and forums are available on LibreCAD website.

2.1.2. QCAD

QCAD is 2D CAD open-source application for Windows, macOS and Linus operation systems, developed by Ribbonsoft and released under the General Public Licence version 3 (GPLv3) license. The OSS, called QCAD Community edition provides a wide range of geometric features, which, however, are severely limited in comparison with QCAD Professional, commercial proprietary software, also developed and distributed by Ribbonsoft.

QCAD's native file format is dxf, but likewise LibreCAD, the program supports pdf and image file formats for import and export. QCAD Professional and QCAD/CAM, at the same time, can export and import in dwg file format. The application provides the indispensable features for 2D representations of projects.

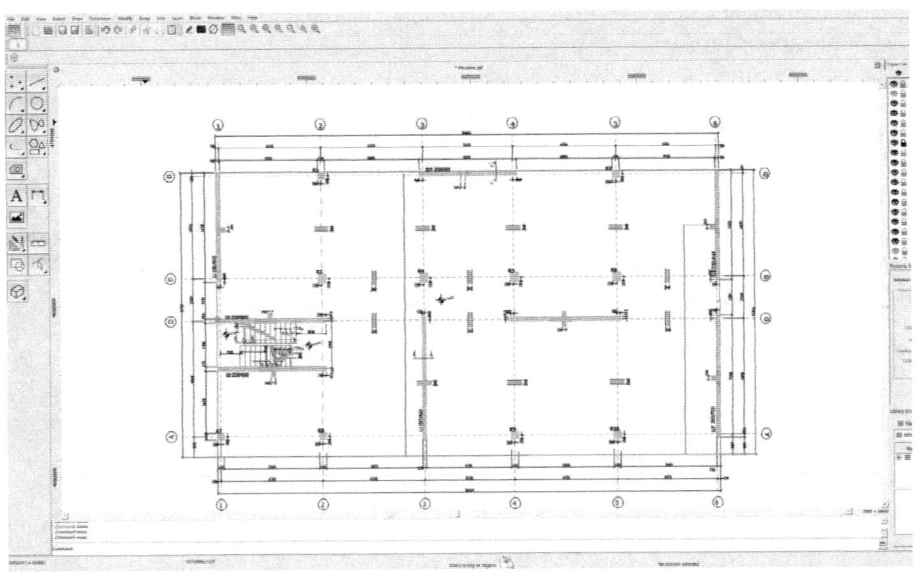

Figure 3. 2D CAD representation of the concrete building in GUI of QCAD

Table 1. Comparing 2D CAD open-source programs LibreCAD and QCAD

	LibreCAD	QCAD
Documentations	3	4
Functions	4	5
Languages	4.	3
Supported units	5	5
Export/Import file format	3 /BMP, JPEG, TIFF, ICO, PPM, XBM, XPM, SGV, PDF, DXF, CXF, and DWG for import only/	2 /BMP, JPEG, TIFF, ICO, PPM, XBM, XPM, SGV, PDF, DXF, CXF/
Print options	3	3
Drawing tools	3	3
Text settings	4	4
Snap options	4	3
Block tools	2	2
Libraries browser	4	3
Dimensions	4	4

The result of the comparison of the LibreCAD, QCAD and the commercial software, widespread in the field of AEC industry has revealed strengths and some critical aspects. One of the major limitations of OSS is the inability to support dwg file format, nevertheless, they have open-source code, which allows the community to contribute to the software. The comparison of the LibreCAD and QCAD also revealed some limitations in both of the applications. In our opinion, the graphical user interface of QCAD is more user-friendly and the geometric features provided more options for the representation of the project. The commercial version of the product, on the other hand, raises an issue for the subsequent development of the application.

2.2. 3D modelling

Building Information Modelling (BIM), a rich, manifold and widespread field of research, remains unexplored territory in experience with OSS. First OSS tools for the implementation of BIM projects, are developed, in an exclusive form, by big IT companies. Nowadays IfcOpenSHell, open-source software libraries, and Open Cascade Technology (OCCT), a platform for 3D, CAM software development, represents an irreplaceable basis for both OSS and proprietary 3D software. [10,12,14]

2.2.1. FreeCAD

FreeCAD, 3D CAD/ CAE parametric modelling application, is free alternatives to proprietary BIM software, developed in 2001 by Jurgen Riegel. The FreeCAD Project was called Graphical Object modeler (GOM), which impalement Open CASCADE, Phyton and Qt multi-platform languages. First inspired by Catia V5, well-known proprietary design software, nowadays FreeCAD is a multidisciplinary software application, due to long term active collaboration of developers and users. Arch, Draft, FEM, Image, Inspection, Mesh, OpenSCAD, Part, Path, Plot Points, Raytracing, Reverse, Robot, Ship, Sketcher, Spreadsheets, Surface, TechDraw are some of the active workbenches, part of the application. FreeCAD is released under LGPv2+ license, which is considered to be a weak copyleft license.

The application allows customers to use Graphic User Interface (GUI), Python scripts in Phyton console and Macros for modelling the building, structure or detail.

Its native file format is FreeCAD Standart file format (FCStd.), nevertheless, it supports dxf, dwg, svg, stl, SCAD, IFC and other file formats. The Arch and BIM Workbenches provide features to create 3D model of building or structure, but on the other hand, FEM Workbench submits complex finite element analysis, which presents the FreeCAD like OSS also for calculation and analysis. [13]

The concrete building was designed in both FreeCAD, open-source software, and Revit, well-known proprietary software, for the ability to consider all the problems, related to the phase of modelling and representing, and proper evaluation.

Figure 4. Model of the concrete building for evaluation

The comparison, during modeling of construction, revealed some critical aspects in FreeCAD, focused on representing AEC objects, but on the other hand, the application provides opportunities to calculate and design complex geometry by using scripts or macros. [9] Libraries, online training, manuals and forums allow the customers to use all features of this multidisciplinary software.

3. Calculation and analysis of building structures and geotechnical problems with complex geometry

The development of more powerful hardware in the last few decades has led to the rapid expanse of software solutions in all engineering areas. The software application allows engineers to design complex structures efficiently and decreases the number of mistakes and errors.

Code Aster is one of the most famous OSS FEM codes for calculating almost every engineering task. Many other also commercial types of software are using the code only as a solver. Calculating the cantilever beam is considered to be designed in Code Aster for the purposes of the report. Modeling and calculating the beam is the fundamental prerequisite for the evaluation. The results are comparable with the results of STAAD Pro for the same section and load, but the design is much more complex and requires knowledge in programming in Python. Nevertheless, Code Aster is one of the widespread OSS for the calculation of complex structures.

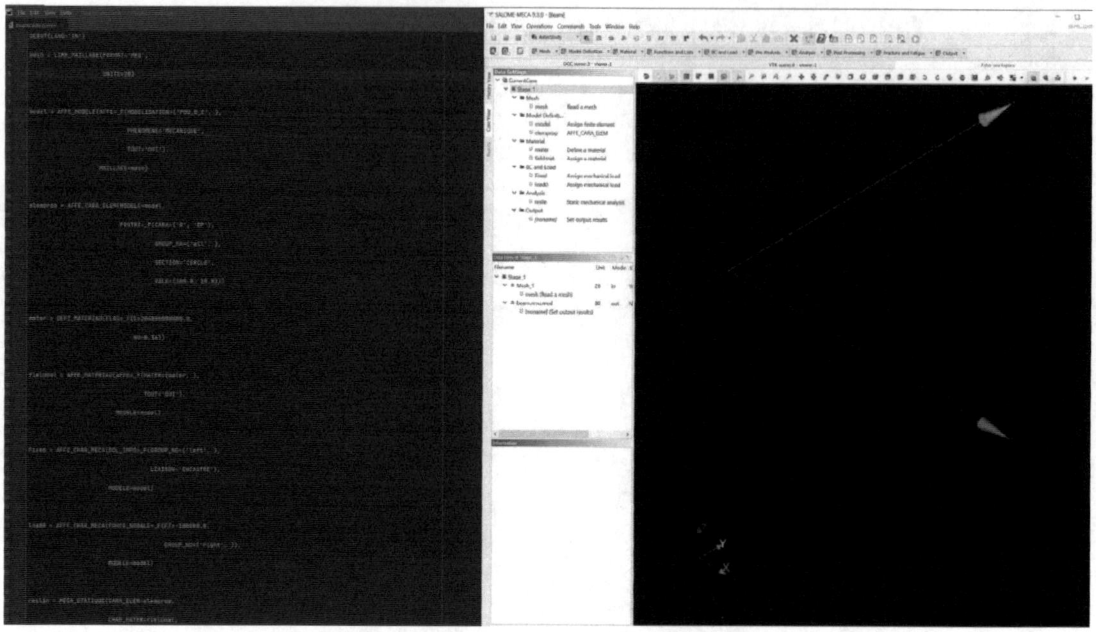

Figure 5. Python code for calculation of cantilever beam in Code Aster

Code-Aster, build in Fortran programming language, is released under the LGPL license. Code-aster combined with the GUI and the powerful pre- and post-processor Salome is released as Salome-Meca. The biggest issue with Code Aster and Salome-Meca was the documentation in French, but lately, there is full English support. Since version 17 Salome-Meca is also available for Windows. Version 19 of the software supports CAD-like Model construction and it is already the most powerful OSS software for FEM Analyses including some aspects in Geotechnics.

There is a vast majority of OSS software for Geotechnics, but almost none of them is general-purpose geotechnical software (solve only one task). Based on the difficulty of the geotechnical field, most of them are either task-oriented or sophisticated code-oriented executable that doesn't support GUI. Barkley University OSS geotechnical software is an example of the last one.

One remarkable OSS, software for Geotechnics is the early developed software with the name ADONIS. The author knows well the already described problem with the geotechnical OSS and tries to implement code, GUI and sophisticated material models in one application.

ADONIS is 2D Finite Element Software for Geo-Engineers. It supports different element types that represent soil body, retaining walls, anchors, geogrids etc. in the model. It supports also geotechnical calculation types as settlement, excavation (also of tunnels), slope stability etc. The software is still in development under LGPv3. Further would be a dynamic simulation, so the contribution to such type of software is always a good idea.

4. Conclusion

Software, considered for the purposes of the report, has advantages and some critical aspects compared to proprietary software in main structural engineering essential parts – modeling, analyses and geotechnical problems. The choice of the right software is determined by the needs of the customer – 2D CAD, 3D CAD or software for calculations and analysis, nevertheless the license's cost and agreements are a substantial part of the decision. OSS provides the comparable functionalities to proprietary software without restrictions and additional costs.

IOP Conf. Series: Materials Science and Engineering **951** (2020) 012003 doi:10.1088/1757-899X/951/1/012003

References

[1] Machado F, Malpica N and Borrmeo S 2019 *Parametric CAD modeling for open source scientific hardware: Comparing OpenSCAD and FreeCAD Python scripts (*Universidad Rey Juan Carlos, Morstoles, Spain*)*
[2] Cariilo K 2008 *The open source movement: A revolution in software development.*
[3] Wade M 2010 *A Comprehensive Review and Synthesis of Open Source Research.*
[4] Martinez-Torres 2015 *A Quantitative Study of a Evolution of Open Source Software Communities*
[5] Stallman R 2010 *Free as in Freedom (2.0): Richard Stallman and the Free Software Revolution* p 123 130.
[6] Ahmed F, Campbell P, Jaffar A and Caprets L 2009 *Defects in Open Source Software: The Role of Online Forums.*
[7] Jang M 2017 *The use of Computer-Aided Design in Small Contractors in a Local Area of Korea.*
[8] Brieri-Cote An, Rivest L and Marazana R 2013 *Comparing 3D CAD Models: Uses, Methods, Tools and Perspectives.*
[9] Al-Badi A and Al-Badi 2014 *Open Source Software in Higher Education: Oman SQU Case Study*
[10] Russel J and Cohn Ronald 2012, *Open Cascade Technology* p70 p100.
[11] Oku O 2015 *FreeCAD: Efficacy of Arch Moduly For Authoring Architectural Content in a BIM Process Using Open-Source Software.*
[12] Junk St and Kuen Chr 2016 *Review Of Open Source and Freeware CAD Systems for Use with 3D-Printing.*
[13] Donato D and Abita M 2019 *Low-cost 4D BIM Modelling A Comparison Between FreeCAD and Commercial software.*
[14] Rynkovskaya M, Elberdov T, Sert E and Ochsner 2020 Study of modern software capabilities for complex shell analysis.

IOP Conf. Series: Materials Science and Engineering **951** (2020) 012004 doi:10.1088/1757-899X/951/1/012004

Constitutive models for FE analysis of pile founded buried arch bridge

K Kazakov[1], L Mihova[2,*] and D Partov[1]

[1] University of Structural Engineering and Architecture "Lyuben Karavelov" (VSU), Sofia, Bulgaria

[2] University of Architecture, Civil Engineering and Geodesy (UACG), Sofia, Bulgaria

* E-mail: l_mihova@yahoo.com

Abstract. The paper presents plane strain FE modeling of a pile founded bridge, made from reinforced concrete, by using plate elements for the structure and the soil body. The soil-structure contact is modeled by interface finite elements. The numerical analysis is performed by the Plaxis 2D software. The mechanical behavior of the bridge is approximated by the constitutive models of Hooke and Mohr-Coulomb. Two sets of values of the Mohr-Coulomb strength parameters (cohesion and internal friction angle) for the concrete are obtained - on the basis of the theoretical uniaxial compressive and tensile strengths and on the basis of relationships in the Eurocode 2. The advanced constitutive HSM model is used for the soil. The construction and service periods of the bridge are simulated by a sequence of representative states. The mechanical response of the bridge undergoing backfill and road loads is analysed.

1. Introduction

Finite element models for the interaction between the structure of a buried arch bridge and surrounding soil body are analysed. The modeling is carried out by the Plaxis 2D software. The purpose of the study is an assessment of the application of different constitutive models for the bridge structure and soil materials. The prototype of the analysed structure is the bridge Tri Voditsi with width of 15,29 m and height of 8,67 m which is built of prefabricated elements of concrete C50/60. It is the structure of the railway-road junction near the villages Tri Voditsi and Novo Selo in Bulgaria (Figure 1). The same bridge superstructure founded on strip footings is considered in previous analyses [1], [2]. In the present research the substructure is a piled-raft foundation of concrete C25/30. This type of foundation reduces significantly displacements of the bridge structure and it has been realized in the design project.

The usual procedure for the finite element modeling of a soil-structure interaction is to introduce one-dimensional (beam) elements for concrete structures within two-dimensional soil elements. For the purposes of the present analysis both the soil body and the concrete structure are discretized by a mesh of 2D plane strain plate elements. This approach allows to describe the complex geometry of the bridge structure more accurate. The stiffness parameters of the discrete piles are adapted to the plane strain model. This approximation gives more conservative results in comparison with a model of 3D discretization of the piled-raft foundation [3].

The mechanical behavior of the bridge structure is modeled by the linear elastic constitutive model of the Hooke and the Mohr-Coulomb elastic-plastic model. Two sets of values for the concrete Mohr-

Coulomb parameters (cohesion and friction angle) are defined - on the basis of the theoretical uniaxial compressive and tensile strengths and on the basis of relationships in the Eurocode 2.

The advanced constitutive Hardening Soil Model is used for modelling of the soil.

Figure 1. Photo of the bridge Tri Voditsi

The loading conditions of the bridge structure during the construction and service periods are simulated by a sequence of representative states. The mechanical behavior of the bridge is analysed undergoing the backfill and road loads. Only static loadings of the bridge are considered. Time history dynamic analysis of the bridge using an accelerogram of the earthquake Pernik-2012 in Bulgaria [4], [5] is performed in reference [6].

2. Finite Element Modeling

2.1. Finite element mesh and boundary conditions

A 2D plane strain finite element model is created considering the symmetry of the geometry and the loading of the bridge structure (Figure 2). 15-node triangular elements are used for discretization of the soil body and the bridge structure and quartic interpolations of the displacements, horizontal, U_x, and vertical, U_y, are conducted. Interface finite elements, developed in the Plaxis software, are used for modeling of the soil-structure contact.

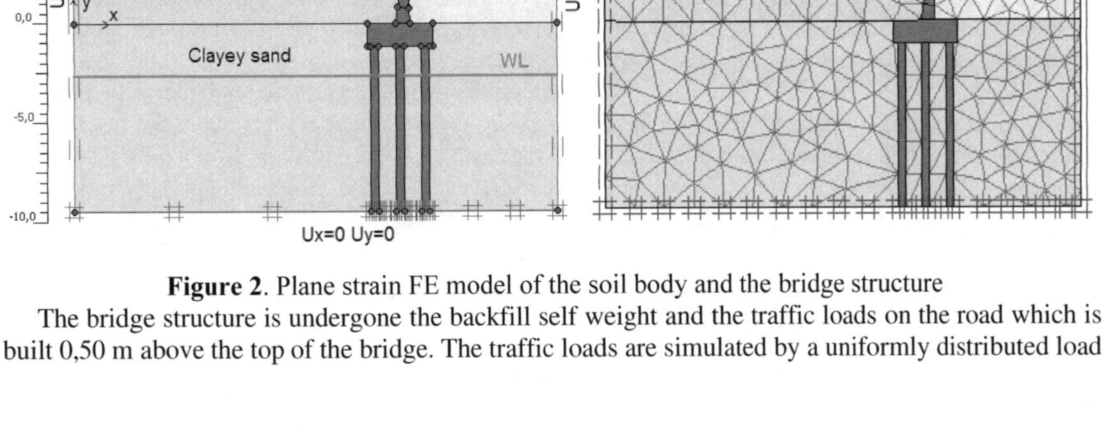

Figure 2. Plane strain FE model of the soil body and the bridge structure

The bridge structure is undergone the backfill self weight and the traffic loads on the road which is built 0,50 m above the top of the bridge. The traffic loads are simulated by a uniformly distributed load

IOP Conf. Series: Materials Science and Engineering **951** (2020) 012004 doi:10.1088/1757-899X/951/1/012004

of value q = 9 kPa which corresponds to the load on the road lane number one according to the UDL system of the traffic model LM1 in the Eurocode 1 [7].

2.2. *Constitutive model for the soils*

The constitutive model (law) is a mathematical description of stress-strain behavior of the material in response to the applied loads. Here the elastic-plastic Hardening Soil Model (HSM), implemented in the Plaxis software [8], is used for modeling of the soil ground and backfill of the bridge. The HSM is a state-of-the-art scientific achievement which integrates many conceptions of the soil mechanics [9], [10]. It uses more than 10 material parameters and approximates the mechanical behaviour of the soil with a high degree of accuracy.

Table 1. HSM parameters used in the FE model

Parameter	Symbol	Dimension	Backfill	Clayey sand
Unit weight of unsaturated soil	γ_{unsat}	kN/m^3	18	18
Unit weight of saturated soil	γ_{sat}	kN/m^3	-	20
Reference stress for stiffnesses	p_{ref}	MN/m^2	0,100	0,100
Tangent stiffness for primary oedometer loading	$E_{oed,ref}$	MN/m^2	50	20
Secant stiffness in standard drained triaxial test	$E_{50,ref}$	MN/m^2	50	20
Unloading/reloading stiffness	$E_{ur,ref}$	MN/m^2	150	60
Poisson's ratio for unloading/reloading	ν_{ur}	-	0,2	0,2
Power for stress-level dependency of stiffness	m	-	0,5	0,6
Coefficient of lateral earth pressure	K_0	-	0,384	0,515
Failure ratio	R_f	-	0,9	0,9
Friction angle	φ	^{0}	38	29
Cohesion	c	kPa	0	10

2.3. *Constitutive models for the concrete*

The Hooke model (law) of linear elasticity is the simplest constitutive law used for structure elements in the design practice. This model requires only two material parameters – Young modulus, E, and Poisson ratio, ν, and assumes unlimited values of stresses. These features of the Hooke model are the reason for receiving of conservative results for stresses that are not relevant to the real mechanical behavior of the concrete.

The Mohr-Coulomb (MC) constitutive law describes elastic-perfectly plastic behavior and this model is appropriate for materials such as geomaterials, concrete, for which the compressive strength far exceeds the tensile strength. According to the MC model, the stresses are proportional to the strains until the yield (fail) point is reached. The stress state corresponding to the yield points is described by a set of six linear equations in effective principal stresses σ_1', σ_2', σ_3', neglecting the effect of the intermediate principal stress. Assuming no order of the principal stresses the MC criterion is graphically represented as a hexagonal pyramid (Figure 3). When the stress state corresponds to the yield (fail), then the stress point lies on the pyramid surface. The stress points corresponding to the elastic behavior lie inside the surface.

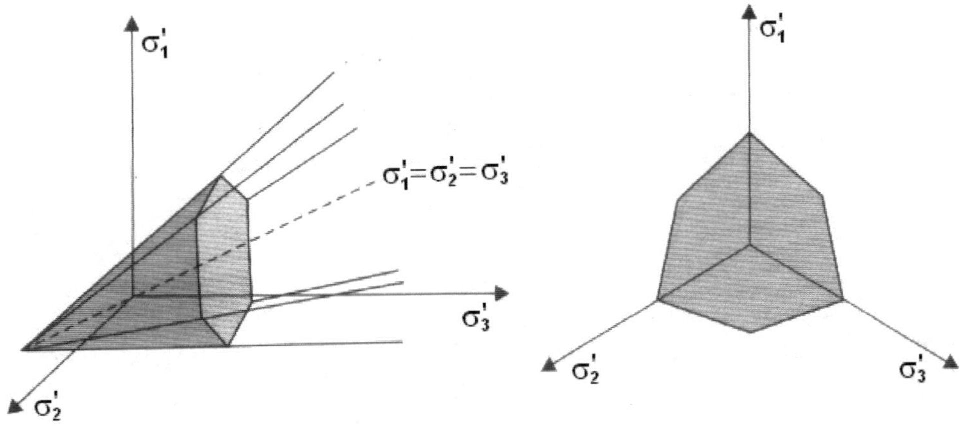

Figure 3. Mohr-Coulomb yield function: a) Surface in 3D space of principal stresses; b) Cross section in deviator plane ($\sigma_1' + \sigma_2' + \sigma_3' = 0$)

For the particular case $\sigma_1' \geq \sigma_2' \geq \sigma_3'$ the MC criterion can be written as a following functions of the major σ_1' and the minor σ_3' principal stresses

$$F(\sigma'_1, \sigma'_3) = (\sigma'_1 - \sigma'_3) - \sin\varphi\ (\sigma'_1 + \sigma'_3) - 2c\cos\varphi\,, \tag{1}$$

and normal σ and shear stresses τ on the failure plane

$$\tau = c + \sigma.\ \mathrm{tg}\ \varphi\ . \tag{2}$$

There are many experimental and theoretical researches on the application of the MC model to concrete. The most important point is the determination of the concrete strength parameters values. A review of the literature shows a very wide range of values. In the reference [11] indirect tensile tests and numerical simulations are used and values c = 1,8-2,7MPa and φ = 31° are obtained. On the basis of equivalence of split test results and wedge theoretical solutions in [12] it is proved that the friction angle of the concrete is equal to φ = 55°-57,5°. Authors of reference [13] carried out direct shear tests and the cohesion and the friction angle are specified between 2,94MPa and 12,34MPa and between 29,8° and 41,7°, respectively. Concrete samples are tested in a large shear box in [14] and values of c = 0,58-2,27MPa and φ = 36°-52° are summarized. Concrete Construction Engineering Handbook [15] recommends for the concrete values of c = 2,75MPa and φ = 30°-60°. The Plaxis Bulletin [16] presents three algorithms based on: Montoya's considerations [17], the Spanish Structural Concrete Code EHE-98 and the Eurocode 2. According to these algorithms the following values of the strength parameters are determined: c = 0,387; 0,365; 0,712MPa for the concrete of compressive strength f_c = 15MPa and c = 0,5; 0,513; 1,186MPa for the concrete of f_c = 25MPa; φ = 9°; 35°; 55° for the both concrete classes.

In the present research the theoretical grounds of three sets of material parameters for concrete, named Model I, Model II and Model III, are developed. They are applied to the bridge structure and a comparative analysis of the results is carried out.

- Model I – validity of the Hooke constitutive law for the concrete.
- Model II – validity of the MC constitutive law for the concrete with parameters c and φ calculated using the theoretical uniaxial compressive, f_c, and tensile, f_t, strengths.

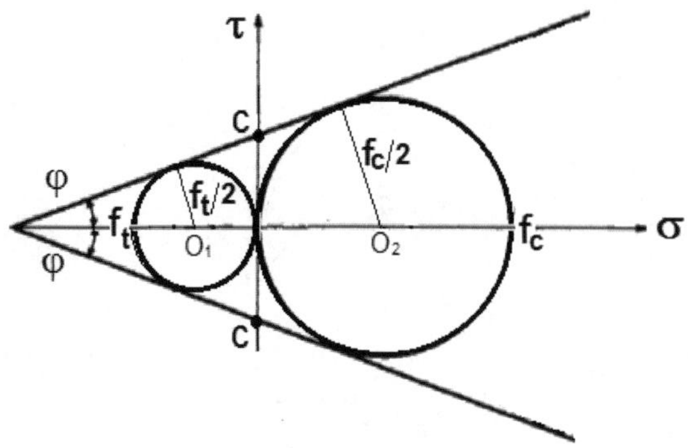

Figure 4. Diagrams of the Mohr-Coulomb criterion

In consequence of the MC relationships (Figure 4) the expressions are valid:

$$f_c = 2.c.\cos\varphi \,/\, (1 - \sin\varphi)\,, \tag{3}$$

$$f_t = 2.c.\cos\varphi \,/\, (1 + \sin\varphi)\,. \tag{4}$$

After the substitution of (3) and (4) in equation $F(\sigma'_1, \sigma'_3) = 0$, the following expressions are obtained for the strength parameters:

$$\sin\varphi = (f_c - f_t) \,/\, (f_c + f_t)\,, \tag{5}$$

$$c = f_c.(1 - \sin\varphi) \,/\, (2.\cos\varphi)\,. \tag{6}$$

The similar methodology is discussed in references [18], [19].

The values of the strength parameters calculated according to the Model II for the concrete classes used in the bridge structure are given in Table 2.

Table 2. Values of c and φ for the concrete according to the Model II

Concrete class	***c* (MPa)**	(°)
C25/30	3,36	60
C50/60	6,04	63

- Model III – validity of the MC law for the concrete with parameters c and φ determined according to the Eurocode 2.

The Eurocode 2 suggests the following formula for the design shear resistance of concrete elements without shear reinforcement

$$\tau_{Rd,c} = v_{min} + k_1\,\sigma_{cp}\,, \tag{7}$$

where:

$v_{min} = 0{,}035 \,.\, k^{3/2} \,.\, f_{ck}^{1/2}$, where f_{ck} is the characteristic compression strength in MPa;

$k = 1 + (200/d)^{1/2} \leq 2$, where d is the effective height of a cross section in mm;

$k_1 = 0{,}15$ – empirical coefficient;

σ_{cp} – design normal stress.

The form of the formula (7) is identical to the expression (2) of the MC criterion. Using the substitutions

$$c = v_{min} \text{ and } \operatorname{tg}\varphi = k_1\,, \tag{8}$$

the design values of the strength parameters c and φ can be calculated.

The values of the strength parameters calculated according to the Model III for the concrete classes used in the bridge structure are given in Table 3.

Table 3. Values of c and φ for the concrete according to the Model III

Concrete class	***c* (MPa)**	**φ(°)**
C25/30	0,495	9
C50/60	0,700	9

2.4. Simulation of the construction and service periods of the bridge

Six loading states in the period of backfilling (State 1 – State 6) and one loading state in the service period (State 7) of the bridge are numerically analysed (Figure 5).

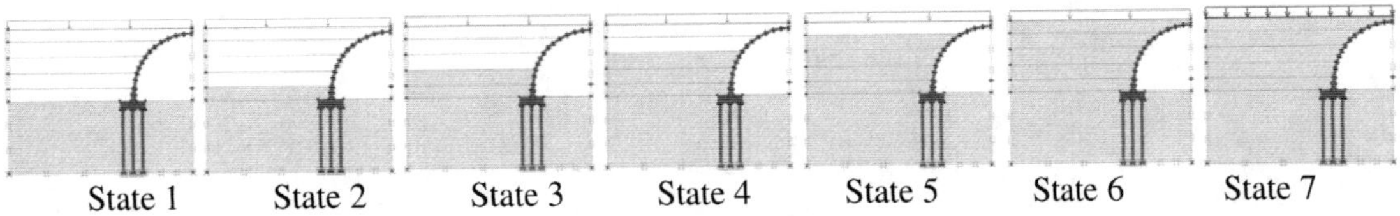

Figure 5. States in construction and service periods of the bridge

3. Results from FE analyses

3.1. Initial stress conditions

The initial stresses ("in situ" stresses) in the ground exist as a result of action of the soil and ground water selft-weight. Diagrams of the "in situ" stresses are shown in Figure 6, where the following symbols are used:

$\sigma_{y,0}$' – normal vertical effective (in the soil skeleton) stress;
$\sigma_{x,0}$' – normal horizontal effective (in the soil skeleton) stress;
p_w – hydrostatic pressure.
The initial ground displacements are equal to zero.

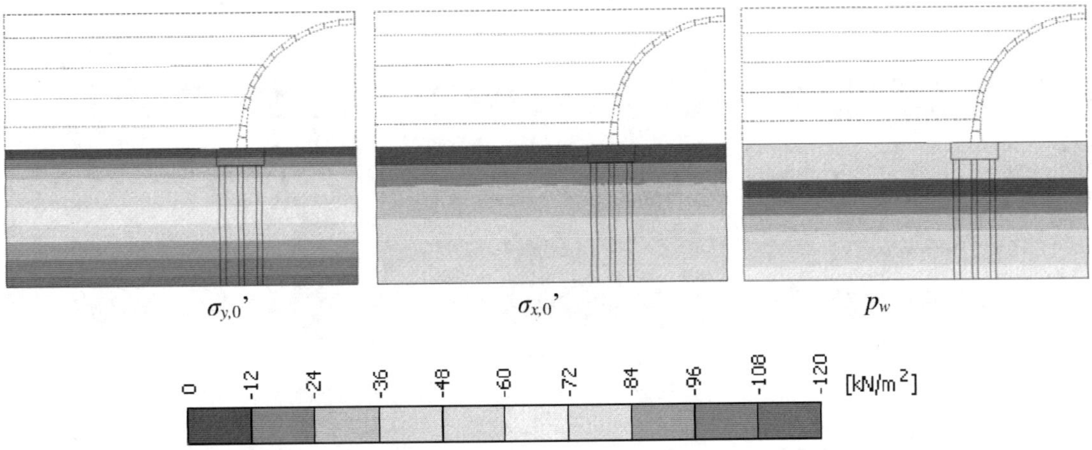

Figure 6. Initial stress conditions

IOP Conf. Series: Materials Science and Engineering **951** (2020) 012004 doi:10.1088/1757-899X/951/1/012004

3.2. Service period analysis

The mechanical response of the bridge, undergoing backfill and road loads (State 7 in Figure 5) is analysed. Results from FE solutions using defined above constitutive models for the bridge structure: Model I, Model II and Model III, are presented in Figures 7-9.

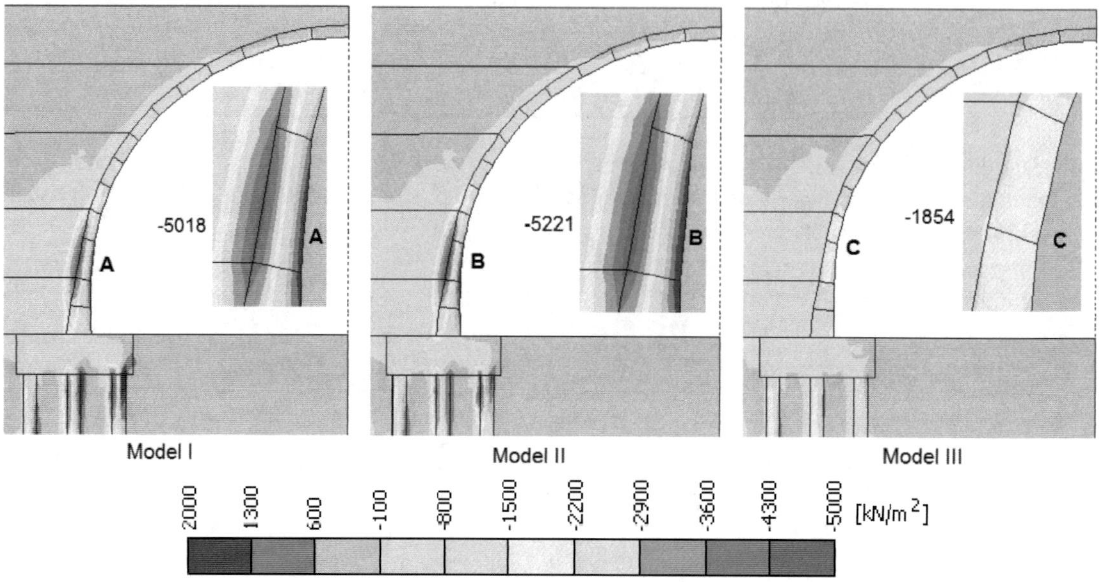

Figure 7. State 7: Diagrams of the normal vertical effective stresses $\sigma_{y,0}$'

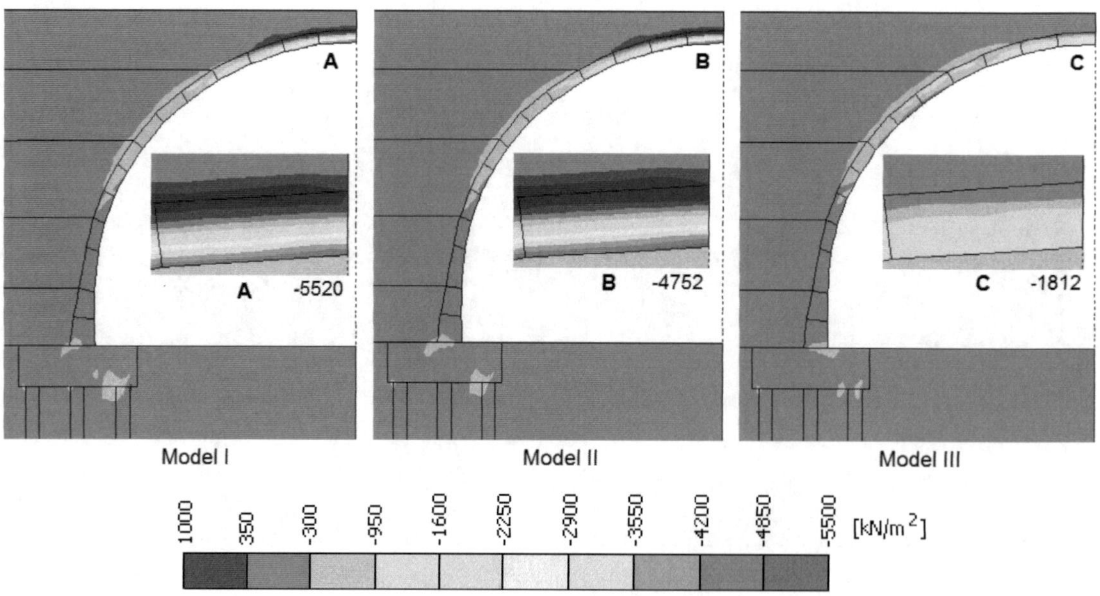

Figure 8. State 7: Diagrams of the normal horizontal effective stresses $\sigma_{x,0}$'

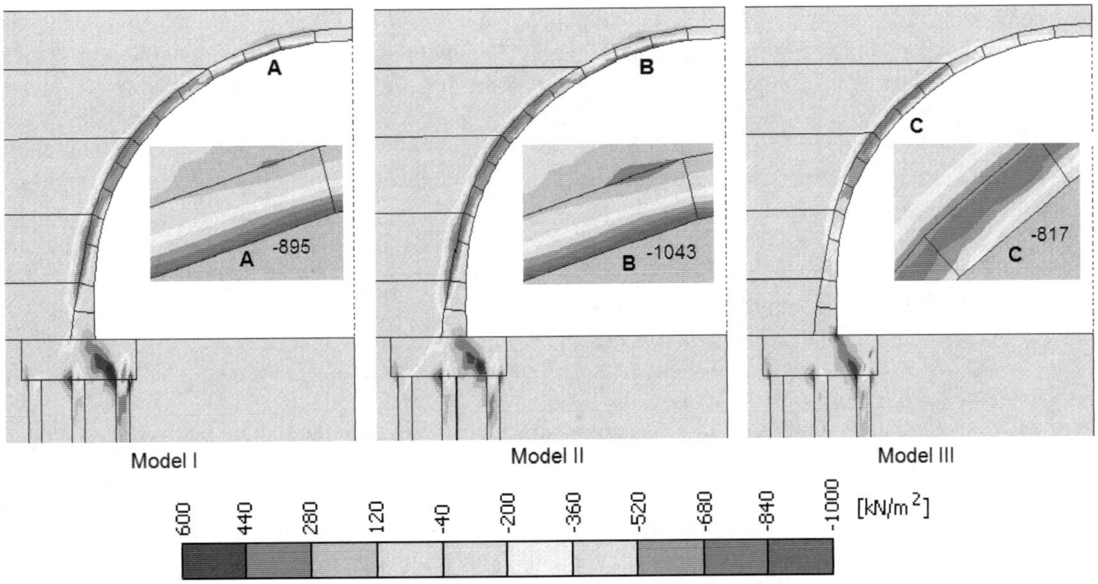

Figure 9. State 7: Diagrams of the shear stresses τ_{xy}

4. Conclusion

The comparative analysis of the results confirms the following conclusions:

- The FE analyses of Model I, Model II and Model III give similar distribution of areas with extreme values of stresses;
- The solution using the Hooke constitutive model for the bridge structure gives the most conservative results for the stresses;
- The values of the Mohr-Coulomb strength parameters – cohesion and angle of internal friction – have significant influence on the results of the stresses in the bridge structure. The Model II uses strength parameters about 7 times larger than the strength parameters in the Model III. And the Model II gives larger values of max stresses in the bridge in comparison with the Model III: 3 times in the vertical normal stress; 2,6 times in the horizontal normal stress; 1,3 times in the shear stresses;
- The rather high values of the strength parameters in the elastic-plastic Model II are the reason for results of this model closed to the results of the elastic Model I: the differences are about 15% and 4% in the normal and the shear stresses, respectively.

The present study shows that the constitutive modeling of concrete structures according to the Mohr-Coulomb law requires precise analysis of the strength parameters – cohesion and internal friction angle. The parameters influence significantly on the results of the mechanical behavior of concrete structures. The literature sources include different experimental and theoretical methods for determination of the concrete strength parameters and the obtained values are in very large ranges. Two methods are applied in the present paper for determination of the strength parameters. The first method is based on the theoretical relationships between uniaxial compressive and tensile strengths of the concrete. The second method is based on the analogy between the equation of the Mohr-Coulomb criterion and the formula for the design shear resistance of concrete elements without shear reinforcement in the Eurocode 2. The application of these methods for a bridge structure shows big differences in the results of the structure mechanical response.

IOP Conf. Series: Materials Science and Engineering **951** (2020) 012004 doi:10.1088/1757-899X/951/1/012004

References

[1] Kazakov K, Mihova L and Partov D 2019 Finite element modeling of the road buried arch bridge "Tri Voditsi" in Bulgaria *Proc. 18th Int. Symp. of Macedonian Association of Structural Engineers MASE* (Ohrid)

[2] Kazakov K, Mihova L and Partov D 2019 Comparative analysis of different finite element models of the soil-buried arch bridge interaction *J. for Research of Materials and Structures* 4 (Serbia) pp 15-28

[3] Tanev T, Kerenchev N, Manolov A and Dimitrov H 2012 Analyzing the displacements of a pile frame to determine its transverse load-bearing capacity *Proc. 22th European Young Geotechnical Engineering Conf. EYGEC2012* (Sweden)

[4] Paskaleva I, Ivanov R, Kaneva A and Ivanchev I 2018 Seismic effects in the Pernik area after the 22, May 2012 earthquake *Sixth Int. Conf. Earthquake Engineering and Engineering Seismology* (Serbia) pp1-12

[5] Ivanov R, Kaneva A, Paskaleva I, Ivanchev I and Gorolomov A 2018 Earthquake-induced ground motion effects from the seismic sequence of 2012, May on the territory of Pernik. *XVIII Jubilee Int. Scientific Conf. „Civil Engineering Design and Architecture"* (Sofia) pp 294-299

[6] Kazakov K., Sulay I, Partov D 2020 Seismic analysis of a buried road bridge *Proc. XIX Int. Scientific Conf. VSU'2020* (Sofia)

[7] Eurocode 1: Actions on structures, Part 2: Traffic loads on bridges - EN 1991-2, 2003.

[8] Schanz T, Vermeer P and Bonnier P 1999 The hardening-soil model: Formulation and verification (In Brinkgreve R B J *Beyond 2000 in Computational Geotechnics*) (Balkema Rotterdam)

[9] Plaxis Material Model Manual 2015

[10] Duncan J and Chang C 1970 Nonlinear analysis of stress and strain in soil. *J. Soil Mechanics and Foundations Division ASCE* 96

[11] Levović S, Vasović D and Stojić D 2019 Determination of the Mohr-Coulomb Material Parameters for Concrete under Indirect Tensile Test *Technical Gazette Scientific Professional Journal of Technical Faculties* Josip Juraj Strossmayer Univ. 26 2 (Croatia) pp 412-419

[12] Fujita Y, Ishimaru R, Hanai S and Suenaga Y 1998 Study on internal friction angle and tensile strength of plain concrete. *Proc. Fracture Mechanics of Concrete Structures FRAMCOS 3* (Japan) pp 325-334

[13] Pul S et al 2017 Experimental determination of cohesion and internal friction angle on conventional concretes *Materials J.* 114 3 pp 407-416

[14] Lindenbach E 2017 *Concrete Shear Strength Parameters Compared to Other Laboratory Determined Properties* (Department of the Interior, Bureau of Reclamation Technical Service Center, Denver Federal Center) p 55

[15] Nawy E 2008 *Concrete Construction Engineering Handbook* (Taylor&Francis Group) 1584 p

[16] Ardiaca D 2009 Mohr-Coulomb patameters for modeling of concrete structures *Plaxis Bulletin* (Spring Issue) pp 12-15

[17] Montoya P 1971 *Hormigón Armado* Tomo 1(Editorial Gustavo Gili S A/Barcelona) p 631

[18] Zhang Y 1992 *Finite element modeling of tornado missile impact on reinforced concrete wall panels* Dissertation in Civil Engineering (Texas Tech. Univ.)

[19] Soutsos M, Robinson D and Chen J 2017 Numerical modeling of UHPFRC beams under impact load *Proc.1st Int. ICT Conference on Cement and Concrete Technology* (Oman) pp 1-6

IOP Conf. Series: Materials Science and Engineering **951** (2020) 012005 doi:10.1088/1757-899X/951/1/012005

Aurubis iron-silicate fines: universal sustainable construction material: a state-of-the-art review

E Manolova

Iron-silicate Business Development Manager, Industrial zone, 2070 Pirdop, Bulgaria

E-mail: e.manolova@aurubis.com

Abstract. This article brings together a lot of research information that deals with iron-silicate fines (ISF), as a basic by-product produced by the copper plant of Aurubis, located in Bulgaria. As a rare material, produced only in couple of plants in the world, ISF has a wide variety of applications due to its specific physical, chemical and mechanical characteristics. Well known for decades by the clinker cements industry, nowadays Aurubis presents a wide variety of new perspectives for creating green and sustainable building materials out of ISF. Sustainability in the construction sector is mostly referred to as zero waste scenarios related to the use of recycled and secondary materials (RSM). Such RSM materials differ significantly and their performance remains to be established separately as well as the properties of the construction products produced by them. Besides this each production unit has its own experience with raw materials for its production and introducing a new one is always causing some initial doubts. ISF not only covers preliminary expectations but even exceeds them.

One of the main purposes of this article is to present opportunities to academics, researchers, design engineers, research and development & production managers in companies, specialists and contractors, for sustainable building products development of this not so well-known material with great potential. Most of the presented applications are prepared on industrial scale with different percentages of ISF for testing the effect of its quantity on the properties of the final products.

1. Introduction or what is iron-silicate fines?

Iron-silicate fines (ISF) is a by-product obtained after flotation of smelted copper slag in the plant of Aurubis located in Bulgaria. The main difference between ISF and other iron-silicates produced worldwide is the grain size. Bigger grain size gives bigger advantages in finding the right applications than smaller one. Although other copper slags are mostly cast in pots, followed by air-cooling and crushing up to the desired grain size, or water-quenching, ISF is formed by consecutively cooling the pots, crushing, milling and floating, as a final step of copper extraction. The type of slag treatment is responsible for the microstructural phases, but the chemical composition depends mostly on the initial blend collected by the plant.

The annual world copper slag production in 2014 was about 40 million tonnes [1]. More than one third of it is generated in China, about 60% of the total production is in Asia, less than 20% is in Europe and more than 15% is in South and North America [1]. Different standards define copper slags as a safe material from environmental point of view, including it in the green list of non-hazardous materials from Japan through Europe to USA. ISF production accounts for a small amount of the abovementioned percentages, and is produced mainly in Europe, Asia, Japan and South America and its characteristics

IOP Conf. Series: Materials Science and Engineering **951** (2020) 012005 doi:10.1088/1757-899X/951/1/012005

and areas of application are not so well defined. One of the reasons is the absence of standards for copper slag utilization globally, except for some attempts in Korea and Japan. This article is a state-of-the-art review of all different applications discovered in Aurubis Bulgaria.

ISF contains mostly Fe_2O_3 in the form of fayalite (Fe_2SiO_4) and magnetite (Fe_3O_4). Iron is in a hard relationship with silicon, only a small amount is in the form of free Fe, which makes magnetic separation very difficult. Although there are some promising studies in this direction in and outside of Bulgaria. Slag from the smelter is cooled down using two different methods – air-cooling after the pots are discharged and cooling in the pots with water streaming on the top surface. Phase formations are in the same range for both methods: about 35% crystalline phase and 65% amorphous phase. These results differ from those of the other copper slags, which are predominantly crystalline for ordinary air-cooled slags and mostly amorphous for granulated ones. According to many scientists, amorphous structure is responsible for the pozzolanic activity of the final copper slag [2,3]. That is why the glassy structure of the ground granulated blast furnace slag and the fly ash have been facilitating the development of cement and concrete production for decades and have been included in building standards. The different cooling technology of ISF forming 2/3 amorphous structure can make this by-product another sustainable alternative as a cementitious component compared to the other already established materials. The average size of 20 μm can be beneficial for production of cement, mortars, concretes, ceramics and asphalts which require fine grained material. The specific density of the material, based on BNS EN 1097-6, is found to be 3.8 g/cm^3, which is close to the densities of 3.91 g/cm^3, reported by other copper slag producers [4,5]. Additional information on the physical, chemical and mechanical characteristics of the ISF can be found in [6,7]. Chemical and physical characteristics are constantly maintained.

The goal of the article is to give a wide overview of all established applications of ISF discovered and tested in and out of Bulgaria. The future of this material is in production of green building materials, which is stimulated by the sustainable policy of the EU and the EU Green Dial strategy [8,9].

2. Where is the bottleneck in by-products material application?

Difficulties in direct application of by-products can be divided into two directions: absence of standards and the waste status of the materials. Each application requires defined procedure for its utilization. This procedure is described in standards and normative documents required by the Institutions. In Bulgaria, due to the waste status of the ISF, each company which utilizes ISF must have a permission by the Regional Inspection of the Environment and Waters for its application in their production and must provide regular reports on the quantity of material they have included in their production. For customers outside of Bulgaria the procedure is the same in accordance with the European Waste Directives. The third aspect is the additional technological efforts for implementation of a new material in production, which requires installation and software upgrades.

Despite the abovementioned difficulties, the construction sector in Bulgaria shows willingness to start ISF utilization on a large scale, due to its benefits and the absence or scarcity in Bulgaria of other alternatives defined in the construction standards, such as fly ashes, ground granulated blast furnace slag, pozzolans and silica fume. Together with the EU directives on wider utilization of wastes, ISF can cover this gap for local effective material. Certified as an aggregate by the Declaration of Exploitation Characteristics, the ISF fulfils the criteria for application in concrete.

3. Applications of iron-silicate fines

The engineering characteristics of ISF affect very positively the properties of the final products where it is applied. This paper gives a wide overview of many years of experiments with ISF in completely different areas – from construction (clinker; cement; alkali-activated binders; dry plasters; concrete; ceramics; glass ceramics; glass fibres; asphalts; mineral wool, rubbers and caoutchoucs; soil embankments) through steel production to the mining sector and the coal industry. Application of ISF in construction, as one of the largest global markets, is the most promising area of application.

3.1. Cement production

Ordinary Portland Cement (OPC) production includes first clinker production and then crushing and milling with additives for regulation of the setting time. ISF has been successfully applied in cement clinker production for many decades. This article provides an overview of other applications of ISF not only in clinker, but also in blended cements, as well as alkali-activated binders, which are one of the most published about new types of binders out of all types of slag during the recent years.

3.1.1. Clinker cement

Production of cement clinker requires adjustments of four basic minerals. For the formation of the one of them (Ca_4AF) is needed a certain amount of iron. If local clays are poor in iron, it is necessary to increase the iron content using additional material. Fe_2O_3 drops the melting point and it gives higher strength to the cement, better corrosion resistance, increases the setting time and drops the exothermic reaction. ISF is one of the best materials for this purpose. Moreover, it has lower melting point and reduces the calcination temperature in the kiln during cement clinker production [10]. Because of this, the need of mineralizer is reduced or eliminated [11]. Some authors claim that the performance of the cement is even better than that produced with traditional limestone, clay and mill scale [12]. ISF improves the grindability of the clinker and does not require additional crushing.

3.1.2. Blended cements

Building materials utilized in cement-based products to reduce the level of cement to a certain extent are called 'Supplementary Cementing Materials'. The pozzolanic properties of copper slags in blended cements have been researched for decades [13, 14, 15]. The potential suitability of copper slag for use as a cementitious component has been examined and its positive effects have been proven. Moreover, studies reported that copper slag does not need to be completely amorphous for a significant hydration process to occur.

It is examined the effect of ISF and 2 different types of fly ashes (FA1 and FA2) substituting 10% and 25% of the cement. Test results show that the mortars prepared using ISF give relatively the same flexural strength at all ages, but the compressive strength differs. The early (2day) and late (3month) compressive strength of those mixtures, which contain ISF, is higher than the compressive strength of those with FA substitution. These results are more obvious at higher percentages of cement substitution (25%). The final strengths at 10% are comparable (>95%) to pure cement mortar. One of the reasons for the lower strength values at higher cement substitution of both FA and ISF is the higher water consumption for one and the same workability. When sand is substituted with ISF, even though the water/cement ratio increases, the compressive strength of the mortar increases by 30%. This means that when appropriate chemical admixture is applied to improve the rheology of the mix, the final strength will be even higher.

3.1.3. Alkali-activated binders

Alkali-activated binders (AAB) have been proclaimed as a real Portland Cement (PC) alternative during recent years. In addition, the use of industrial by-products like fly ash, high furnace slag, blast furnace slags, fayalitic slags, etc, in AAB has been declared the most promising area of their application. This new type of binder can contain a certain amount of PC, but it is not even necessary if the by-product contains enough SiO_2 and Al_2O_3 and has high reactivity with the precursor. Some researchers [16, 17] suggests that AAB have a higher resistance to fire and improved performance at elevated temperatures and even against acids.

Production of AAB from ISF has been proven through activation with potassium silicate. The fayalite phase within the ISF is reactive during alkali activation, and the addition of a small quantity of either PC, ground granulated blast furnace slag, or steel slag, brings advantages in strength development [18]. The properties of fresh mortar made out of AAB have been improved by adding water and chemical admixture. The reaction products formed are not only CSH gel, but also new alumo-silicate hydrates, made by the combination of the potassium silicate and the CSH gel. Curing conditions also significantly affect the properties of AAB. One of the most interesting applications of ISF in AAB is for anti-skid surfaces (ASC). Additional information can be found in [19].

3.2. Plasters based on different binders

Plasters and different types of coatings and polishes are another huge segment in construction chemistry. They have developed very fast during the past several decades with worldwide trading names having local plants in every country. Most plasters are prepared in sacks and buckets for direct application after adding water (for cement based) or applied in air (for air-hardening). This type of products is fast growing due to its precisely developed composition containing a wide range of materials and chemicals.

3.2.1. Cement plasters

Cement plasters are everywhere around us: inside and outside of buildings, infrastructural facilities, transportation lines, industrial areas, etc. Applied on horizontal and vertical surfaces they need to provide comfortable living and working environment, sustainable construction and effective material resource utilization in changing weather. ISF not only makes plasters more sustainable, but also improves their properties first in fresh state for easier handling and enhance the mechanical characteristics in hardened state for plasters where this is required. ISF give volume and linear stability and decrease shrinkage, which is one of the most important parameters in cement-based plasters.

3.2.2. Polymer plasters

Polymer plasters are made out of a matrix of special types of fillers, bonded together by a polymer to achieve the desired properties. The know-how of the producers of these products is related to choosing the correct type of polymer for different applications in many industries, including marine, aerospace, automotive and electrical. The use of ISF with an average particle size of 20µm as a filler increasec the adhesion properties and thermal conductivity of the resultant composites, as proven by Aurubis. In combination with glass-fibres, according to some authors [20], small grain size copper slag like ISF improves flexural and tensile strength, tensile modulus and hardness.

3.3. Concrete production

Application of ISF in concrete can substitute partially cement and/or sand. The following paragraphs discuss ISF as an alternative material to sand. ISF increases the consistency of concrete, but it also retards the setting time. Nevertheless, initial strength is higher than reference mixtures with ordinary sand. The water-cement ratio is adjusted by adding chemical admixtures and thus the final strength and modulus of elasticity are improved. ISF compacts the structure by filling the pore structure both mechanically and chemically, enhancing concrete durability and agent resistivity, improving abrasion and wear resistance. Concretes do not show any shrinkage after 1-year examination and show lower level of heat of hydration. ISF can be implemented in many different types of concrete presented below.

3.3.1. Ready-mixed concrete

Ready-mixed concrete (RMC) is used as a reference for the stage of growth of a unit (city; country; union; continent). ISF can be applied as an aggregate which additionally improves the properties in fresh and hardened state. Due to the small grain size and the limits by the standards ISF cannot be applied in large amounts. Nevertheless, even 50÷100kg/m^3 increase compressive strength, especially at early age (2nd and 7nd day) and after 90 days. There are two reasons for this: the small grain size of 20microns immediately mechanically compacts the structure at early age, compared to evaporated water and bigger open porosity in ordinary concrete, and the late formation of additional chemical products, which require time for the activation of the reactive amorphous phase, which is followed by additional CSH gel formation. This process is supported by appropriate composition and suitable chemical admixture with cement and ISF. The other great advantage of ISF is that there is no risk of alkali-silica reaction due to the very high thermal treatment of the material and absence of reactive agents.

3.3.2. Concrete products

Vibro-pressed concrete products are widely distributed in urban areas. The technology of their production requires a concrete mix design with specific workability and flowability of the fresh concrete mix and at the same time high mechanical properties and durability of the hardened products. ISF, which in general improves the properties of hardened concrete, is effectively utilized in paving products by covering the requirements for low water-cement ratio and formability of elements. ISF can be used for a wide range of concrete products like tiles, blocks, panels, curbs, etc. One of the greatest benefits of ISF application in concrete products is the improvement of mechanical properties by more than 25%. Additional information can be found in [7].

3.3.3. Special concretes

There is a wide range of special concretes, but here we will focus on those, in which the use of IFS can have special benefits due to its physical and chemical characteristics.

3.3.3.1. Heavy-weight concrete

Heavyweight concrete is a material which mostly uses aggregates like barite and magnetite, which are rare in Bulgaria. Heavyweight concretes are used where high density and heavy weight required by standards. BNS EN 206 defines heavyweight concrete as a material which has an oven dry density of more than 2600kg/m^3. The specific density of ISF is about 3700kg/m^3, which is about 50% higher than ordinary aggregates used in construction, and fulfils this criterion with higher aggregate/cement ratios, higher volume of chemical admixtures and the same cement and water/cement ratios for the desired strength class. According to detailed Laboratory tests by Aurubis (both mechanical and durability) heavyweight concrete with ISF can be applied into: heavy foundations for special structures; underground and tunneling facilities; marine structures (coast protection, retaining walls, tetrapods); bridges; anchorages; backfilling of pipelines, LegoBlocks, etc. Improved durability, especially waterproofing (Cw0,8 & Cd50 (BNS EN 206:2013 & NA.N)) and protection against the aggressive impact of chlorides, sulphates and salts in a combination with increased frost resistance, abrasion resistance and volume stability make it a perfect material for all these uses.

3.3.3.2. Radioactive concrete

ISF shielding concrete is a building material used for the construction of protection shields. These shields guarantee the work safety of the people inside the structures, subjected to different types of radiation and attacks. This type of concrete is used for: walls, ceilings and floors against ionizing radiation; multi-threat protection from electromagnetic radiation; radiation protection in hospitals, laboratories, oncology clinics and waste treatment facilities; protection against the hazard of nuclear radiation leakage; protection against terrorist attacks and protection against extreme natural disasters. The high density of ISF and its small grain size compact the structure and make it resistant to external attacks.

3.3.3.3. Ultra high performance concrete

Ultra High Performance Concrete (UHPC) is characterized by a very high compressive strength (above 150 MPa) and extreme durability, even when exposed to highly aggressive environmental impacts. All these characteristics make it an alternative to other construction materials (e.g. steel) for the construction of transport and hydraulic facilities, high-rise buildings, as well as for special mine structures, structures for military application and high aesthetic architectural elements and details. ISF is one of the best materials for making UHPC, which brings plenty of benefits in the structures and elements starting from micro- and macro- structural optimization, improving workability and creating long-lasting products. Aurubis has developed UHPC products with high aesthetic vision for outdoor application.

3.3.3.4. De-icing concrete

De-icing concrete is a special mixture containing ingredients, which are electrically conductive. Ordinary concrete does not have this property. The composition of de-icing concrete consists of high quantities of ISF in combination with other geological materials with high iron content and a cocktail of building materials. When de-icing concrete is connected to a power source, its ingredients are heated and prevent ice formation on the concrete surface. This special type of concrete is used on several US bridges and sidewalks and was invented by Prof. Christopher Tuan and assigned by the University of Nebraska with Patent No.: US 6,825,444 B1 [21]. The concrete surface is heated up to +20° Celsius above the ambient temperature with minimum electrical supply. Aurubis applied a sidewalk in its plant in order to prove the electroconductive properties of ISF and successfully melts snow during the winter period.

3.4. Asphalt mixtures
Asphalt mixtures are mixtures that combine different grain size aggregates bound by bitumen. Mineral filler is one of the important aggregates, because it not only fills voids in mixes, but also improves the cohesion of the asphalt binder. The most frequently used type of filler is limestone powder, containing over 90% of $CaCO_3$, together with waste materials from crushing coarse aggregates. ISF, as extremely fine aggregate, is successfully applied in asphalt mixtures, once it is dried. Due to its large specific surface (about 0,92m^2/g), ISF sticks on the bitumen and increases the specific surface of the binder by itself by creating better cohesion with the other coarser aggregates. As a result, the compactness of the mixtures is increased and the void content is reduced, which is important for the lifetime of the asphalt layer and its durability. ISF also improves the Marshall's stability and the Marshall flow test results.

3.5. Ceramic products
Ceramic is one of the most ancient building materials on our planet, dating from earlier than 10 000 years BC. Nowadays, scientists and the industry have been testing a wide range of alternative materials for ceramic production, which improve the properties of the final products on the one hand, reduce the extraction of natural resources on the other and drop energy consumption in total. ISF is a material, which utilization leads to several benefits in different types of ceramic products (bricks, ceramic and clinker tiles & plates, roof tiles, etc.). Aurubis investigations in this direction show that the use of ISF contributes greatly to the ceramic products' properties both is plastic and hardened state. ISF first improves the plasticity of the clay by decreasing the decomposition of the clay, then decreases plasticity during drying and linear shrinkage, followed by increasing the compressive strength of the products and improving their thermal and sound insulation capacity and acid resistance. Moreover, due to the high iron content in ISF, which is threated at high temperatures during copper extraction, the sintering temperature drops and therefore the energy consumption is reduced. The application of ISF in ceramics has great potential.

3.6. Other applications
The other possible applications of ISF are also interesting and can be implemented in large scale if ISF covers the requirements of the final customers in respect to their production peculiarities.

3.6.1. Soil embankments and stabilization
ISF is a product that has been tested *in-situ* in combination with soil for different applications to enhance the physical and mechanical properties of the soil. ISF increases the shear strength and can control the shrinkage and swelling properties. This improves the load bearing capacity of the subgrade of roads and other types of construction works. According to the type of soil and application different percentage of ISF can be applied for stabilization of the soil.

3.6.2. Mineral and glass wool
The incorporation of slags, mainly blast furnace and steel slags, in mineral (or stone) wool production is not a new practice. The goal of their utilization, besides the environmental aspect, is lowering the melting temperature and supporting the flow of the molten rock. ISF can efficiently do the same, as the iron component is much higher compared with the slags from the ferrous industry. Mineral wool

production utilizes about 25% by-products in the total mix and 10% off-cuts from the plants' own production process. Each company has its own requirements, mostly related to chemical composition to keep the flux, solid content to act as a fuel and particle size to avoid interference with the kiln operation. Copper slags are successfully applied in USA for mineral wool production.

Glass wool utilises a high percentage of post-consumer recycled glass as recycled content. Industrial by-products are also recycled to a limited extent, because the slightest variation in feedstock composition can affect the fibre length of the product.

3.6.3. Glass ceramics and glass fibre
Iron-rich phases of copper slag impart magnetic, electrical and thermal properties, as well as a brownish shade to glass-ceramics [22]. Each copper slag, based on its specific chemical composition and percentage of clay substitution, must be examined for its sintering peaks in a range between 750^{0}C and 1100^{0}C. According to some researchers [23, 24] the optimum values for high density and hardness in a combination with low porosity and water absorption were discovered at 950^{0}C with 61,5% copper slag. Other researchers [25] achieved the best properties after sintering at 1025°C for 1 hour. The tiles have a bending strength of 57 MPa, 2wt% water absorption and excellent resistance to mineral acids, suitable for production of chemical resistant floor tiles. ISF addition in glass ceramics leads to an unusual sintering process in low temperature internals between 1150^{0}C and 1200^{0}C. These temperatures provide significant technological advantages such as great reduction of the firing shrinkage, as well as low porosity and high crystallinity, followed by good mechanical characteristics of the final products. Another interesting application of copper, which has been reported by [26] is for glass fiber production.

3.6.4. Rubbers and caoutchoucs
Fillers are the second most important component in rubber and caoutchouc composites by volume used. These additives are commonly used either as diluents, process improvers, aesthetic improvers or as reinforcers. Like fillers in asphalts, here they are used to reduce the polymeric matrix and to reinforce the composite by physical and chemical interactions. The fine grain size, high specific surface area, structure and density of the ISF make its application beneficial. The complex interactions allow material flexibility to be maintained, while enhancing strength and resistance to deformation. ISF with moisture content below 1% has been successfully tested in production of heavy rubbers.

4. Conclusion
The following article gives a wide overview of ISF application in more than 15 different directions. This is the first detailed presentation of this kind of special copper slag produced only in a couple of plants in the world. The article gives summarized information of the several years of investigations made by Aurubis. Material peculiarities, together with absence of EU standards do not discourage Aurubis from finding new directions for its product. On the contrary, these peculiarities make it more attractive, more interesting and, since it is available in Bulgaria, it fulfils the criteria of the EU commission for sustainable utilization of by-products, as prescribed by the Green Deal of construction products producers. In this respect, ISF, with its constant chemical composition and physical properties, can maintain the high quality of products and support producers in making their industries greener.

References
[1] ICSG 2015 Lisbon The World Copper Factbook. International Copper Study Group 64 pp.
[2] Ravindra K. Dhir OBE, Jorge de Brito, Raman Mangabhai, Chao Qun Lye 2017 Sustainable construction materials: Copper slag Elsevier.
[3] Walker R, Pavиa S 2011 Physical properties and reactivity of pozzolans, and their influence on the properties of lime-pozzolan pastes. Materials and Structures 44 (6) 1139–1150.
[4] Sankh A. C, Biradar P. M, Naghathan S. J. and Ishwargol M. B. 2014 Recent trends in replacement of natural sand with different alternatives IOSR J. Mech. Civ. Eng. 59–66.
[5] Velumani M, Nirmalkumar K 2014 Durability and characteristics of copper slag as fine aggregate

and fly ash as cement in concrete, Proceeding of 2nd Int. Conf. Current Trends Eng. Technologies 222–227.
[6] Manolova E, Stoyanova V 2017 Leuven Characteristics of iron-silicate fines as replacement material in cement concrete, Proceedings of the 5th Slag Valorization conference 279-283.
[7] Manolova E, Stoyanova V 2017 Leuven Utilization of iron-silicate fines as replacement material of fine aggregates in cement concrete pavements, Proceedings of the 5th Slag Valorization conference 235-239
[8] https://ec.europa.eu/growth/sectors/raw-materials/eip/commitments/recycling-secondary-raw-materials-sustainable-optimization-construction-processes-civil_en
[9] https://ec.europa.eu/info/strategy/priorities-2019-2024/european-green-deal_en#actions
[10] Tan Y, Zhou Q, Wei R. 2000 The initial study on the use of copper slag as a additional materials for cement production Journal of Xinjiang Institute of Technology 21(3) 236–239.
[11] Shia C, Meyer Ch, Behnood A 2008 Utilization of copper slag in cement and concrete Elsevier
[12] Liu, et al. 2007 Using more tailings and copper slag as cement raw materials, blast furnace slag and phosphorus slag as blending components to produce blended cements Chinese Building Materials No. 5. 98–100.
[13] Al-Jabri KS, Taha R, Al-Hashmi A, Al-Harthy AS 2006 Effect of copper slag and cement bypass dust addition on concrete properties Construction and Building Materials 22 322–31.
[14] Douglas E, Mainwaring PR 1986 Pozzolanic properties of Canadian non-ferrous slag, SP-91-75. Farmington Hills, MI: American Concrete Institute 1525–50.
[15] Tixier R, Devaguptapu R, Mobasher B 1997 Effect of copper slag on the hydration and mechanical properties of cementitious mixtures Cement and Concrete Research 27(10) 1569–1580.
[16] Abdulkareem, O. A. et al. 2014 Effects of Elevated Temperatures on the Thermal Behavior and Mechanical Performance of Fly Ash Geopolymer Paste, Mortar and Lightweight Concrete Construction and Building Materials 50 377–387.
[17] Ariffin, M.A.M. et al. 2013 Sulfuric Acid Resistance of Blended Ash Geopolymer Concrete Construction and Building Materials 43 80–86.
[18] Provis J, Angel L. Gomez M, H. Hussein O, Koma G, Manolova E 2019 Rovijn Production of alkali-activated binders from iron-silicate fines, Proceedings of the SMSS 353-361
[19] Wilkinson A, Eadie R, Woodward D, Magee B, Manolova E 2017 Initial investigation into the effect of iron-silicate fines addition to metakaolin-based geopolymers. ECI.
[20] Biswas S and Satapathy A 2010 Use of copper slag in glass-epoxy composites for improved wear resistance Waste Management and Research 28 (7) 615–625.
[21] Tuan Ch, Nguyen L, Chen B 2012 Patent No.: US 6,825,444 B1 Concrete mix for electromagnetic wave/pulse shielding.
[22] Chinnam R, Francis A, Will J, Bernardo E and Boccaccini A 2013 Review: functional glasses and glass-ceramics derived from iron waste and combination of industrial residues Journal of Non-Crystalline Solids 365 63–74.
[23] Yang Z, Lin Q, Xia J, He Y, Liao G and Ke Y 2013 Preparation and crystallization of glass ceramics derived from iron-rich copper slag Journal of Alloys and Compounds 574 354–360.
[24] Mohamed E, Shahsavari P, Eftekhari-Yekta B and Margussian V K 2015 Preparation and characterization of glass ceramic foams produced from copper slag Transactions of the Indian Ceramic Society 74 (1) 1–5.
[25] Marghussian, V. K, Maghsoodipoor A 1999 Fabrication of unglazed floor tiles containing Iranian copper slags Ceramics International 25 617-622.
[26] Chung C. H, Mizuno T, Mackenzie J. D 1980 Iron recovery and glass fiber production from copper slag Mines and Mineral Resources 123-128.

IOP Conf. Series: Materials Science and Engineering **951** (2020) 012006 doi:10.1088/1757-899X/951/1/012006

Alkali-activated geopolymers based on iron-rich slag from copper industry

A Nikolov

Institute of Mineralogy and Crystallography, Bulgarian Academy of Science (IMC-BAS, Acad. G. Bonchev Str., bl.107, 1113 Sofia, Bulgaria

E-mail: drsashko@imc.bas.bg

Abstract. In the present study geopolymers were prepared using fayalite slag (iron-silicate fines) from copper producing plant (Aurubis Bulgaria AD). The influence of the different concentrations of alkali activator on properties and structure was examined. The prepared geopolymers were characterized by maximal compressive strength of about 27 MPa. The microstructural examination revealed that only small amount of the precursor reacted, but certain amount of ferric iron (Fe^{3+}) participate in the structure of newly formed geopolymer gel.

1. Introduction

Geopolymers, also referred as inorganic polymers or alkali activated materials, are alumino-silicate materials which possess superior physical and chemical properties and a wide range of potential applications. The synthesis of geopolymers involves reactions between powder precursor and concentrated solution yielding at ambient temperature tridimensional polymer structure [1]. There are two routes of synthesis: in alkaline media using alkali hydroxides and silicates; and second: in acid media – using phosphoric acid [2]. The raw material is homogenized with activator solution to initiate the geopolymerization process. Based on existing model for aluminosilicate weathering in aggressive media, the model of geopolymerization includes dissolution, reorientation and reprecipitation stages including an autocatalytic polycondensation step [3]. Initially kaolinite (and calcined version – metakaolin [2]) were used as geopolymer precursors but last decades wide range of natural [4] and industrial waste materials were intensively studies. Most of the studies are oriented to fly ash [5], [6] and blast furnace slag [7]. Both are utilized in concrete industry as mineral additive to Portland cement according EN 206-1. On other hand other important wastes, in terms of volume, are coal mine and floatation tailings, red mud from the production of alumina and several types of iron-rich slags from metallurgical plants [8]. Usually those industrial waste contain heavy metals as Cu, Zn, Cr, Pb, Mo, Cd, etc [9], thus are considered toxic and disposal without any proper management can cause serious environmental issues related to contamination of soil and water. Geopolymer technology has already been patented and used to immobilize and stabilize toxic or wastes materials [10]. One of the potential raw material is iron-rich slag from copper industry [11]. The role of iron in the geopolymerization process is not well studied and the results are often contradictory. Some of the authors report that iron could be part of the polymer structure partly replacing the aluminum in poly ferro-sialate framework [12].

The aim of the present study is to examine the influence of different concentration of the activator solution to the mechanical properties and to characterized prepared geopolymers based on fayalite slag (iron-silicate fines) from copper producing plant (Aurubis Bulgaria AD). The fayalite slag is hardly marketable product due to its specific composition, fineness and presence of heavy metals. Huge amount remains unutilized and disposed in landfills in the territory of Bulgaria.

2. Materials and methods

2.1. Materials

The geopolymer precursor in the present study was copper by-product - fayalite slag from Aurubis (Pirdop, Bulgaria) with the following chemical composition (%): Fe_2O_3 – 58.42; SiO_2 – 29.34; Al_2O_3 – 4.40; CaO – 2.66; MgO – 0.89; K_2O – 0.71; Na_2O – 0.58; CuO – 0.49; PbO – 0.37; TiO_2 – 0.30; MoO_3 – 0.26, measured by XRF. The mineral composition of the slag was presented by fayalite, magnetite, minor quantity pyroxene, and absence of significant amorphous phase [13].
The chemical reagents used for the preparation of the activator solutions are: sodium silicate (water glass) derived from local industry with molar ratio of SiO_2/Na_2O=2.98 and density respectively – 1.49 g/cm^3; potassium hydroxide pellets and tap water.

2.2. Methods of analysis

The morphology and the chemical analysis of the prepared geopolymer was studied by scanning electron microscopy JEOL 6390 with energy dispersive spectrometer (SEM-EDS) . Thin section of samples was prepared by polishing and covered with gold. The Mössbauer spectra were obtained at room temperature by a Wissel electromechanical spectrometer working in a constant acceleration mode. A 57 Co/Rh (activity @ 10 mCi) source and α-Fe standard were used. FTIR spectra were collected using a Tensor 37 spectrometer (Bruker) with a 4 cm^{-1} resolution after averaging 72 scans on KBr pallets in the spectral region 400-4000 cm^{-1} at room temperature.

2.3. Specimens preparation

The geopolymer mixtures were prepared by mixing oven dried fayalite slag with tempered hardener solution. The composition design is presented at Table 1. Series 15F1 and 15F2 share same Fe_2O_3/M_2O and Al_2O_3/M_2O ratio but former contains twice more sodium silicate solution. Series 15F5 was prepared only with potassium hydroxide solution. Each series have water to solid ration equal to 0.15 (w/w). The fresh geopolymer mixtures were homogenized with mechanical stirrer for 90 seconds and then poured into cylindrical moulds (50x30mm). The specimens were sealed in polyethylene and cured at laboratory conditions (23 °C, w=65%) for 14 days, then 7 days in 80 °C. On 21^{-th} day samples were demoulded and placed in laboratory conditions.

Table 1. Composition design of the prepared geopolymers based on fayalite slag

Series	Activator solution per 100 g fayalite, (g)			Total molar ratios		
	Water glass	KOH	water	H_2O/M_2O	Fe_2O_3/M_2O	Al_2O_3/M_2O
15F1	10	10.2	8.7	7.59	2.50	0.32
15F2	20	8.5	2.7	7.64	2.50	0.32
15F3	10	5	8.8	11.40	4.00	0.50
15F4	10	16.3	8.5	5.46	1.75	0.22
15F5	-	17.9	14.4	5.36	1.75	0.22

3. Results and discussion

3.1. Compressive strength

The compressive strength was measured at 28^{-th} day and the results (with relative deviation) are presented in figure 1. The samples 15F2 and 15F3 showed the highest compressive strength of 27.96 MPa and 27.53 MPa, respectively. The compressive strength increased with the increase of Fe_2O_3/M_2O ratio. Increased alkalinity of the activator solution induces more dissolution and hinder the polycondensation step which result in lower compressive strength. The results from previous study where geopolymers with natural zeolite were obtained, also precursor with low reactivity, it was observed that the increase of alkalinity lead to slower strength gain rate - the 28^{-th} day compressive strength was lower, but 180^{-th} higher [2]. The samples with higher concentration alkali in the activator showed bleeding during curing and were characterized by very low strength ~3 MPa.

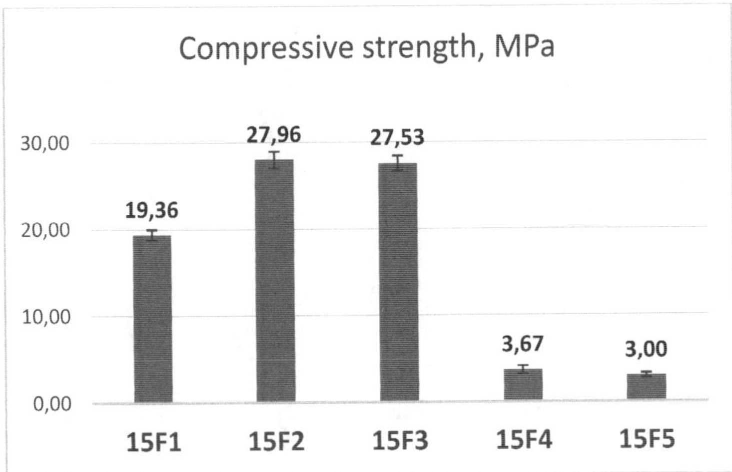

Figure 1. Compressive strength of the prepared geopolymer series based on fayalite slag

However, sample 15F3 is of a greater practical and economical interest because it was prepared with less water glass and potassium hydroxide. Compared to other geopolymers based on similar waste materials the compressive strength was about 30 MPa for geopolymers based on Romanian lead slag, and more than twice less than geopolymer based on Greek ferronickel slag (56 MPa) [14], or 67 MPa – Polish ferronickel slag [15]. It should be noted that mentioned ferronickel slags contain significant amount amorphous phase and reactive minerals –like cristobalite, tridymite, which are not presented in the slag used in the present study.

3.2. Mössbauer spectroscopy

The Mössbauer spectrums are presented in Figure 2. The parameters of hyperfine interaction such as isomer shift (δ), quadrupole splitting (ΔEq), effective internal magnetic field (B), line widths (Γexp), and relative weight (G) of the partial components in the spectra were determined (Table 2). The Mössbauer spectrum of RAW sample is composed of sextet and doublet components. The presented model comprises of 3 sextets and 3 doublets is used for spectrum fitting, as results for calculated component parameters. The parameters of the sextet components correspond to the mineral magnetite: Sx1 - tetrahedrally coordinated Fe^{3+}; Sx2 - octahedrally coordinated $Fe^{2.5+}$ (actually Fe^{3+} and Fe^{2+}, but due to fast electron exchange between them, the spectral effect is one sextet component), Sx3 - also octahedrally coordinated $Fe^{2.5+}$ ions, which have non-iron cationic neighbors in the spinel structure (e.g., Mg, Mn, Cu, Al, etc.). The calculated parameters of the doublet components (Db1 and Db2) correspond to the two different positions of the Fe^{2+} ions in the structure of the mineral fayalite (Fe_2SiO_4). The doublet component Db3 cannot be related to the iron position in the crystalline phases of magnetite nor fayalite. Isomer shift values above 1.00 mm/s are typical for iron in second oxidation state. In the absence of a third crystalline phase in the sample (powder XRD data [13]), it can be assumed that Db3 is due to Fe^{2+} ions in an amorphous phase.

The spectrum of sample 15F3 was identical with the raw material. Not much changes in the oxidation states or coordination of the iron were observed after geopolymerization, which means that both magnetite and fayalite stayed predominantly inert. Still, part of the doublet component Db3 related to amorphous phase decreased and new weak doublet named Db4 was observed. Calculated parameters of Db4 correspond to Fe^{3+} ions in amorphous state. Similar doublet was reported by Pyes et. al. after geopolymerization synthetic CaO-FeOx-SiO_2 slag. The observed doubled was significantly more distinct which could be explained by the superior reactivity of the laboratory prepared slag contained 97% amorphous phase [16]. It is believed that tetrahedral Fe^{3+} shows a quadrupole doublet of room temperature Mössbauer spectrum with its high-energy line at ~0.4 mm/s [17]. On the contrary, Mysen stated that isomer shifts below 0.3 mm/s is typical for tetrahedral Fe^{3+}, whereas the isomer shifts of octahedral Fe^{3+} occur above 0.4 mm/s. However, it could be stated that minor quantity of the iron during geopolymerization oxidizes to amorphous Fe^{3+}.

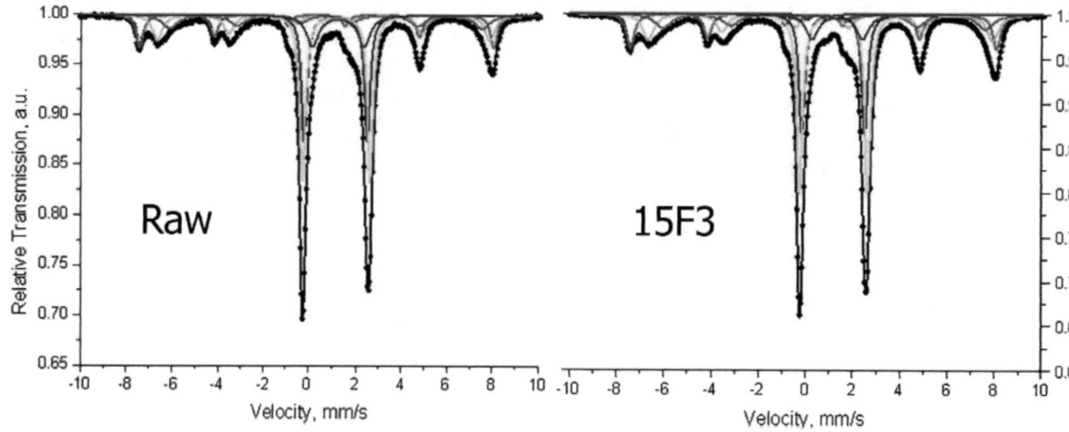

Figure 2. Mössbauer spectrums of geopolymer precursor (RAW) and prepared geopolymer (15F3)

Table 2. Mössbauer parameters of investigated samples: fayalite slag (RAW) and geopolymer (15F3)

Sample	Components	δ, mm/s	ΔE_q , mm/s	B, T	Γ_{exp}, mm/s	G, %
RAW	Sx1-Fe_3O_4, Fe^{3+}_{tetra}	0.3	0	48	0.36	13
	Sx2-Fe_3O_4, $Fe^{2.5+}_{octa}$	0.62	-0.05	45.1	0.54	15
	Sx3-Fe_3O_4, $Fe^{2.5+}_{octa}$	0.72	-0.05	42.3	0.78	13
	Db1-Fe_2SiO_4, Fe^{2+} - M1	1.14	2.68	-	0.3	21
	Db1-Fe_2SiO_4, Fe^{2+} - M2	1.17	2.88	-	0.3	27
	Db3- Fe^{2+}	1.24	2.19	-	0.65	11
15F3	Sx1-Fe_3O_4, Fe^{3+}_{tetra}	0.31	0	48.1	0.4	14
	Sx2-Fe_3O_4, $Fe^{2.5+}_{octa}$	0.62	-0.04	45.3	0.58	17
	Sx3-Fe_3O_4, $Fe^{2.5+}_{octa}$	0.74	0	42.4	0.83	13
	Db1-Fe_2SiO_4, Fe^{2+} - M1	1.15	2.67	-	0.3	19
	Db2-Fe_2SiO_4, Fe^{2+} - M2	1.17	2.91	-	0.3	28
	Db3- Fe^{2+}	1.31	2.16	-	0.64	8
	Db4- Fe^{3+}	0.45	0.83	-	0.3	1

3.3. FTIR

Figure 3 presents the FTIR spectra of the raw material and geopolymer series 15F3. The spectra of the slag (Raw) correspond to the in-depth study of another batch of the present slag [18]. Most of the observed bands are correlated with fayalite. The bands at 873 cm^{-1} and 827 cm^{-1} arises due to asymmetric stretching vibration of the υ3 mode of Si-O in olivine group. The band at 827 cm^{-1} was due symmetric stretching (υ1), while features between 460-610 cm^{-1} can be attributed to splitting of the degenerate υ4 asymmetric bending vibration. The peaks at 639 cm^{-1} and 1047 cm^{-1} was assigned to magnetite and pyroxene which showed overlapping peaks in this range [19].

The features of mineral phases presented in the precursor was still observed after geopolymerization. A significant peak appeared at about 1010 cm^{-1} associated with asymmetric stretching vibration T-O-Si (T – Si or Al). A strong band around 1000 cm^{-1} is a fingerprint of the inorganic polymer matrix [20]. Usually, this band is used for determination of the degree of geopolymerization. The band about 1642 cm^{-1} and 3431 cm^{-1} are bending (H-O-H) and stretching vibration (-OH) of water molecules, respectively. The band about 1456 cm-1 is related to C-O carbonate stretching vibration – a product of the efflorescence and contact of unreacted activation solution with atmospheric CO_2.

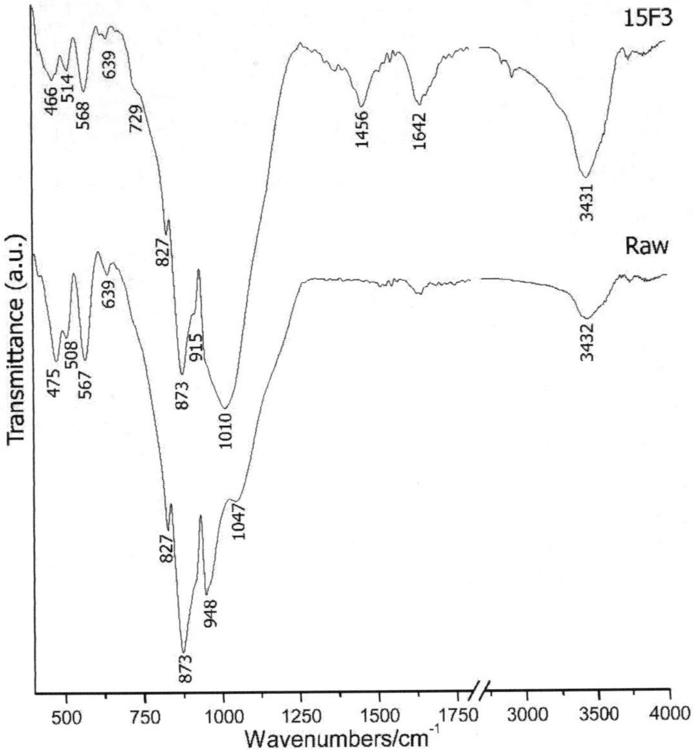

Figure 3. FTIR spectrums of raw fayalite (Raw) and the prepared geopolymers (15F3)

3.4. SEM-EDS

The SEM images are presented in Figure 4. Individual fayalite and magnetite particles were observed on the cross section of geopolymer sample 15F3 (Figure 4 – a and b). The spectroscopic data above showed that those mineral phases did not react during geopolymerization. Thus, the non-reacted particles act as fillers bonded together by geopolymer gel. It was visible that the binder phase contains significant amount of capillary pores. The solid to binder ratio is relatively low (0.15 w/w), so probably the porous structure was due to insufficient binder phase to fill the volume between non-reacted particles. On higher magnification Figure 4-c) it was observed a particle embedded in geopolymer gel. The elemental analysis (EDS) in point 2 showed that the chemical composition corresponds to fayalite. The surrounding geopolymer binder phase was characterized by higher potassium, sodium and aluminum content, which is typical for geopolymer gel. Also, it contains iron. Based on elemental content we could calculate the molar ratio: (Na+K+½Ca) / (Fe+Al) = 0.95. Theoretically the cations (Na^+, K^+, Ca^{2+}) should be equal to Al^{3+} and Fe^{3+} to balance their negative charge in the tetrahedral coordination [4]. The presence of Fe in the binder (~17% w.) and (Na+K+½Ca) / (Fe+Al) equal to about 1 could be a sign that part of the Fe is structurally integrated in the geopolymer binder.

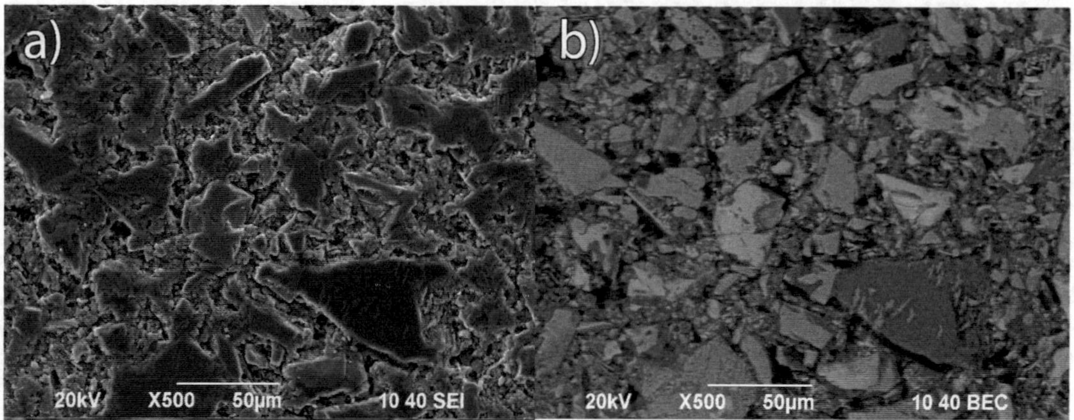

EDS analysis of image c) by elements

Elements/ wt%	Point 1	Point 2
O	41.05	31.91
Na	1.6	-
Mg	-	-
Al	3.48	-
Si	22.56	16.61
K	12.66	0.46
Ca	1.52	1.43
Fe	17.01	48.59

Figure 4. SEM images of geopolymer (15F3): a), c) - secondary electrons SEI; b) - backscattered electrons BEC, results from EDS analysis

4. Conclusion

The results from present study showed that the fayalite slag (iron-silicate fines) from copper producing plant (Aurubis Bulgaria AD) is low reactive geopolymer precursors. The first step of geopolymerization – dissolution could not provide sufficient binder phase to produce dense matrix. The concentration of the activating solution had great influence on properties of the prepared geopolymers. The optimal strength was obtained by preparing geopolymer with H_2O/M_2O=11.40, Fe_2O_3/M_2O=4; Al_2O_3/M_2O=0.50. With the increase of the alkali concentration - the compressive strength decrease, also bleeding of activator solution was observed on higher alkali dosages. The microstructural examination revealed that the main phases fayalite, magnetite and pyroxene stayed unreacted after geopolymerization. However, part of the amorphous phase presented in small amount react during geopolymerization and certain amount of ferric iron Fe^{3+} was incorporated in the structure of the geopolymer binder phase.

Acknowledgments

The results in this work has been achieved in fulfillment of a project financed by the National Science Fund of Bulgaria under contract No. DM17/3 from 12.12.2017. The publication of the present article was financed the K2-2020 of IMC-BAS.

References

[1] Duxson P., A. Fernández-Jiménez, J. L. Provis, G. C. Lukey, A. Palomo, and J. S. J. van Deventer (2007) Geopolymer technology: the current state of the art, J. Mater. Sci., vol. 42, no. 9, 2917–2933.
[2] Davidovits, J. (2008) Geopolymer chemistry and applications. Institut Géopolymère, 14
[3] Provis, J. L. (2006) Modelling the formation of geopolymers, PhD Thesis, University of Melbourne.
[4] Nikolov A., I. Rostovsky (2017) Sodium-silicate geopolymers based on natural zeolite – clinoptilolite, Comptes Rendus L'Academie Bulg. des Sci., vol. 70, no. 12., 1655-1662
[5] Gollakota, A., V. Volli, and C.-M. Shu (2019) Progressive utilisation prospects of coal fly ash: A review, Science of the Total Environment, vol. 672, 951–989.
[6] Nikolov A., B. Barbov (2018) Lightweight geopolymer based fly ash, Review of the Bulgarian Geological Society, volume 79, part 3, Geosciences, ISSN 0007-3938, 23-24.
[7] Davidovits, J. (2011) Application of Ca-based geopolymer with blast furnace slag, a review, in 2nd International Slag Valorisation Symposium, 33–49.
[8] Komnitsas, K., D. Zaharaki (2007) Geopolymerisation: A review and prospects for the minerals industry, Minerals Engineering, vol. 20, no. 14, 1261–1277.
[9] Nikolov, A. (2020) Alkali and acid activated geopolymers based on iron-silicate fines - by-product from copper industry, International scientific journal "Machines, Technologies, Materials", Year XIV, Issue 1, p. 37-39.
[10] Davidovits, J. (1994) Method for obtaining a geopolymeric binder allowing to stabilize, solidify and consolidate toxic or waste materials, Google Patents.
[11] Nikolov, A. (2019) Geopolymers based on Bulgarian raw materials – preliminary studies, International scientific journal "Machines, Technologies, Materials", vol. XIII, no. 2, 197–199.
[12] J. Davidovits and R. Davidovits, (2020), Ferro-sialate geopolymers, Technical papers # 27, Geopolymer Institute Library, www.geopolymer.org. DOI:10.13140/RG.2.2.25792.89608/2
[13] Nikolov, A., R. Titorenkova, N. Velinov, and Z. Delcheva (2018) Characterization of a novel geopolymer based on acid-activated fayalite slag from local copper industry, Bulgarian Chemical Communications, vol. 50, no. F, 54–61.
[14] Komnitsas, K., Zaharaki, D. and Perdikatsis, V. (2009) Effect of synthesis parameters on the compressive strength of low-calcium ferronickel slag inorganic polymers, Journal of Hazardous Materials, 161(2-3), 760-768.
[15] Komnitsas, K., G. Bartzas, V. Karmali, E. Petrakis, W. Kurylak, G. Pietek, J. Kanasiewicz (2019) Assessment of alkali activation potential of a Polish ferronickel slag, Sustainability, 11(7), 1863.
[16] Peys, A., A. P. Douvalis, V. Hallet, H. Rahier, B. Blanpain, and Y. Pontikes (2019) Inorganic polymers from CaO-FeOx-SiO2 slag: the start of oxidation of Fe and the formation of a mixed valence binder, Frontiers in Materials., vol. 6, 212.
[17] Rancourt D., M. Z. Dang, and A. E. Lalonde (1992) Mössbauer spectroscopy of tetrahedral Fe3+ in trioctahedral micas, American Mineralogist., vol. 77, no. 1–2, 34–43.
[18] Mihailova I.,D. Mehandjiev (2010) Characterization of fayalite from copper slags, J. Univ. Journal of Chemical Technology and Metallurgy, vol. 45, no. 3, 317–326.
[19] Pisciella P., M. Pelino (2005) FTIR spectroscopy investigation of the crystallisation process in an iron rich glass, Journal of the European Ceramic Society 25.11: 1855-1861. https://doi.org/10.1016/j.jeurceramsoc.2004.06.012
[20] Zaharaki D., K. Komnitsas, and V. Perdikatsis (2010) Use of analytical techniques for identification of inorganic polymer gel composition, Journal of Materials Science., vol. 45, no. 10, 2715–2724, doi: 10.1007/s10853-010-4257-2.

Influence of sulphate ions on the crystal chemistry of white cement mortars with a high content of marble filler powder

V A Stoyanov[1,2]*, V Petkova[3], B Kostova[4]

[1] Department of Technology and Construction Management, Faculty of Construction, USEA (VSU) "L. Karavelov", 175 Suhodolska St., 1373 Sofia, Bulgaria;
[2] Academy of the Ministry of Interior, Department of Safety and Prevention, 17 Pirotska Str., 1309 Sofia, Bulgaria
[3] Institute of Mineralogy and Crystallography "Acad. Ivan Kostov", Bulgarian Academy of Sciences, Acad. Georgi Bonchev Str., Block 107, 1113, Sofia, Bulgaria;
[4] Department of Natural Sciences, New Bulgarian University, 21 Montevideo St., 1618 Sofia, Bulgaria

* E-mail: vensy.stoyanov@vsu.bg

Abstract. Objects of this research are white cement mortars, characterized by both of high content of addition of marble powder and reduced water-cement ratio. The hardened cement mortars, formed after 28-and 120-day curing under standard conditions, are studied. The newly formed phases, containing $[SO_4]^{2-}$, $[CO_3]^{2-}$, $[OH]^-$, etc., are identified using X-ray powder diffraction, infrared spectroscopy and thermal analysis (coupled with analysis of the outgoing gas mixture by mass spectrometry). Based on of the formed calcium silicate hydrates, calcium aluminate hydrates, the influence of sulphate ions is analysed and the mechanism of the thermal decomposition reaction at high temperatures in an oxidizing gas environment is studied. This allows us to establish that the hydration of Portland cement depends on the addition of marble powder (technically calcium carbonate $CaCO_3$), as well as mono- and hemicarboaluminates are formed instead of monosulphoaluminates.

1. Introduction

Modern building materials have a low footprint on the environment and lower energy costs for their production [1, 2, 3]. One of the main directions for their production is the replacement of some of their components with recycled materials, as this effect is most significant for the expensive and polluting production of cement, mortar and concrete [4, 5]. Limestone quarry waste is often used as inert additives and aggregates for the production of cement-based mortars, but this component increases the content of sulphate and carbonate ions in cement mortars. Knowledge of the microstructure and crystal chemistry of such mortars is important to achieve good performance indicators and durability [6].

The aim of this work was to study the amounts of new formed solid phases with $[SO_4]^{2-}$, $[CO_3]^{2-}$, $[OH]^-$, formed at different curing times of mortars based on white Portland cement. Mortars with a high content of inert mineral aggregate (marble powder) and a low water-to-cement ratio were studied. The phase composition was examined by powder X-ray diffraction (PXRD), Fourier transform infrared spectroscopy (FTIR) and thermal analysis (TG/DTG-DSC).

2. Materials and methods

2.1. Components

The samples were prepared with white Portland cement CEM I 52.5 N (produced by Devnya Cement; Bulgaria), aggregate - marble powder type (produced by White Marble Products AIAS SA; Greece), polycarboxylate-based high range water reducer (HRWR - Sika ViscoCrete 5-800) and distilled water. The chemical admixture was chloride-free, soluble in water, without any retarding effects and with a density of 1.07 g/cm^3 (at 20°C).

The chemical composition of white Portland cement was (in wt%): SiO_2 – 24.3; Al_2O_3 – 2.1; Fe_2O_3 – 0.2; CaO – 68.3; MgO – 0.3; Na_2O – 0.13; K_2O – 0.02; Free lime – 1.9. Thus, the mineral composition, calculated by Bogue method was (in wt%): C_3S – 72.13; C_2S – 15.28; C_3A – 5.23; C_4AF – 0.61. The chemical composition of marble powder (in wt. %) was: CO_2 + H_2O – 45.7; SiO_2 – 0.12; Al_2O_3 – 0.38; Fe_2O_3 – 0.14; CaO – 32.9; MgO – 20.0; Na_2O – 0.05; K_2O – 0.19; MnO – 0.01. The decarbonization at 1000°C of this aggregate was divided to two stages (DTA peaks – 806.6°C and 821.2°C) with total mass loses of about 45% and, thus the mineral composition is defined as dolomite and Mg-rich calcite. The polydispersity of the aggregate was determined by the following parameters: the maximal size of grains – 2 mm; grains with sizes < 0.125 mm – 50.0 wt % and grains with sizes < 0.063 mm – 35.0 wt %.

2.2. Samples preparation

The sample composition was characterized by the following ratios:

- cement-to-aggregate ratio – 1:2 (in mass parts);
- water-to-cement ratio – 0.40;
- dosage of HRWR – 2.0% by the weight of white cement.

Based on these data the water-to-fines mass ratio was 0.235, where fines were all particles with sizes below 125 μm. After the procedure of mixing, based on EN 196-1:2016, the samples were moulded without any compacting treatment in 6 prisms (two prisms for each sample), 40×40×160 mm, for 1 day in a moist atmosphere (> 95% RH and 20°C) [7]. After the samples were stored under water (20°C) until strength testing (1, 3, 7, 14, 28 and 120 days). The names of the samples were based on the ageing time of the samples – C001, C003, C007, C014, C028 and C120, respectively.

2.3. Experimental Methods

The density and adsorption after immersion were measured according to ASTM C642-13 [8]. Due to the different densities of the samples, the values of adsorption were adjusted to comparable values. The compressive strengths at 28 and 120 days of water curing samples were measured according to EN 196-1:2016 [7].

To measure the porosity of broken parts samples with a mass of 2.0±0.3 mg were analysed by mercury intrusion porosimetry method using Carlo Erba, Porosimeter Mod. 1520 (pressure range 1-150 atm corresponding to pore size range 50-15000 nm. The PXRD patterns were performed on X-Ray powder diffractometer D2 Phaser BrukerAXS, CuKα radiation (λ = 0.15418 nm) (operating at 30 kV, 10 mA) from 5 to 80 °2θ with a step of 0.05° (1.0±0.1 mg grinded sample; particle sizes below 0.075 mm). FTIR measurements were performed by FT-IR Spectrometer, Varian 660-IR, Austria, 2009, covering the range of 400-4000 cm^{-1}, KBr tabletting. TG/DTG-DSC analyses were carried out on a Setsys Evolution 2400, SETARAM, France, combined with an OmniStar mass-spectrometer (heating rate of 10°C.min^{-1}). The operational characteristics of the TG/DTG–DSC- system were: sample mass of 18.0±2.0 mg; ceramic sample crucible and static air atmosphere.

3. Results and Analysis

3.1. PXRD phase analysis

Based on PXRD phase analysis several phases were identified (table 1). They can be divided into the following two groups of minerals:

- relict minerals from the raw composition: belite, albite, anorthite, Mg-rich calcite, quartz, dolomite;
- newly formed minerals: portlandite, ettringite, mono- and hemicarboaluminate, scolecite, hillebrandite, nonotlite, spurrite, artinite and $Ca(HSO_4)_2$.

Table 1. Results from PXRD analysis.

No	Description	Sample	Identified phases
1	Non-hydrated relict cement phases	C001, C003, C007, C014, C028, C120	Belite (C_2S), 49-1673 – $2CaO.SiO_2$
			Alite (C_3S), 11-0593 – $(Na,Ca)Al(Si,Al)_3O_8$
			Anorthite (CAS_2), 41-1486 – $CaO.Al_2O_3.2SiO_2$
2	Non-hydrated phases relict of aggregate phases	C001, C003, C007, C014, C028, C120	Dolomite, 36-0426 – $CaCO_3.MgCO_3$ Mg-rich Calcite, 47-1743 – $CaCO_3$
3	New formed C-S-H gel phases	C001, C003, C007, C014, C028, C120	Portlandite (CH), 44-1481 – $Ca(OH)_2$
3.1	containing OH^- and CO_3^{2-}	C120	Monocarboaluminate, 41-0219 – $Ca_4Al_2(OH)_{12}(CO_3).5H_2O$
		C028, C120	Hemicarboaluminate, 41-0221 – $Ca_4Al_2(OH)_{12}(OH)(CO_3)_{0.5}.4H_2O$
		C003, C007, C014, C028, C120	Artinite, 72-1320 – $Mg_2(CO_3)(OH)_2.3H_2O$
3.2	containing OH^- and SO_4^{2-}	C001, C003, C007, C014, C028, C120	Ettringite, 41-1451 – $6Ca(OH)_2.Al_2(SO_4)_3.26H_2O$ Gypsum, 21-0167 – $CaSO_4.2H_2O$
		C014, C028, C120	$Ca(HSO_4)_2$, 85-1271
3.3	hydrosilicates formed from main oxides: CaO, Al_2O_3, SiO_2	C001, C003, C007, C014, C028, C120	Hillebrandite, 29-0373, 42-0538 – $CaSiO_3.Ca(OH)_2$
		C120	Scolecite, 41-1355 – $CaO.Al_2O_3.3SiO_2.3H_2O$ $CaAl_2Si_3O_{10}.3H_2O$
		C028, C120	Xonotlite, 23-0125 – $5CaO.6SiO_2.Ca(OH)_2$ $Ca_5Si_6O_{17}.Ca(OH)_2$
4	Water-free non-hydrated newly formed phases	C007, C014, C028, C120	Spurrite, 13-0496 – $Ca_5(SiO_4)_2CO_3$

3.2. FTIR Spectroscopy

The spectra, obtained by FTIR spectroscopy demonstrate a wide variety of streaks and are evidence of the formation of multiple hydrate phases in the samples (figure 1). This occurs in three areas – 1750-1240 cm^{-1}, 1300-900 cm^{-1} and 800-400 cm^{-1} in which the overlap of the lines is strongly expressed. At these intervals, the vibrational fluctuations of the functional groups, which are essential for the identification of the cement hydration products, are displayed.

The analysis of the results obtained from the FTIR spectroscopy studies proves the vibration bands of the minerals formed by the hydration of the cement minerals. In the analysis, several groups of new-formed phases can be identified:

- phases, containing chemically bonded water $[OH]^-$;
- phases, containing carbonate ions $[CO_3]^{2-}$, structurally bonded water $[OH]^-$ and chemically bonded water (H_2O);
- phases, containing sulphate $[SO_4]^{2-}$ ions, structurally bonded water $[OH]^-$ and chemically bonded water (H_2O).

IOP Conf. Series: Materials Science and Engineering **951** (2020) 012007 doi:10.1088/1757-899X/951/1/012007

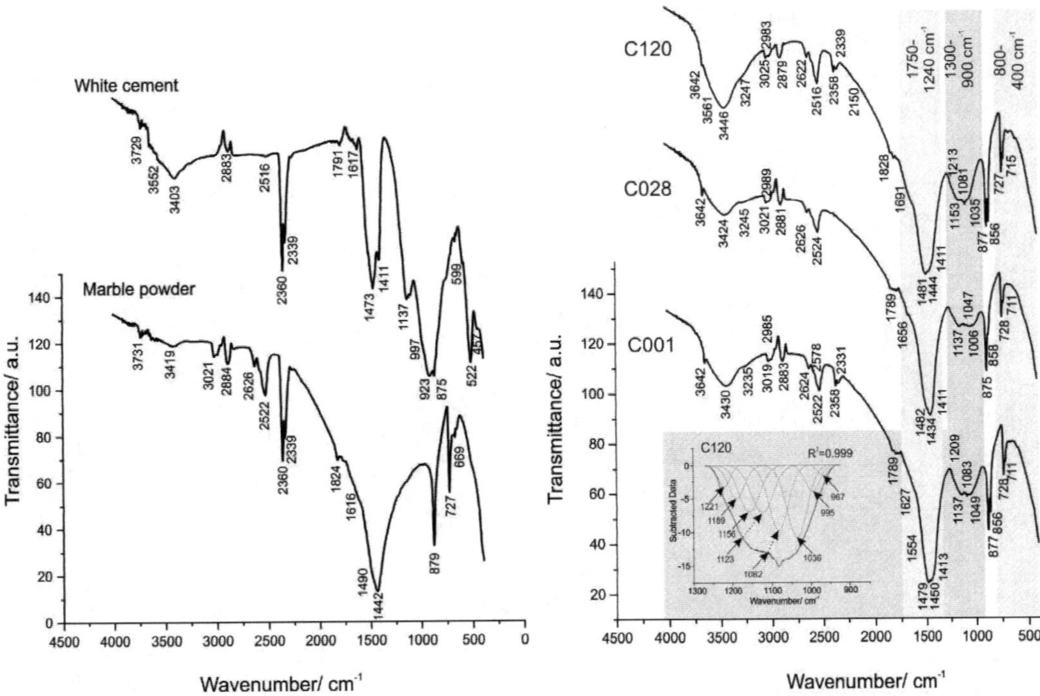

Figure 1. FTIR spectra of raw marble powder, white cement (left), C001, C028 and C120 (right). *Insertion:* peak decomposition via Gauss function within the range 900-1300 cm^{-1} of sample C120.

The analysis of the results obtained from the FTIR spectroscopy studies proves the vibration bands of the minerals formed by the hydration of the cement minerals. In the analysis, several groups of new-formed phases can be identified:

- phases, containing chemically bonded water $[OH]^-$;
- phases, containing carbonate ions $[CO_3]^{2-}$, structurally bonded water $[OH]^-$ and chemically bonded water (H_2O);
- phases, containing sulphate $[SO_4]^{2-}$ ions, structurally bonded water $[OH]^-$ and chemically bonded water (H_2O).

The components, used for the production of cement-based mortars, produce closely related new-formed hydrate phases, some of which are the same for all samples – portlandite, ettringite and hillebrandite. Mono/hemicarboaluminate and nonotlite were formed after 28 days, Scolecite - after 120 days of curing.

The results of the PXRD analysis and FTIR spectroscopy justify the chemical reactions during hydration of samples to be represented by the reactions forming the hydrate silicate phases, Mg-hydrogen carbonate and Ca-hydrogen sulphate phase, and water-free non-hydrated phases as spurrite. Compared to conventional mortars, prepared with quartz sand as fine aggregate, the use of marble powder increases the content of $[CO_3]^{2-}$-ions in cement mortars, leading to an isomorphic substitution of ettringite during hydration. These reactions can be described by the following equations:

$$3Ca^{2+} + HSiO_3^- + CO_3^{2-} + SO_4^{2-} + 5OH^- + 12H_2O \rightarrow 3Ca(OH)_2.Si(SO_4.CO_3).12H_2O \quad (1)$$

$$6Ca^{2+} + 2AlO_2^- + 3CO_3^{2-} + 3SO_4^{2-} + 6OH^- + 26H_2O \rightarrow 3Ca(OH)_2.3CaCO_3.Al_2(SO_4)_3.26H_2O \quad (2)$$

3.3. Thermal analysis

The data from the thermal analysis carried out on the solid components (white Portland cement and marble powder) is shown in figure 2, and the TG-DTG-DSC results for the tested cement compositions in figure 3. Some of the results are supplemented by the analysis of gas phases (figure 4). Processing of these data yields to obtaining both the thermal decomposition temperatures (it is assumed that this is the temperature in the inflexion point T_{infl}) and mass losses (ML), which are then arranged in tabular form (table 2).

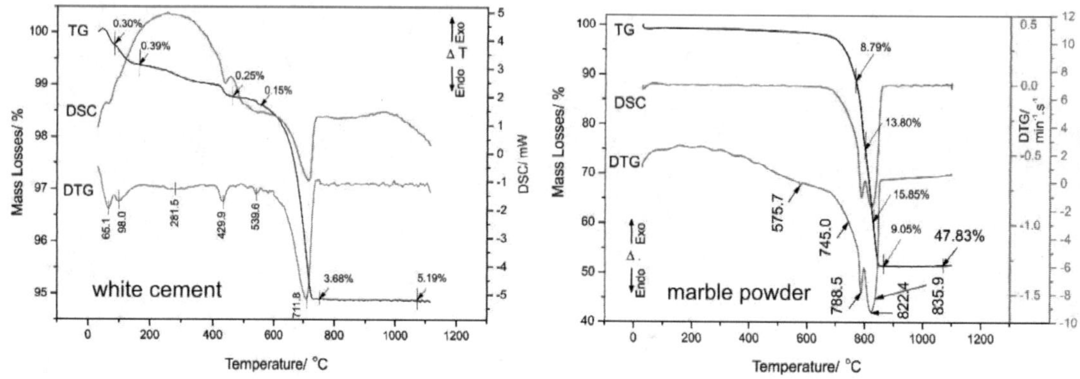

Figure 2. TG-DTG-DSC- curves of raw white cement (left) and raw marble powder (right).

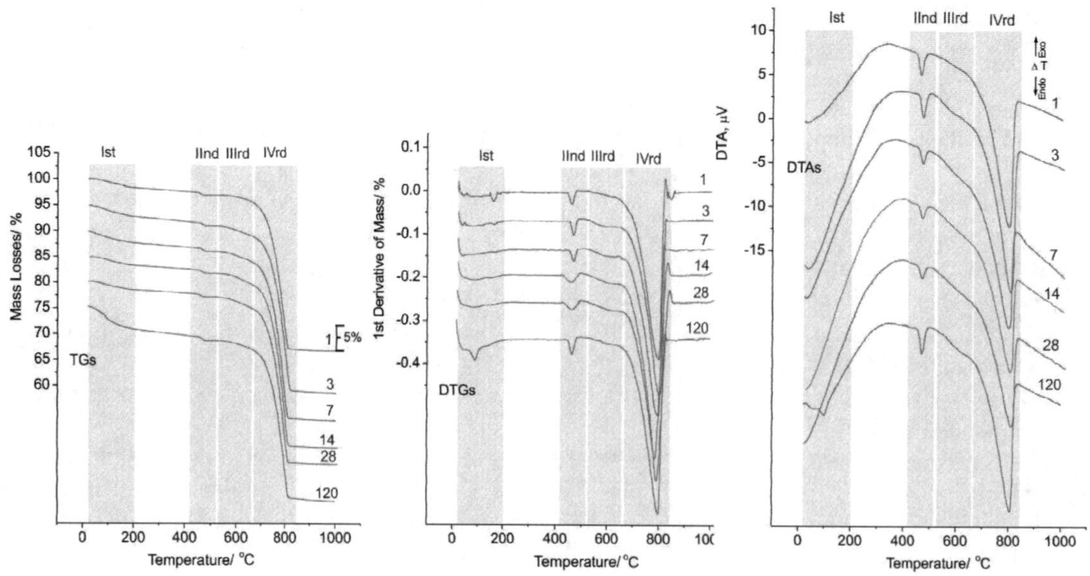

Figure 3. TG, DTA and DTG curves of samples C001-C120.

The mass losses are divided into four ranges: 1st range - 1.00-1.53%, 2nd range – 0.46-0.70%, 3rd – range 2.60-4.36% and 4th range – 3.30-30.10%. Such a distribution of mass losses is also related to four temperature intervals (stages) of samples' decomposition (table 2):

- 1st stage– dehydration (40-200°C);
- 2nd stage – dehydroxylation (420-470°C);
- 3rd stage – dehydroxylation, decarbonation and desulphurization (520-700°C);
- 4th stage– decarbonation and desulphurization (700-830°C).

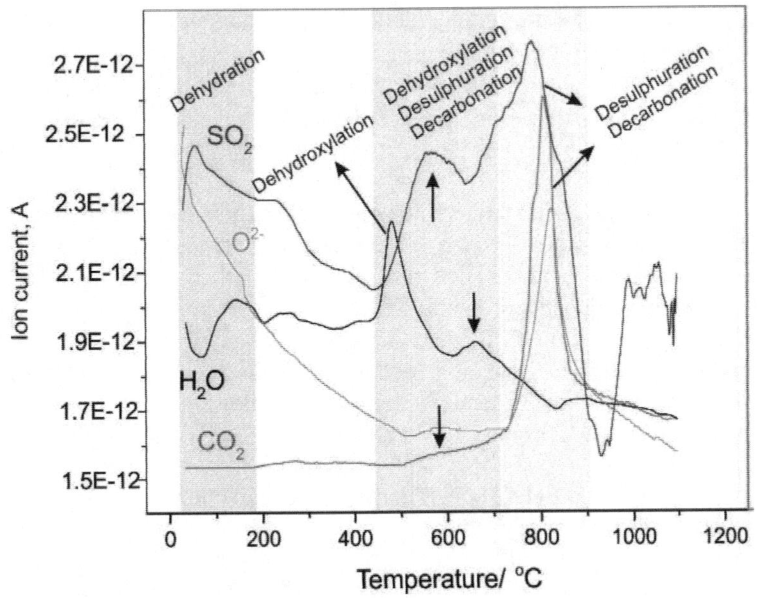

Figure 4. MS analysis of evolving gases of thermal decomposition of C120.

Table 2. Temperatures of thermal decomposition (T_{infl}) and mass losses (ML)

Sample	I stage 40-200°C Dehydration		II stage 420-470°C Dehydroxylation of $Ca(OH)_2$		III stage 520-700°C Dehydroxylation of C-S-H		IV stage 700-830°C Decarbonization of $[CO_3]^{2-}$ phases		Total mass losses, %
	T_{infl}, °C	ML, %	T_{infl}, °C	ML, %	T_{infl}, °C	ML, %	T_{infl}, °C	ML, %	
White cement	65.1 98.0	0.30 0.39	429.9	0.25	539.6	0.15	711.8	3.68	5.19
Marble powder	-	-	-	-	575.7	7.70	806.6	30.02	47.83
C001	84.8 161.1	1.67	459.5	0.57	609.3	0.84	795.8	29.02	33.30
C003	82.4 167.6	2.07	469.4	0.69	574.6	1.81	804.1	29.68	35.83
C007	169.7	2.17	470.8	0.69	589.8	2.40	786.4	29.45	35.94
C014	90.7 160.3	2.75	457.9	0.64	609.9	2.42	790.0	29.43	35.19
C028	135.3	0.76	426.7	1.23	566.5	0.94	721.5	28.98	34.38
C120	88.1 139.8	4.11	464.2	0.82	593.7	1.41	798.6	29.12	37.71

These four stages are referred to both the raw materials and the prepared mortars, taking into account differences in 3rd stage (dehydroxylation, decarbonation and desulphurization) as well as in 4th stage (decarbonation and desulphurization) [9, 10, 11]. In these two stages, the changes in the amount of crystal and bound water, decarbonation of carbonate and hydro-carbonate ions, as well as desulphurization of sulphate ions associated with the raw materials, their ratio and the resultant microstructures are defined. The detailed description of the mass losses (ML) and the related processes are as follows:

1st stage –an ongoing process of dehydration in the temperature range 40-200°C

In the low-temperature range up to 200°C, the crystal water is released from cement hydrate phases – gypsum and C-S-H gel. The results from PXRD analysis prove the existence of phases, forming gypsum and C-S-H gel (hydro-silicate minerals with fine structure, usually PXRD amorphous [9, 12]). The registered low ML are a result of their low amount in the samples – 0.76-4.11% (table 2).

2nd stage – an ongoing process of dehydroxylation in the temperature range 420-470°C

The thermal effects in this temperature range are weak, but well developed for all studied samples. Based on our previous research [9, 12, 13] and according to the other authors [10, 11, 14, 15], it can be assumed that the temperature dependences and ML at this stage are results from dehydroxylation of portlandite. The obtained lower ML (0.57-1.23%) is due to lower water-to-cement ratio as well as to water bonding in hydro-silicates, forming C-S-H gel. Despite the registered low ML, the dihydroxylation has been established by the decreasing of portlandite amount, namely decreasing crystal water in portlandite and increased amount of water included in C-S-H gel.

3rd stage – an ongoing process of dehydroxylation, decarbonation and desulphurization in the temperature range 520-700°C

The result of thermal analysis for this stage was supplemented by evolving gas analysis in the order to clarify and correct interpretation of the thermal decomposition reaction mechanism. A slight increase in ML up to 0.84% (C001) and 2.42% (C014) was measured, which can be explained by the incorporation of larger amounts of hydroxyl, carbonate and/or sulphate ions in these samples. Besides, in this temperature range, the dihydroxylation of bonded water in C-S-H gel minerals (ettringite, mono/polycarboaluminate, monosulphoaluminate, etc.) occur [10, 11, 14, 15, 16]. Decomposition processes in this temperature range are more complex and involve stages of overlapping thermal reactions in which hydroxyl, carbonate and/or sulphate ions are also released. This is defined by the mass spectrometric analysis of the evolving gases, which established that the dehydroxylation process is carried out parallel with decarbonization and desulphurization. The combined separation of carbonate and hydroxyl ions implies both their independent emission from the solid phase and the emission as hydroxycarbonate ions.

The ongoing reactions refer to dihydroxylation-decarbonation to release of carbonate ions from hydrogen-carbonate phases. For example, $CaHCO_3$ and monocarboaluminates, formed during the initial hydration of cement minerals [15], are metastable and they are decarbonated at temperatures, lower than that of calcite. This fact explains the inserted low-intensity peaks in the main decarbonation peak in the next temperature range (700-830°C). The presence of HCO_3^- explains the registered emissions of hydroxyl and carbonate ions [17, 18], while the O_2^- is related to equilibrium $CO_3^{2-} \leftrightarrow CO_2 + O^{2-}$.

The obtained results show that the equilibrium between HCO_3^-/CO_3^{2-} depends on cement components and parameters of the environment: humidity, temperature, the content of sulphur, nitrogen, carbon, gases, pH, etc. The change in the ratio of the phases containing HCO_3^-/CO_3^{2-} ions takes place at the solid-gas boundary (solid phases-atmosphere). These processes lead to changes in the chemical activity and solubility of the components of cement mortars and determine their durability, which is important for construction practice.

4th stage – an ongoing process of decarbonation and desulphurization in the temperature range 700-830°C

The decomposition of carbonate phases in the samples is carried out at a higher temperature in comparison to those of white Portland cement and marble powder (figures 2 and 3). Significant differences in mass losses were registered (3.68% for white Portland cement, 30.02% for marble powder and 29.12% for C120), because of different quantity of carbonate minerals (calcite and dolomite) in the raw materials and studied samples. The stepwise decarbonation is an indication of thermal decomposition of different carbonate-containing phases: dolomite, Mg-calcite, calcite-hydrated calcium-aluminium-carbonate silicates. Their decomposition occurs in a narrow temperature range, resulting in splitting and intensity changes of the DTG/DSC peaks [19, 20]. The curve, describing the CO_2 release in the evolving gas analysis corresponds to the change in the DTG/DSC curves and proves the existence of overlapping thermal reactions in this stage.

4. Findings and conclusions

Traditionally, the marble powder is recognized as an inert mineral additive without taking part in processes of cement hydration. Nevertheless, this study found that the use of a marble powder addition as an aggregate and substitute for part of Portland clinker leads to:

- increasing of variety of hydrate phases;
- reducing the amount of portlandite;
- increasing the content of structurally bound water;
- densifying macro- and microstructure of mortars.

The water-to-cement ratio is an essential factor for the proper hydration of cement minerals, but also other parameters control the hydration process (quantity and pollution of mineral fillers, type and dosage of chemical admixtures, etc.). The small amounts of mass losses during the dehydration of the samples prove that studied cement-based mortars are characterised with an insufficient quantity of water for complete hydration of the cement minerals.

These results can be used to study the structure and properties of building artefacts, related to the discovery of ancient composite production technologies and environmental knowledge in various archaeological epochs.

Acknowledgements

This work was supported by the Operational Program "Science and Education for Intelligent Growth", co-financed by the European Union through the European Structural and Investment Funds under grant BG05M2OP001-1.001-0008 of National Centre for Mechatronics and Clean Technology (V.P.), by the Bulgarian Science Research Found under grant KP-06-N39/9 (V.S, B.K.) and by Laboratory of Geology - BF at NBU (B.K.).

References

[1] Aïtcin P-C 2019 *Lea's Chemistry of Cement and Concrete*, ed P Hewlett and M Liska, (Oxford: Butterworth-Heinemann) chapter 17 pp 806-820
[2] Hamad B S 1995 Investigations of chemical and physical properties of white cement concrete *Advn Cem Bas Mat* 2 161
[3] Ling T-C and Poon C-S 2011 Properties of architectural mortar prepared with recycled glass with different particle sizes *Mater. Des* 32 2675
[4] Aïtcin P-C 2007 *Binders for Durable and Sustainable Concrete* (London and New York: CRC Press) p 528
[5] Aïtcin P-C and Mindess S 2011, *Sustainability of Concrete* (Oxon and New York: CRC Press) p 328
[6] Taylor H F W 1997 *Cement Chemistry* (London: Thomas Telford Publishing) p 480
[7] EN 196-1:2016 2016 *Methods of testing cement - Part 1: Determination of strength* p 38
[8] ASTM C642-13 2013 *Standard Test Method for Density, Absorption, and Voids in Hardened Concrete* p 3
[9] Stoyanov V, Kostova B, Petkova V and Pelovski Y 2012 Structure of white cement mortars with high content of marble powder, *J Therm Anal Calorim* **110** 405
[10] Tavangarian F and Emadi R 2011 Mechanism of nanostructure bredigite formation by mechanical activation with thermal treatment, *Mater Lett.* **65** 2354
[11] Pacewska B, Wilinska I and Nowacka M 2011 Studies on the influence of different fly ashes and Portland cement on early hydration of calcium aluminate cement *J Therm Anal Calorim* **106** 859
[12] Petkova V, Stoyanov V and Pelovski Y 2012 TG–DTG–DTA in studying white self-compacting cement mortars *J Therm Anal Calorim* **109** 797

[13] Kostova B, Petkova V, Stoyanov V and Serafimova E 2016 *Proc. XV Int. Conf. on Thermal Analysis and Calorimetry in Russia (Saint-Petersburg)* vol. I (Saint-Petersburg: Peter the Great St. Petersburg Polytechnic University) p 435
[14] Olszak-Humienik M and Jablonski M 2015 Thermal behavior of natural dolomite *J Therm Anal Calorim* **119** 2239
[15] Frankeová D and Slížková Z 2016 Determination of the pozzolanic activity of mortar's components by thermal analysis, *J Therm Anal Calorim* **125**, 1115
[16] Huang Y, Wang S, Hou P, Chen Y, Gong C and Lu L 2015 Mechanisms and kinetics of the decomposition of calcium barium sulfoaluminate *J Therm Anal Calorim* **119** 1731
[17] Kamruddin M, Ajikumar P K, Dash S, Tyagi A K and Baldev R 2003 Thermogravimetry-evolved gas analysis-mass spectrometry system for materials research *Bull. Mater. Sci.* **26** 449
[18] Ghorab H Y, Zahran F S, Kamal M and Meawad A S 2018 On the durability of Portland limestone cement: Effect of pH on the thaumasite formation *HBRC Journal* **14**, 340
[19] Ximena G, Borrachero M V, Payá J, Monzó J M and Tobón J I 2018 Mineralogical evolution of cement pastes at early ages based on thermogravimetric analysis (TG), *J Therm Anal Calorim* **132** 39
[20] Djordje M, Paulin I, Kržmanc M M and Škapin S D 2018 Physical and chemical treatments influence on the thermal decomposition of a dolomite used as a foaming agent, *J Therm Anal Calorim* **131** 1125

IOP Conf. Series: Materials Science and Engineering **951** (2020) 012008 doi:10.1088/1757-899X/951/1/012008

Recycling of concrete construction and demolition waste in alternative binders: Part 1 – Technical study

B Y Petrov and R A Zaharieva*

Dept. of Building Materials, Faculty of Civil Engineering, University of Architecture, Civil Engineering and Geodesy, 1 Hristo Smirnenski Blvd., 1046 Sofia, Bulgaria

* E-mail: roumiana.zaharieva@gmail.com

Abstract. The concrete construction and demolition waste (CDW) is usually recycled as coarse aggregate. The fine fraction remains unused – if recovered at all, it is for downcycling in backfilling. Part 1 of this article is dedicated to a technical study on recycling of the waste concrete powder (WCP) as supplement in blended cements. Part 2 presents assessment of the environmental impacts of those binders. XRD and DTA/TG analyses of WCP are performed. WCP from two types of concrete CDW is used. The behaviour of fresh and hardened pastes with replacement of CEM I 52,5 by WCP (15%, 30% and 45%) is investigated. Standard methods of testing with some modifications are used. The influence of WCP content on the setting, consistency, strength and water absorption is evaluated. Four different curing regimes are applied. The results confirm the initial assumption that WCP can be consider as cementitious supplement. The replacement rate up to 15% practically does not change the behaviour of the binder, while 45% seems to be the maximum, because of binders performances degradation. The replacement up to 30% allows to keep satisfactory technical properties, but offers greater economic and environmental advantages.

1. Introduction

The Portland cement concrete is still the most used construction material. Cement production generates more than 30% of CO_2 [1] emissions, because of high fossil energy demand and chemical transformations of raw materials to get the clinker. Replacement of clinker by mineral additives is one of the ways to reduce the negative environmental impact of Portland cement, but also a path for cost optimisation. Moreover, some mineral additives, such as ground-granulated blast furnace slag (GGBFS), fly ash, silica fume, burnt shale, natural pozzolana are active, i.e. they modify the properties of cement and lead to increased durability in aggressive environment or exposure to high temperature. Other additives such as limestone filler, quartz mill, etc. are rather inert (at normal hardening conditions) and contribute mainly to the costs optimisation.

Blended cements become increasingly used because of their technical and economic advantages. However, it shall be considered that some mineral supplements coming from industrial waste are going to be in a shortage in Europe because many of the metallurgical factories have been closed and the coal thermal power plants are going to be decommissioned in the next 2-3 decades as per the Grean deal [2]. Natural pozzolana such as volcanic ashes and pumice are not always available at

IOP Conf. Series: Materials Science and Engineering **951** (2020) 012008 doi:10.1088/1757-899X/951/1/012008

reasonable distances from cement production plants. For example, in Bulgaria all 3 cement factories are situated in the Northern part, while the deposits of natural pozzolona are concentrated in a small region in the southern part, which means transport distances of more than 400 km. Thus, alternative binders such as geopolymers [3] or alternative supplementary cementitious materials are being explored [4] with increasing interest.

Notwithstanding the foregoing, the use of recycled waste concrete powder (WCP) from construction and demolition waste (CDW) possesses several advantages: a) environmental: it contributes to closing the loop of concrete by increasing the CDW recovery degree, the recycling process of concrete CDW is not sophisticated and is low-energy consuming, the WCP is available at a short distance and CDW generation is relatively constant; b) technical – depending on the original concrete and the recycling technology the old cement stone can have higher or lower hydraulic activity, due to non-hydrated clinker, while the hydrated Portland cement stone might serve as germs of crystallization, i.e. WCP is expected to be an active supplement. The main source of WCP is the fine fraction (crushed sand and powder) which results from the recycling of CDW into crushed stone and aggregate. At present in Bulgaria, this residue, if recovered at all, is downcycled as backfilling material.

There are a couple of studies on the utilization of WCP as a substitution material for cement. The results are not consistent and often contradictory, but all report a significant decrease in compressive strength when WCP is used. To overcome this problem, some researchers suggest a thermal activation at 800°C to be performed in order to compensate the strength reduction [8]. There are even more radical approaches where the WCP replaces the clinker after being treated at 400–1450°C and is further blended with supplementary cementitious materials to produce cement [9-11]. Although technically feasible and chemically justified, such approach does not resolve the major environmental drawbacks of cement production – CO2 emissions and huge energy consumption.

The purpose of the first part of the presented study is to assess the limits in clinker replacement by WCP from technical perspective – presence of hydraulic activity, influence on the structure of the cement stone, impact on strength and sourcing of WCP.

2. Materials and methods

2.1. Materials

Portland cement CEM I 52,5 R having fineness of 3420 cm2/g measured as per Blaine method (BDS EN 196-6 [12]) is used for the preparation of alternative binders with different amount of WCP. Two different sources of concrete CDW to produce WCP are explored:

- "young concrete" waste, at age of 2 months, originating from concrete cubes prepared in the laboratory compressive strength conformity assessment for a high-rise building in Sofia; the binder content in concrete is expected to be relatively high and cement class is most likely to be CEM I 52,5, so a great WCP hydraulic activity is theoretically probable; however, the real activity might be lower, because this is a common practice in Bulgaria to replace the Portland cement by fly ash and to achieve the higher classes of cement by a greater fineness, resulting in reduction of the amount of non-hydrated cement particles;
- old concrete" waste at age of more than 35 years, originating from CDW from large panel buildings; it is expected that the majority of concrete CDW of interest to be recycled (incl. as WCP) in Bulgaria, will be of that type; the used cements were coarser at that time, so the amount on non-hydrated part of cement stone is expected to be big; the amount of carbonates is also expected to be higher due to the carbonation process.

A laboratory recycling of CDW was performed. A preliminary crushing of concrete elements was done to reduce their maximum size to approximately 10 cm. The CDW pieces were oven dried at +105°C to a constant mass. After jaw crushing of concrete waste, the coarse fraction (above 4 mm) was removed and the WCP was obtained by grinding the sand fraction in a ball mill to a fineness of 3000-3500 cm^2/g, measured as per the Blaine method. Particles bigger than 0.125 mm were removed by sieving. The WCP was stored in airtight PE bags for not longer than 2 days, before being incorporated in alternative binders.

A (DTA/TG) and (XRD) analyses of WCP (from the young concrete) were performed at Bulgarian Academy of Sciences, in the partner Institute of Physical Chemistry. The DTA results show two main endothermic effects related to the chemical dissociation of hydrated products and the release of chemically bound water: the peak at 450-480°C is associated with the calcium hydro-silicates and calcium hydro-aluminates (Figure 1). The second, more pronounced peak (i.e. representative for a bigger amount of the compound) at 700-800°C is related to the dissociation of carbonates resulting mainly from limestone/dolomite aggregate, present in the primary concrete [4].

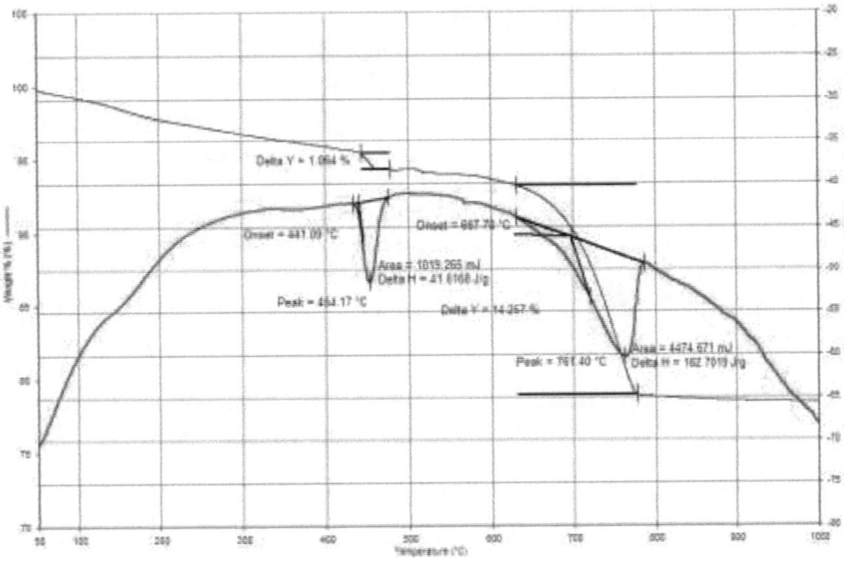

Figure 1. DTA and TG curves of WCP from the "young" concrete waste.

The XRD results confirms the findings (Figure 2) on the dominant presence of hydration products and carbonates.

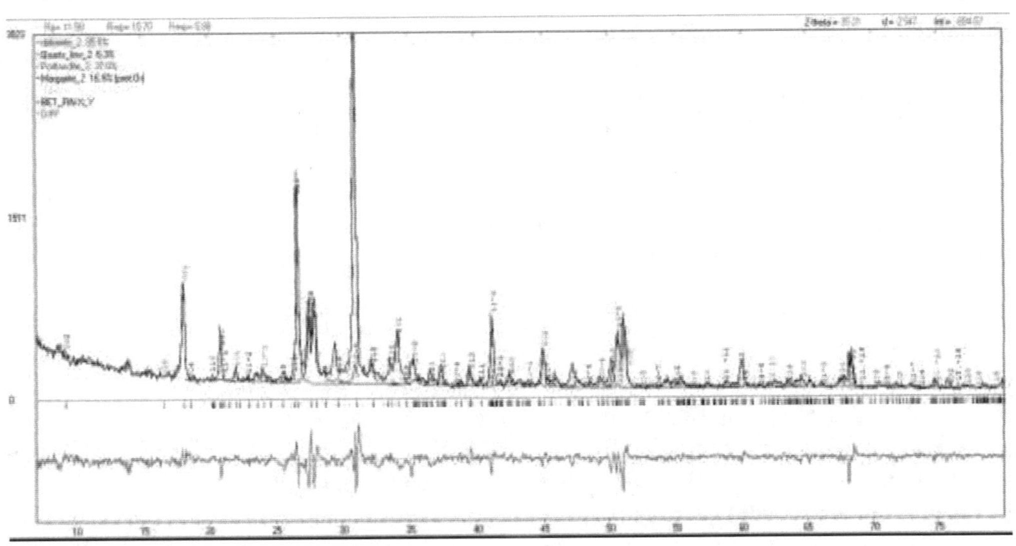

Figure 2. XRD diffractogrames of WCP from the "young" concrete waste.

IOP Conf. Series: Materials Science and Engineering **951** (2020) 012008 doi:10.1088/1757-899X/951/1/012008

The amorphous phase, due to the non-hydrated cement and calcium hydro-silicates (looking amorphous because of their nano-size structure) is around 30% of total mass. The crystalline phase consists of calcium magnesium carbonate (ca. 25% of total mass), portlandite (ca.26% of total mass), calcium aluminum silicate hydroxide (ca. 12% of total mass), hillebrandite (ca. 3% ot total mass) and silicon dioxide (ca. 4% of total mass). The last two components and the carbonates are due to the presence of aggregate (dolomitic coarse aggregate and river sand) in the primary concrete, which are unavoidable in the WCP.

No chemical admixtures or other mineral supplements are used, in order to avoid some possible side effects on the properties of the blended binders.

2.2. Blended binders composition and mix design

Three types of so-called alternative binders are investigated, designated as WCP15, WCP30, WCP45. They represent dry mixes of Portland cement CEM I 52,5 and three varieties of amounts of WCP blend- 15%, 30% and 45% of the total mass of the dry mix, respectively. Smaller amounts might have insignificant technical effects. Bigger amounts would have a greater environmental effect, but the technical properties are not likely to be satisfactory.

When mixed with water, the same water/binder ratio is applied (0,27), because the standard consistency does not change much (Table 1). At that water/binder ratio the workability of WCP15 and WCP30 was practically identic to that of the control paste, while the workability of WCP45 was slightly lower but convenient for the samples preparation. No sand is added to produce mortars, in order to clearly distinguish the effect of WCP on the binder properties. Also, a control mix (designated as C100) is prepared based only on CEM I 52,5.

Table 1. Mix design and properties of fresh mixes with WCP

Binder designation	Substitution by WCP, %	Standard consistence, %	Initial setting time, min	Final setting time, min
C100	0%	27	105	200
WCP15	15%	28	132	205
WCP30	30%	28	135	230
WCP45	45%	29	140	255

2.3. Samples preparation and curing

Samples of size 40x40x160 mm were produced from each binder. For each testing 3 specimens were tested and the average value is reported.

The specimens were kept in molds for 24 hours (at 20 ± 3oC and R.H.>95%), wrapped in PE foil.

After demolding of samples, 3 storage regimes were applied for mixes based on C100 and WCP30: in water at 20 ± 3°C, moist curing (wrapped in PE foil, stored at 20 ± 3°C and R.H.>95%), air conditions (20 ± 3°C, R.H.=60±5%). One series of samples was autoclaved at the age of 2 days (100±5°C and 1,3 MPa for 8 hours) and was stored at air conditions until testing. WCP15 and WCP45 samples were stored only in moist conditions.

2.4. Testing methods

Standard methods and modified standard testing methods were applied:

- Standard consistence and setting time as per BDS EN 196-3 [13];
- Compressive and flexural strength as per BDS EN 196-1 [14] on binder pastes (not mortars);
- Water absorption as per BDS EN 12808-5 [15] with a longer time of immersion in water, because of the fine porous structure;
- Water saturation after 5 hours of treatment in boiling water; it is representative for the open porosity, accessible to water in the hardened paste;

IOP Conf. Series: Materials Science and Engineering **951** (2020) 012008 doi:10.1088/1757-899X/951/1/012008

- Coefficient of saturation is calculated as ratio of water absorption and water saturation and is representative for the volume of empty voids, accessible to water, then the hardened paste is immersed in water in normal conditions.

3. Results and discussion

3.1. Properties of fresh pastes
The properties of fresh mixes do not differ much in terms of workability, assessed by the Vicat plunger penetration 5±1 mm. Both initial and final setting time are influenced by the substitution of cement with WCP – the bigger the amount, the longer the time required for setting, up to 30% in average, measured for WCP45 (Table 1). Such behaviour is similar to other ordinary blended cements of type CEM II.

3.2. Hydro-physical properties of hardened pastes
The water absorption and the water saturation increase when WCP is more than 15% (Table 2) which can be attributed to the fine internal capillary porosity of WCP particles due to the presence of old cement stone. The incorporation of WCP at 15% does not cause any changes in the absorption process (even the absorption is marginally lower), because the WCP particles remain "wrapped" by the new cement paste and the blended binder properties are governed by the new cement paste. When the amount of WCP is increased, its porous particles can absorb water more easily and this explains the very high value of the coefficient of saturation of WCP45 (Table 2) – 93% of voids, accessible to water are filled during 120h of immersion in water, while for C100 and WCP15 this percentage does not exceed 80%. The materials are considered vulnerable to freeze-thaw cycles when the coefficient of saturation is more than 80% but this negative result for WCP 30 and WCP45 can be attributed also to the presence of more structural defects due to the relatively lower workability of this mix.

Table 2. Hydro-physical properties of hardened blended binders (WCP from "young" concrete).

Binder designation	Water absorption, %			Water saturation, %	Coefficient of saturation
	After 24h	After 36 h	After 120 h		
C100	2,18	2,47	2,52	3,21	0,79
WCP15	2,12	2,42	2,48	3,27	0,76
WCP30	3,00	3,67	3,73	4,59	0,81
WCP45	5,11	5,96	6,08	6,50	0,93

The coarser porosity of WCP45 is best illustrated by the capillary absorption (Figure 3). The initial capillary absorption (usually measured after 1 hour of partial immersion) is more than 3 times higher than that of the control mix, but the coarse porosity continues to be filled-in up to the 4th hour and the absolute value is 4,6 times higher. The sorptivity of WCP45, representative for the fine porosity, is also significantly higher, while that of WCP15 and WCP30 is practically the same as that of C100. The initial absorption of WCP15 and WCP30 is also more significant than that of the control paste.

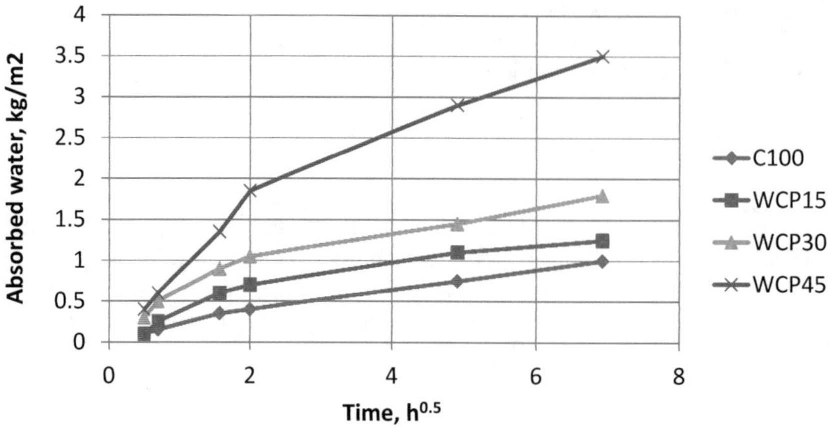

Figure 3. Capillarity of the hardened pastes of different binders (28 days).

3.3. Strength of hardened blended binders.

The replacement of cement by WCP originating from old concrete waste leads to a certain strength decrease. At 28 days the decrease for WCP15 and WCP30 is marginal (only 2 and 3% respectively), within the limits of test precision and can be therefore neglected. The replacement by 45% has also non-proportional reduction of only 10%.

It shall be noted that at early age (7 days) WCP30 and WCP45 have lower strength by 10 and 17% respectively, but the hardening process seems to be more intensive and the difference for WCP30 with the control paste is practically eliminated at 28 days and significantly reduced for WCP45 (Figure 4a). There is practically no difference in compressive strength of blended binders when WCP from the given "young" and "old" concrete waste were used in the same proportion for cement replacement – Figure 4 (a) and (b). This finding needs to be further investigated.

(a) (b)

Figure 4. Compressive strength development of different binder pastes (moist curing): (a) WCP from "old" concrete waste; (b) WCP from "young" concrete waste.

The curing seems to play more important role on the strength development of blended binders. The strength of C100 follows the usual trends (autoclaving accelerates the hardening, but exhausts the strength gain capacity; the air conditions are less favourable; at 28 days practically there is no difference in strength between the moist curing and curing in water), while the curing impact on WCP looks somehow surprising – the lowest strength is determined after the autoclaving and the moist curing is more favourable than the curing in water for 28 days strength (Figure 5).

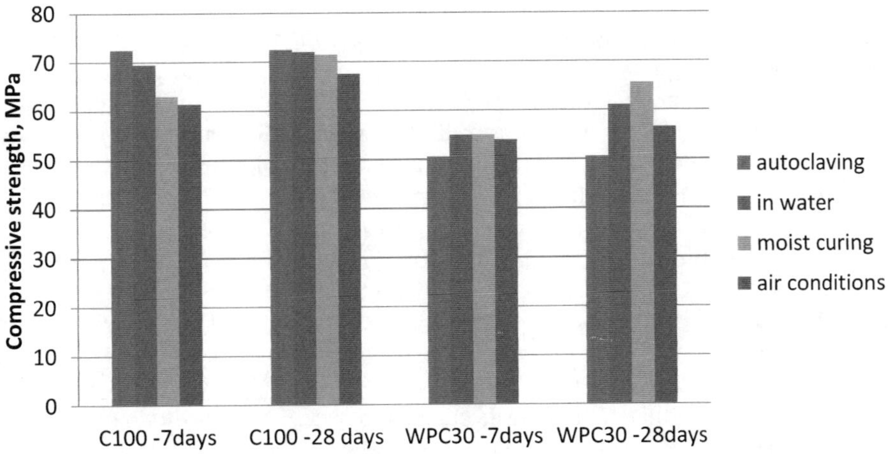

Figure 5. Impact of curing regime on control hardened binder and WCP30 ("young" concrete waste).

It seems the autoclaving has induced self-destroying processes, while a mobilization of quartz mill in pozzolanic type reactions was expected. This result requires further research. Because of the more porous structure of WCP30 it is normal that the air conditions lead to more intensive evaporation and overdrying of concrete surface, which results in lack of water for hydration processes.

4. Conclusion

This first technical part of the study confirmed the possibility to use recycled waste concrete powder as a cementitious supplement in blended binders, without the need of additional thermal activation of the recycled material. When the replacement of cement by WCP is not more than 15%, the main properties of the blended binder – workability of fresh paste, compressive strength, values of water absorption and water saturation, remain practically the same as those of cement without supplements.

The replacement degree of 30% has no significant effect on strength at 28 days but leads to an increased coarse porosity of hardened paste and, thus, to higher water absorption. The early age strength is about 13% lower. The initial setting time is extended by almost 30%. The use of this binder would be then limited by certain ambient conditions and a longer moist curing will be required.

WCP of 45% in the blended binder can be considered as a maximum, because the water absorption and water saturation are up to 3 times higher and the 28-days compressive strength is reduced by 10%. The setting of the binder is even more postponed than in WCP-30. The use of WCP in a such big proportion can be motivated by higher positive environmental effect but is limited to a few applications.

It can be therefore concluded that WCP is rather an active supplement able to compensate the reduction of clinker in the blended binder than just quasi-inert filler. Further investigations are to be made on the dependency of this effect on the amount of non-hydrated primary cement, or on a complex of factors (such as alcalis, products from primary clinker minerals hydration, etc., carbonates and quartz dust) in order to control the performance of blended binders. Another significant challenge to be resolved is the unavoidable presence of impurities in WCP that result from CDW.

Acknowledgments
The research is performed within the Project BG05M2OP001-1.002-0019:„Clean technologies for sustainable environment – water, waste, energy for circular economy“ (Clean&Circle), for development of a Centre of Competence, financed by the Operational programme "Science and Education for Smart Growth" 2014-2020, co-funded by the EU through the European structural and investment funds.

References

[1] Andrew R M, Global CO_2 emissions from cement production, 1928–2018, Open Access Earth System Science Data, https://doi.org/10.5194/essd-2019-152 (last accessed July 5th, 2020)
[2] A new circular economy action plan for a cleaner and more competitive Europe, Communication from the Commission to the European parliament, the Council, the European economic and social committee and the Committee of the regions, COM(2020) 98 final, 2020
[3] Hassan A, Arif M, Shariq M, Use of geopolymer concrete for a cleaner and sustainable environment – A review of mechanical properties and microstructure Journal of Cleaner Production Volume 223, 20 June 2019, pp 704-728 https://doi.org/10.1016/j.jclepro.2019.03.051
[4] Paris J M, Roessler J G, Ferraro C C, DeFord H D, Townsend T G, A review of waste products utilized as supplements to Portland cement in concrete, Journal of Cleaner Production, Elsevier, Volume 121, 10 May 2016, pp 1-18, https://doi.org/10.1016/j.jclepro.2016.02.013
[5] Kim Y J, Choi Y W, Utilization of waste concrete powder as a substitution material for cement, Construction and Building Materials, Volume 30, May 2012, pp 500-504
[6] Berndt M L, Properties of sustainable concrete containing fly ash, slag and recycled concrete aggregate. Construction and Building Materials, Volume 23, Issue 7, July 2009, pp 2606-2613.
[7] Jaroslav T, Prošek Z, Plachý T, Influence of increasing amount of recycled concrete powder on mechanical properties of cement paste. IOP Conference Series: Materials Science and Engineering 236 012094, 2017.
[8] Sui Y, Ou C, Liu S, Zhang J, Tian Q, Study on properties of waste concrete powder by thermal treatment and application in morta, Applied Sciences 10(3):998, doi:10.3390/app10030998.
[9] Wang J, Mu M, Liu Y, Recycled cement. Construction and Building Materials Volume 190, 30 November 2018, pp 1124-1132
[10] Xuan D X, Shui Z H, Rehydration activity of hydrated cement paste exposed to high temperature, Fire and Materials 35(7), pp 481–490, 2011
[11] Shui Z, Xuan D, Chen W, Yu R, Zhang R, Cementitious characteristics of hydrated cement paste subjected to various dehydration temperatures. Construction and Building Materials Volume 23, Issue 1, January 2009, pp 531-537
[12] BDS EN 196-6: Methods of testing cement - Part 6: Determination of fineness
[13] BDS EN 196-3: Methods of testing cement - Part 3: Determination of setting times and soundness
[14] BDS EN 196-1: Methods of testing cement - Part 1: Determination of strength.
[15] BDS EN 12808-5: Grouts for tiles - Part 5: Determination of water absorption
iately following the last numbered section of the paper.

IOP Conf. Series: Materials Science and Engineering **951** (2020) 012009 doi:10.1088/1757-899X/951/1/012009

Recycling of concrete construction and demolition waste in alternative binders: Part 2 – Environmental footprint

Y D Kancheva[1]* and R A Zaharieva[2]

[1]Dept. of Applied Geodesy, Faculty of Geodesy, University of Architecture, Civil Engineering and Geodesy, 1 Hristo Smirnenski Blvd., 1046 Sofia, Bulgaria
[2]Dept. of Building Materials, Faculty of Civil Engineering, University of Architecture, Civil Engineering and Geodesy, 1 Hristo Smirnenski Blvd., 1046 Sofia, Bulgaria

* E-mail: kancheva_fgs@uacg.bg

Abstract. The concrete demolition and construction waste (CDW) is usually recycled as aggregate. The fine fraction (crushed sand and powder) obtained from the recycling process remains less recovered – for downcycling purposes such as backfilling, if recovered at all. The feasibility study on the recycling of concrete waste as a component of a blended binder is presented in Part 1 of the study. Part 2 is dedicated to a preliminary assessment of the environmental impacts of the partial replacement of Portland cement binder by a waste concrete powder (WCP). The assessment is focused on the binder production stage as per BDS EN 15804 and does not consider any further scenarios. The environmental impacts of alternative binders with a different amount of WCP is compared to the impacts of conventional cement types CEM I, CEM II/A and CEM II/B. The results confirm that there is a significant reduction of the environmental impacts when WCP is used for cement replacement, because of the more efficient use of material resources and energy. The effect is more pronounced with the higher replacement rate. Therefore, the use of WCP in cements leads to more eco-friendly binders and can contribute to closing the concrete loop.

1. Introduction

Waste recycling and recovery is one of the main directions for achieving the sustainable goals and circular economy in EU. Construction and demolition waste (CDW) forms 25% to 30% of all generated waste in EU [1] and is estimated as the largest waste stream in the EU. CDW includes various materials that differ in type, composition, treatment needs and recovery options. The significance of CDW management is reflected in the European policies by the Waste Framework Directive [2], the EU Construction & Demolition Waste Management Protocol [3], Strategy for the Sustainability of construction sector [4], Circular Economy Action Plan [5], etc.

From environmental point of view, the recovery of demolition waste is considered a way to recapture the resources that are used originally for the manufacturing of building products. In this situation, both energy and resource efficiency can be achieved and, with advancing of technologies for reclaiming raw materials from CDW, the regulatory requirements by the EU legislation are becoming stringent and mandatory. The Waste Framework Directive [2] requires that by 2020 a minimum of 70% (by weight) of non-hazardous CDW (excluding excavated soil) must be prepared for re-use, must be recycled or must undergo other material recovery. This requirement was transposed into Bulgarian legislation by the Waste Management Act (WMA) in 2012 [6].

When aiming at the high recovery degree, the technical and economic feasibility of treatment solutions is usually considered, but the decisions have to be taken after assessing the environmental impact of the various solutions.

The present part of the study (Part 2) is dedicated to a preliminary assessment of the environmental impacts of the partial replacement of Portland cement binder by a waste concrete powder (WCP). It is interesting to explore the recovery of concrete CDW not only in terms of reducing the waste for disposal (i.e. closing the loop for concrete), but also from the perspective of the potential positive effects on the environment from using construction materials with recycled concrete content.

2. Specifics of CDW from concrete

The demolished concrete waste is one of the main CDW streams – for Bulgaria its share is estimated to be around one quarter to one third [7] of the overall amount of CDW. According to WMA [6], at least 85% of concrete CDW shall be recovered. This percentage is calculated as quantity diverted from landfilling and treated for recovery purposes.

Preparation to reuse, recycling and backfilling are recognised by the legislation. However, due to undeveloped market for recycled construction products in Bulgaria, lack of appropriate legislation and lack of awareness, only a few options from the vast recovery possibilities are explored – recycling as crushed stone for road construction purposes (base, sub-base and drainage) and backfilling. Some additional barriers for higher recycling rate of concrete CDW are related to the availability and affordability of good quality natural aggregates and low taxes for landfilling of inert CDW (less than 2.00 EUR per ton of waste), which makes the selective demolition and at source separation of the mineral fraction of CDW not economically attractive. Many illegal or not regulated landfills still exist.

Even when recycled into aggregate, a large part from concrete CDW remains as a fine fraction (sand and filler), for which there is no market interest and this fraction is usually used for backfilling (downcycling). The purpose of our research study is to identify more valuable applications of that fraction, in accordance with the requirements of waste management hierarchy and circular economy principles. The technical feasibility study on recycling concrete waste as a component of a blended binder is presented in Part 1 of the research study. The interest in this approach arises also from the possibility to reduce the environmental footprint of concrete by using a blended binder with a smaller amount of Portland cement. So, there are opportunities for improving both energy and resource efficiency associated with the manufacturing of construction products with recycled contents. This paper explores whether and to what extents these expectations are justified by developing a preliminary theoretical assessment of some environmental indicators of alternative binders with varying content of waste concrete powder (WCP)

3. Methodology for environmental assessment

The effect of substitution of CEM I by WCP on environmental impacts is evaluated by comparing the results on predefined environmental indicators of a set of binders – CEM I, CEM II/A and CEM II/B where the amount of mineral additives is respectively up to 20 and up to 35% (very often these mineral additives are active and come from industrial waste such as fly ashes, ground granulated blast furnace slag (GGBFS) and silica fume) and 3 types of so-called alternative binder (AB), which are a blend of CEM I and various amount of WCP (5%, 15% and 30%). The substitution percentage is justified in Part 1 of the present study. Thus, the assessment includes the most used "competitors" of the WCP-blended binders – CEM II/A and CEM II/B which also have some recycled content and the amount of clinker is reduced by the same order as in the ABs.

For providing a reliable calculation of the environmental impacts of alternative binders with varying WCP content and ensuring an unbiased comparison by outlining the key environmental indicators that are influenced from the substitution of conventional cement binder by WCP, a life cycle assessment (LCA) approach in accordance with ISO 14040 [12] and ISO 14044 [13] is used to model a preliminary product system for the alternative binders. EN 15804+A1:2012 [14] serves as a basis for development of the model and the calculations. Considering the production processes of the binders do not differ in general, the following assumptions are made:

- transport distances and the means of transport for cement raw materials, additives (limestone, GGBFS, burnt shale and WCP) are similar;
- the additives and WCP are delivered in bulk, the ready binder product is also delivered in bulk (packaging is not necessary);
- the grinding of clinker and mineral additives to the suitable fineness requires the same machinery and energy per unit mass.

3.1. Definition of the LCA goal and scope

The goal of this study is to outline some important aspects of the composition of binders that can significantly influence the environmental performance of the product. The results from this preliminary assessment are intended to be included in a more comprehensive research on the feasibility of alternative binders with recycled CDW and to justify the ecological benefits from substitution of cement binder.

The assessment is focused on the production stage ("cradle-to-gate") and does not take into account any further scenarios that can release environmental impacts during the next stages of the life cycle according to EN 15804 (installation, use and end-of-life).

3.2. Life cycle inventory analysis

3.2.1. Functional unit

Binders for concrete are sold and used in construction by mass measurements –tons. Thus, the results are calculated for the functional unit of 1 kg binder.

3.2.2. Collection of data

Usually in practice the collection of information about the manufacturing process of a specific product is done within the context of the factory. In this case, the study is analytical and is based on referent data from the ecoinvent v.3.4 database. Certain datasets describing the production of cement type CEM I are modified in order to model the substitution of the cement with WCP. Table 1 summarises the set of the studied binders.

Table 1. Binders considered in the study.

Binder designation	Mineral additives, %	Type of mineral additives	Recycled content
CEM I	< 5%	None	None
CEM II/A	< 20%	Limestone, Burnt shale, GGBFS	0.2% Burnt shale 1.0% GGBFS
CEM II/B	20–35%	Limestone, Burnt shale, GGBFS	0.4% Burnt shale 2.1% GGBFS
WCP-5	5%	WCP	5%
WCP-15	15%	WCP	15%
WCP-30	30%	WCP	30%

Based on the referent data, the following varieties of conventional binders are considered: CEM I is a cement with more than 90% clinker by mass; CEM II/A and CEM II/B are limestone Portland cements (the mineral additive is mainly a limestone filler) with recycled content of GGBFS. The clinker content in CEM II/A is 83% by mass, the clinker content of CEM II/B is 68%. It is assumed that the binders cover the requirements of the same strength class and can be used for the same purpose.

3.3. Life cycle impact assessment

3.3.1. Characterisation Model and Impact Categories (Indicators)

For this preliminary assessment, the EN 15804 list of indicators is reduced to the aspects that are considered most relevant for the development of alternative binders. This includes all environmental indicators, consumption of energy and water and disposed waste (table 2).

Table 2. Assessed environmental indicators.

Indicator	Unit
Environmental impacts	
Global Warming potential (GWP)	kg CO_2-eq.
Ozone depletion potential (ODP)	kg CFC-11-eq.
Depletion of abiotic resources-elements (ADPE)	kg Sb-eq.
Depletion of abiotic resources-fossil fuels (ADPF)	MJ
Acidification potential (AP)	kg SO_2-eq.
Eutrophication potential (EP)	kg PO_4-eq.
Photochemical ozone creation potential (POCP)	kg ethylene-eq.
Use of resources	
Use of non-renewable energy resources	MJ-eq.
Use of fresh water	m^3
Use of secondary materials	kg
Generated waste	
Non-hazardours waste disposed	kg
Hazardous waste disposed	kg

3.3.2. Calculation

The results are calculated with the openLCA software. The ecoinvent v.3.4 database is used to build the model. CML method and characterisation factors [15] are applied for calculation of the assessed indicators.

4. Results and discussion

The addition of WCP is expected to reduce the environmental impacts of ABs, compared to CEM I and result confirm this assumption. The reductions per binder type are presented in table 3.

Table 3. Summarised reductions of the environmental impacts.

Product	Reduction of impacts compared to CEM I, %
CEM I	0%
CEM II/A	5-10%
CEM II/B	15-20%
WCP-5	5-15%
WCP-15	15-20%
WCP-30	30-35%

It shall be noted that the environmental impacts of both CEM II types are also lower. The reduction of their environmental impact logically depends on the amount of the mineral additives and is more pronounced for CEM II/B than for CEM II/A. This is mainly due to the use of limestone (10% in CEM II/A and 22% in CEM II/B) rather than to the use of recycled industrial waste – the clinker production process is very energy intensive, so bigger decrease of clinker amount in binders results in higher reduction of the environmental impacts, associated to the use of fuels – figure 1 to figure 7. However, the replacement of clinker by WCP leads to a bigger positive effect than the replacement by limestone and industrial waste – compared to CEM I, the WCP-30 has up to 35% smaller values of certain environmental indicators, while for CEM II/B the reduction does not exceed 20% (table 3).

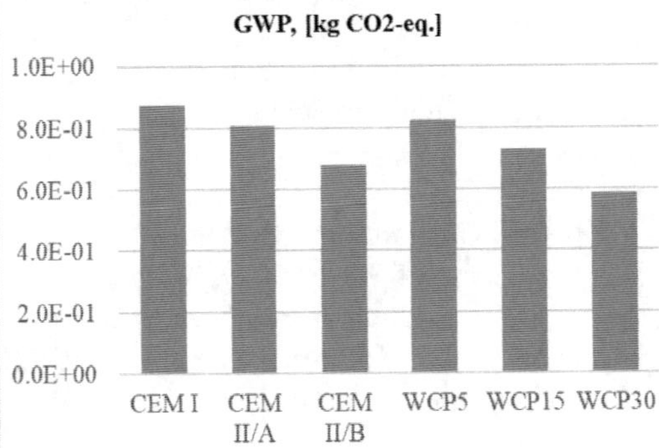

Figure 1. Global warming potential (GWP) of the binders under study

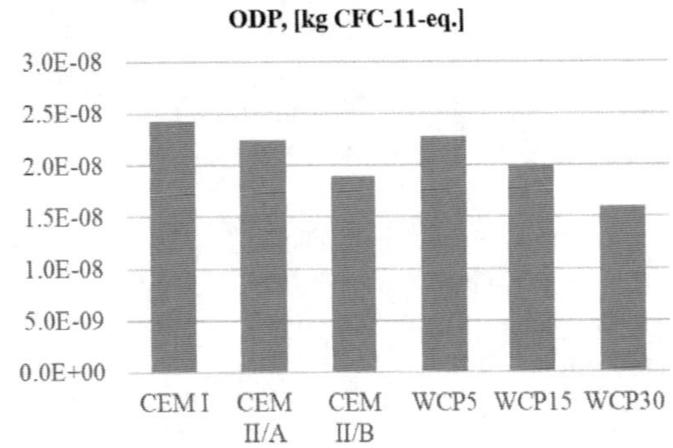

Figure 4. Ozone depletion potential (ODP) of the binders under study.

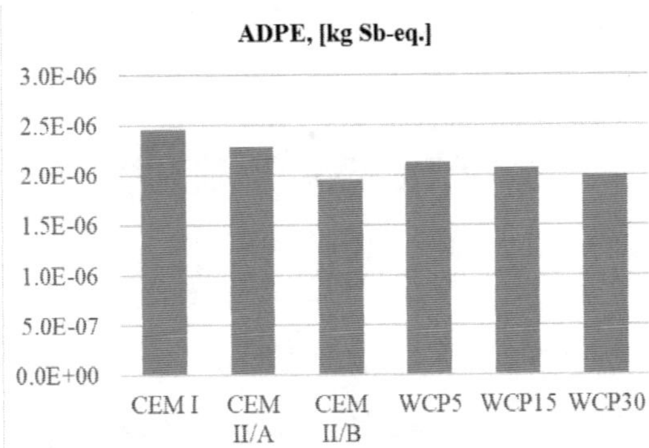

Figure 2. Depletion of abiotic resources-elements (ADPE) of the binders under study.

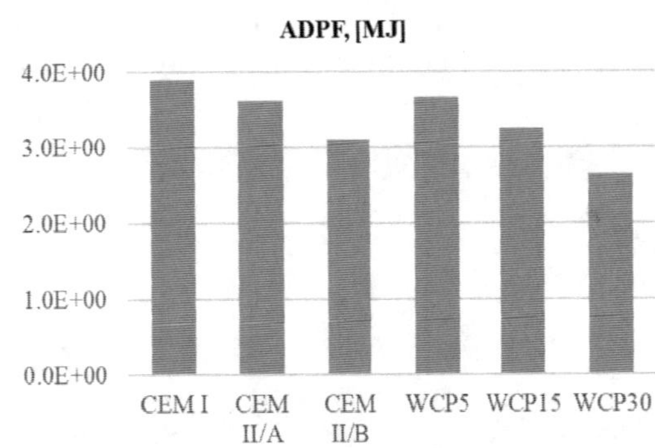

Figure 5. Depletion of abiotic resources-fossil fuels (ADPF) of the binders.

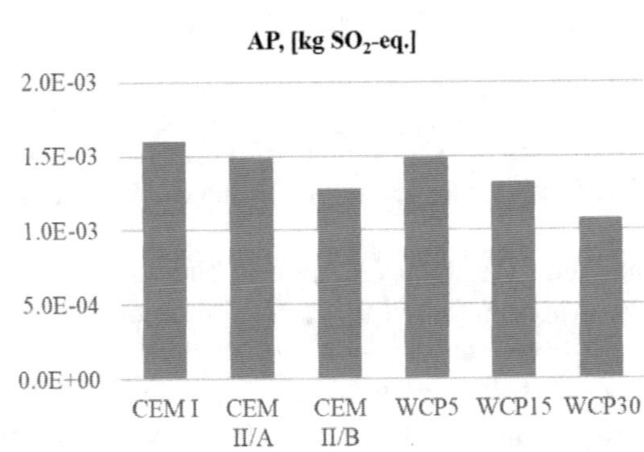

Figure 3. Acidification potential (AP) of the binders under study.

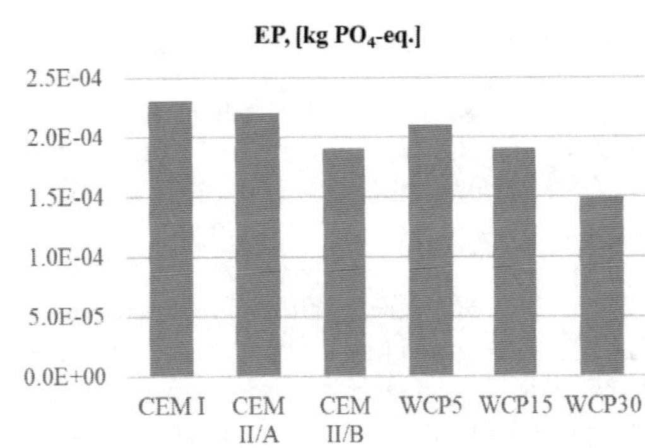

Figure 6. Eutrophication potential (EP) of the binders under study.

IOP Conf. Series: Materials Science and Engineering **951** (2020) 012009 doi:10.1088/1757-899X/951/1/012009

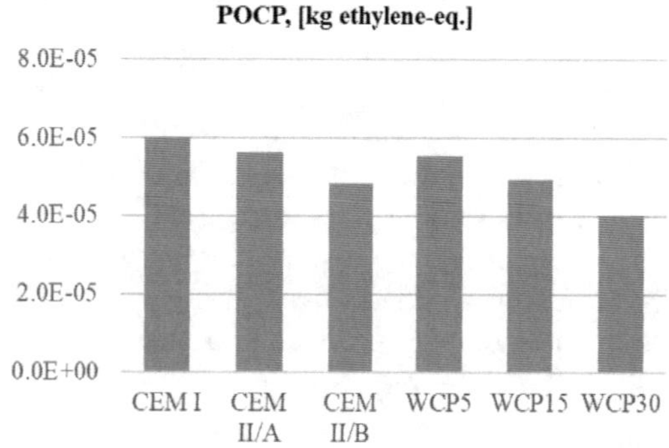

Figure 7. Photochemical ozone creation potential (POCP) of the binders under study.

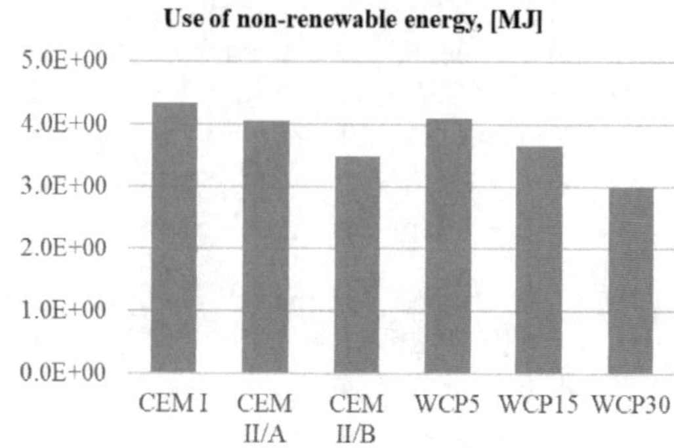

Figure 8. Use of non-renewable energy resources.

The environmental impacts of WCP5, where the WCP is only 5%, are comparable to CEM II/A with a higher clinker replacement degree (up to 20%) and the results of WCP15 (WCP of 15%) are systematically lower than those of CEM II/A. The studied environmental impacts of WCP15 are comparable to those of CEM II/B, while the clinker replacement in WCP15 is twice smaller. WCP30 is superior to CEM II/B by having the smallest values of the environmental indicators, except for the depletion of abiotic resources-elements (ADPE).

The results for the indicators about resource use (figure 8 to figure 10) mostly reflect the same trend except for the use of secondary materials. Use of energy depends on the energy for the acquisition and processing of raw materials, so binders with less clinker content use less total energy for production. The same applies to the water use, related mainly to the processing of raw materials and the energy consumption for clinker production.

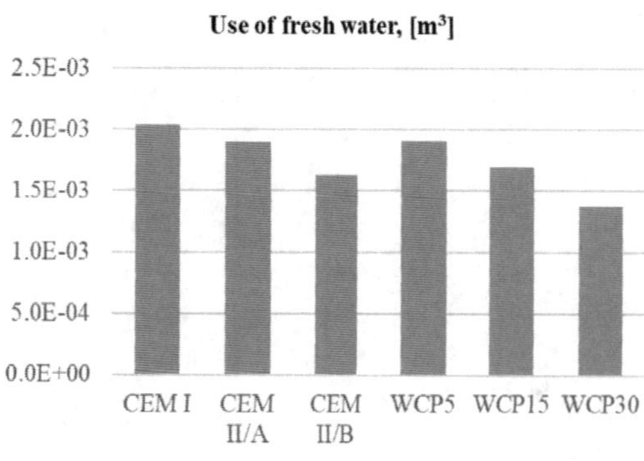

Figure 9. Use of fresh water.

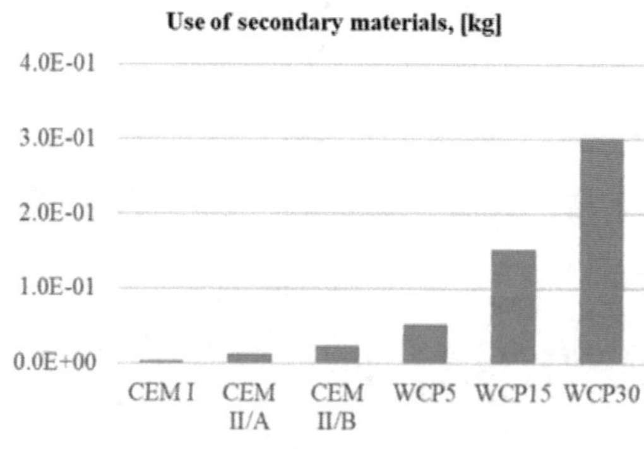

Figure 10. Use of secondary materials.

The use of secondary materials makes the biggest difference of the ABs, because it depends on the recycled material content as a substitute of the primary constituents. In CEM II the amounts of GGBFS are minor and it does not differ much from CEM I, while in the blended binders the share of recycled material from WCP is considerable.

It is also interesting to pay attention to the waste generated form the production of binders although the quantities are very small. In case that concrete CDW is not recovered for backfilling purposes but landfilled, the recycling of that waste into WCP is a possible way for significant reduction in the indicator for non-hazardous waste disposed. The non-hazardous waste (figure 11) generated from the production of CEM I and CEM II is similar in magnitude, though the absolute values decrease from CEM I to CEM II/A and CEM II/B, due to the use of recycled materials from industrial waste. The values of non-hazardous waste for ABs are negative, which indicates that more waste is recovered than generated, which is positive for the environment. The higher the contents of WCP, the more waste is diverted from disposal to recovery.

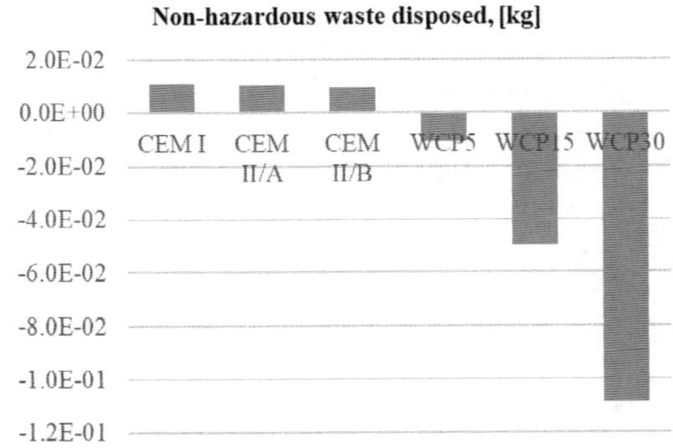

Figure 11. Non-hazardous waste disposed.

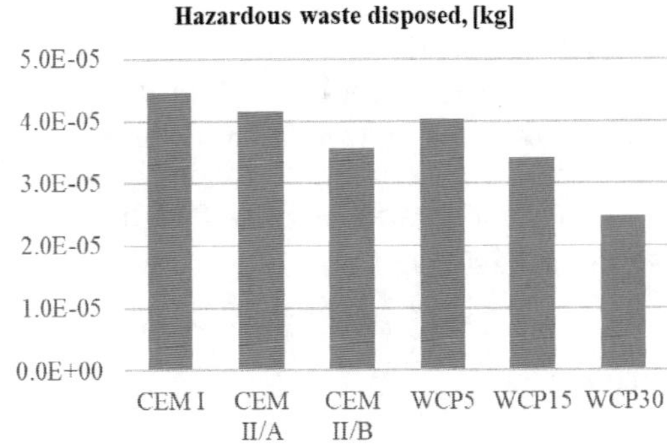

Figure 12. Hazardous waste disposed.

The result on hazardous waste generation (figure 12) is linked mostly to the production of clinker (contribution of more than 90%). Figure 12 shows the expected result that decreasing the quantity of clinker by substitution with WCP leads to reduced hazardous waste. However, the production of clinker is a complex process regarding the waste aspects. One source of hazardous waste is the high demand for energy and fuels for acquisition, transport and processing of raw materials (lime, clay, marl). On the other hand, the cement factories in Bulgaria are permitted to implement recovery operation R1 (Use principally as a fuel or other means to generate energy) in accordance with the WMA [6], so part of the fuels used for clinker production might be RDF including hazardous waste from petrol products. Another interesting aspect is that cement binder can be used for treatment before disposal of hazardous waste by stabilization/solidification process through chemical interactions, physical absorption or encapsulation of contaminated waste [16]. In the case of WCP use, further investigations on the permeability of ABs are necessary in order to assess the possibility to use the ABs for such treatment.

5. Conclusion

The recycling of concrete CDW into powder for being used in blended binders seems to be a prominent perspective from technical and environmental point of view. The recycled concrete powder can be considered as mineral additive to cement, having intermediate properties between the hydraulic additives such as GGBFS and burnt shale and inert additives such as limestone filler. Taking into consideration that the amounts of industrial waste as GGBFS are decreasing (many metallurgical factories have been closed, including the biggest ones in Bulgaria), the alternatives for clinker replacement in cement production are also decreasing. Moreover, some other common additives such as fly ashes will also disappear when the coal thermal power plant are closed. Depending on the degree of clinker substitution by WCP, different effects can be achieved: when modest compressive strength of the binder is required (for example class 32,5 or lower) and no specific requirements related to the use of concrete based on alternative binders exist, the incorporation of up to 30% WCP can be applied and this leads to a significant decrease of the environmental footprint of the concrete. If a smaller amount (5 to 15%) of WCP is used, because of technical reasons, the replacement still has a noticeable positive environmental effect, similar to that of CEM II.

However, further research on the alternative binders has to be performed on both technical aspects and environmental footprint. The specific features of alternative binders such as concrete CDW recycling technology, way of mixing the binders constituents (clinker, WCP, gypsum), etc. are to be modelled and the environmental effects are to be evaluated. The scope of LCA might also be extended by precisely considering the source of concrete CDW, transportation activities, work of machinery used for CDW fractioning and WCP grinding. Since end-of-life aspects are important for assessing the re-use and recycling options of building materials, the multiple recycling of WCP can also be investigated.

Acknowledgments
The research is performed within the Project BG05M2OP001-1.002-0019:„Clean technologies for sustainable environment – water, waste, energy for circular economy“ (Clean&Circle), for development of a Centre of Competence, financed by the Operational programme "Science and Education for Smart Growth" 2014-2020, co-funded by the EU through the European structural and investment funds.

References

[1] https://ec.europa.eu/environment/waste/construction_demolition.html (last visited May 13, 2020)
[2] Directive 2008/98/EC on waste of the European Parliament and of the Council of 19 November 2008 on waste (Waste Framework Directive), 2008
[3] EU Construction & Demolition Waste Management Protocol, European Commission Directorate-General for Internal market, Industry, Entrepreneurship and SMEs, September 2016
[4] Strategy for the Sustainable competitiveness of the construction sector and its enterprises, COM (2012) 433, http://eurlex.europa.eu/procedure/EN/201859 (last visited July, 2020)
[5] A new circular economy action plan for a cleaner and more competitive Europe, Communication from the Commission to the European parliament, the Council, the European economic and social committee and the Committee of the regions, COM(2020) 98 final, 2020
[6] Waste Management Act (WMA), State Gazette No. 105/30.12.2016 (last amendment)
[7] MOEW, National Strategic Plan on CDW management in Bulgaria for the period 2011-2020, 2011, Sofia, Bulgaria.
[8] Lahner T E, Brunner P H, Buildings as reservoirs of materials—their reuse and implications for future construction design, Demolition and Reuse of Concrete, ed E K Lauritzen, RILEM, Published by E & FN Spon, 1994, pp. 548-555
[8] Winkler G, *Recycling Construction & Demolition waste: a LEED-based toolkit*, USA, McGraw Hill, 2010.
[10] Garbarino E, Blengini G A, *Handbook of Recycled Concrete and Demolition Waste,* Chapter 6 The economics of construction and demolition waste (C&DW) management facilities, ed F Pacheco-Torgal et al., Cambridge, UK, Woodhead Publishing Limited, 2013, pp. 108-138
[11] Coelho A, de Brito J, Traditional vs. Selective Demolition - Comparative Economic Analysis Applied to Portugal, Conf.: Portugal SB10–Sustainable Building Affordable to All, Algarve, Portugal, March 2010
[12] ISO 14040:2006 Environmental management — Life cycle assessment — Principles and framework
[13] ISO 14044:2006 Environmental management — Life cycle assessment — Requirements and guidelines
[14] EN 15804:2012+A1:2014 Sustainability of construction works - Environmental product declarations - Core rules for the product category of construction products
[15] ecoinvent v.3.4 database, The ecoinvent Association, 2017
[16] CML Impact Assessment method version 4.1, CML - Department of Industrial Ecology, Leiden University, the Netherlands, last update August 2016

IOP Conf. Series: Materials Science and Engineering **951** (2020) 012010 doi:10.1088/1757-899X/951/1/012010

Experimental study of flat reinforced concrete slabs created in the method of early striking

S Bakardzhiev

Department of Technology and Construction Management, Faculty of Construction, USEA (VSU) "L. Karavelov", 175 Suhodolska St., 1373 Sofia, Bulgaria

E-mail: st.bakardzhiev@mail.bg

Abstract. This report examines the behaviour of an experimental two-storey monolithic construction during the construction of its flat slabs with a formwork system suitable for the application of early striking. The experimental results of the deformations in the slabs are compared with those from the numerical model of the construction. The peculiarities in the findings of the strength characteristics of the concrete used for the monolithic structure in the tests are analysed. Conclusions are made regarding the production of reinforced concrete slabs through formwork systems suitable for early striking from these results and comparisons.

1. Introduction
The method of Early striking is becoming more and more popular in modern building. The principle of early striking is based on the technology for propping or re-propping the main field of a slab with shores, which divide it into smaller fields, which in turn requires overcoming smaller bending moments and shear forces in the partly stroked slab. When using formwork systems for Early striking, it is possible to striking a large part of the formwork elements as early as 2-3 days after laying the concrete in the slab. Only part of the shores with the adjacent dropheads remain. Technically and economically, this is a good solution for the construction of buildings and structures [1], [2]. This technology, in turn, has its peculiarities. There are many studies on this issue worldwide. Some of these studies research the issue of the transfer of loads to reinforced plates by shores to others below them also propped by shores and which have not yet reached their full strength characteristics. As well as to these plates, which are completely striked and have reached their full strength characteristics. Some of the authors who have studied this problem numerically and experimentally are [3], [4], [5], [6]. Other authors [7], summarize the research done in the second half of the XX century and those in the XXI century. Other studies have examined the conditions that a reinforced concrete slab must meet to be produced using the Early Striking method [8], [9]. Some studies also address globally the problem of early loading of reinforced concrete slabs by loads arising from the production of buildings and structures, [10]. Generally, several main peculiarities that need special attention when applying the method of early striking.

2. Object and tasks in the study
The object of the report is to present an experimental study on the behavior of an experimental construction with two flat reinforced concrete slabs produced with a frame formwork system with dropheads for Early striking. Also to identify the main issues and peculiarities in the application of the method, which should be taken into account when using the technology of early striking in the production of reinforced concrete slabs. The tasks in the report are - measuring the deflections of the

IOP Conf. Series: Materials Science and Engineering **951** (2020) 012010 doi:10.1088/1757-899X/951/1/012010

two slabs of the experimental construction in the different stages of striking and loading, comparing the deflections in the slabs with the elastic ones obtained from a numerical model of the structure and tracking the cracks in the slabs in the stages of construction, and analyzing the results.

3. Experimental construction

The report examines an experimental two-storey reinforced concrete construction with two beamless slabs. The dimensions of the slabs are 125 cm wide and 550 cm long. The height of the construction is 232 cm. The foundations are 160x60cm in size and 30cm high Figure 1. The concrete used for the construction is C20/25. The shores used between the two slabs in the formwork system are adapted for small inter-floor heights (about 1m). The shores with adjacent dropheads, under the two slabs, are placed longitudinally in the middle of the slabs, and the axial distances between the shores and columns are shown in Figure 4. For clearer measurement results, the thicknesses of the reinforced concrete slabs of the structure are 16 and 14 cm, respectively, and the reinforcement embedded in them is close to the minimum reinforcement coefficient. The loads in the slabs are of their weight, the load from the formwork system for the second slab and simulated loads during construction used in some of the stages of construction. The additional simulated loads are from formwork panels weighing 200 kg, which sit on a second plate, and simulate the load of pallets weighing 1422 kg, which steps linearly on a second slab Figure 4. The deflections in the slabs are measured by dial gauges attached to a metal frame between the slab of level +1.21 (first slab) and the slab of level +2,32 (second slab) separate from the construction.

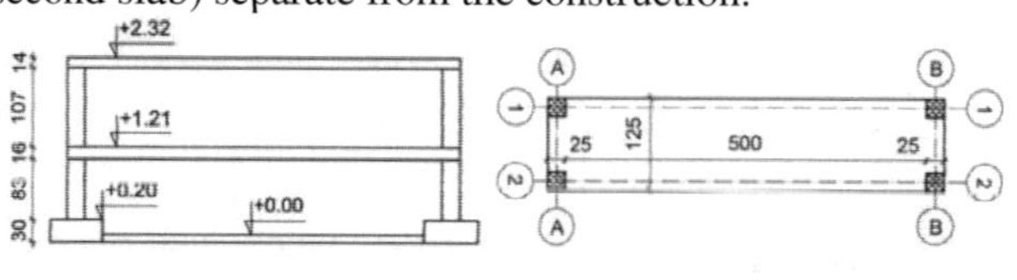

Figure 1. Formwork plan of experimental structure.

Figure 2. Stage of building structure.

Figure 3. Stage of simulating loading.

4. FE model of experimental construction

The numerical model of the construction is made with software based on FEM. For the modeling of the slabs, a two-dimensional slab finite elements are used, and the reinforced concrete columns and metal shores - are modeled with beam finite elements [11]. The shores were modeled with a composite section with dimensions such as those used in the experiment – two metal tube sections (inner and external) with a length of 50 cm, the external tube with a diameter of 60 mm and a thickness of 2 mm, the inner with a diameter of 50 mm and a thickness of 1.7 mm. The adjustable thread, ring, and pin for shore height adjustment are not taken into account and are not modeled. The column-slab node and the shores-slab node are modeled as simple joint connections. The load from the formwork system and the temporary loads placed on the slabs are modeled as point and linear loads. For each of the days, the actual values of the elastic modulus obtained from laboratory tests are entered.

5. Experimental study

The study traced the deflections and cracks in the slabs. The study was performed in 15 steps shown in Figure 4. In each step, the level of casting and striking, the corresponding additional load, the age of the concrete of the first slab (t_1), axial distances (cm) between shores, columns and additional loads, and the elastic modules for the respective age of the concrete of each slab(E_1, E_2) are shown. There are 9 stages of the construction, and the steps included in the separate stages are listed in Table 1.

5.1. Steps:

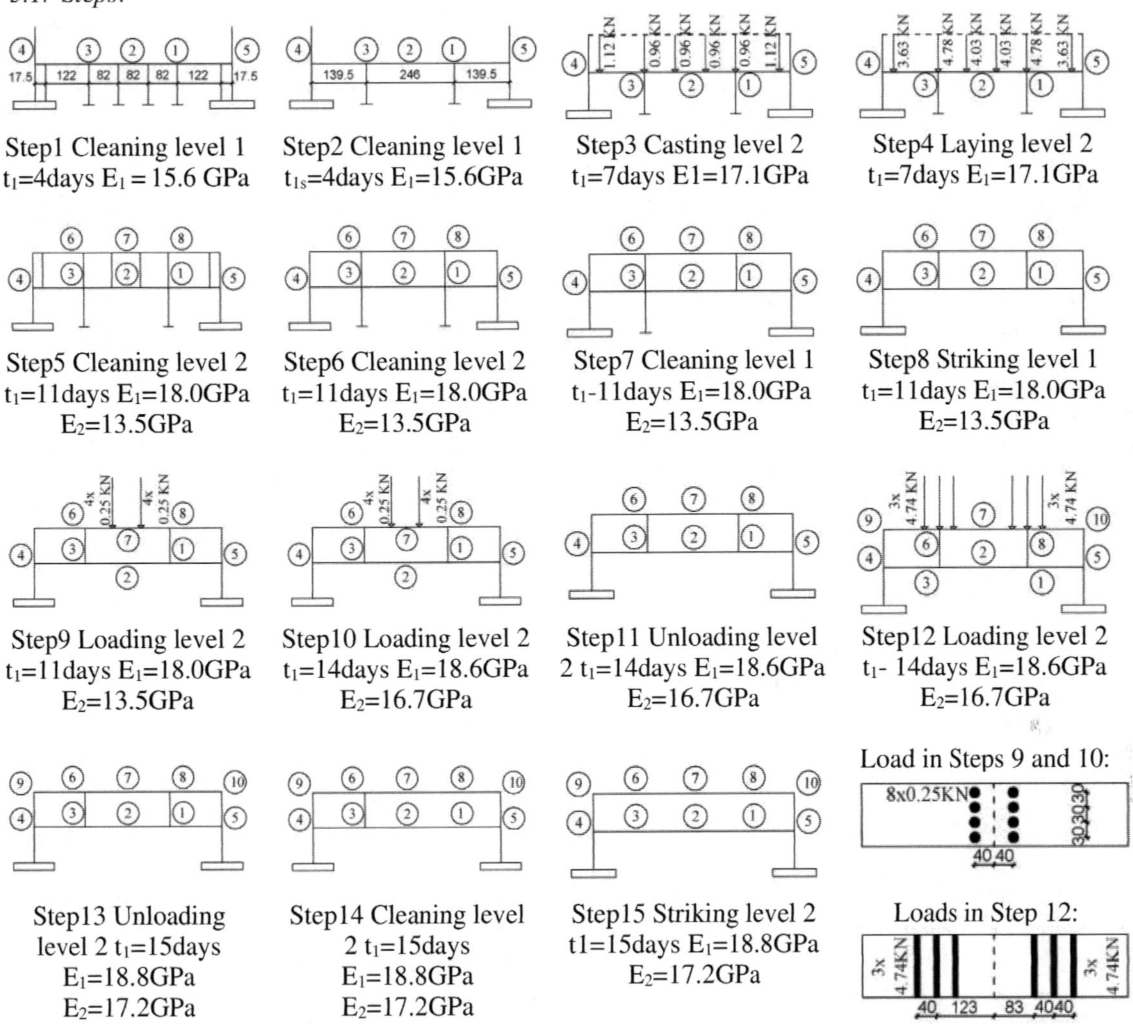

Figure 4. Constructive process.

5.2. Stages:

Table 1. Stages of the constructive process.

Stage	Step	Period (days)	Time (days)	Actions
1	1-2	4th	1	Cleaning 1st slab
2	3	4th – 7th	3	Casting 2nd slab
3	4	7th	1	Laying 2nd slab
4	4-5	7th – 11th	4	-
5	5-8	11th	0,5	Cleaning 2nd slab and striking 1st slab
6	9-10	11th – 14th	3,5	Loading 2nd slab with temporary load
7	11	14th	0,5	Unloading the temporary additional load
8	12	14th	0,5	Loading 2nd slab with temporary load
9	13-15	15	1	Removing the additional load and striking 2nd slab

6. Results and discussion

The values of the elastic deflections (Elastic Zo), the experimental measurements of the vertical deflections (Exp. Zo), and the cracks in the slabs are presented in Table 2 and Table 3.

6.1. Elastic deflections from numerical model (mm)

Table 2. Elastic deflections in each of the points.

Day, t1st slab	Step	Deflection in point (mm)									
		1	2	3	4	5	6	7	8	9	10
4th (1) day	1	0.08	0.06	0.08	0.00	0.00	0	0	0	0.00	0.00
4th (2)day	2	0.15	0.31	0.15	0.00	0.00	0	0	0	0.00	0.00
7th (1)day	3	0.19	0.38	0.19	0.00	0.00	0	0	0	0.00	0.00
7th (2)day	4	0.32	0.67	0.32	0.00	0.00	0	0	0	0.00	0.00
11th (1)day	5	0.27	0.52	0.27	0.00	0.00	0.37	0.54	0.37	0.00	0.00
11th (2)day	6	0.27	0.46	0.27	0.00	0.00	0.41	0.71	0.41	0.00	0.00
11th (3)day	7	1.63	1.47	0.47	0.00	0.00	0.67	1.70	1.67	0.00	0.00
11th (4)day	8	6.07	8.14	6.10	0.00	0.00	6.14	8.27	6.11	0.00	0.00
11th (5)day	9	6.52	8.75	6.56	0.00	0.00	6.62	8.97	6.58	0.00	0.00
14th (1)day	10	5.92	7.96	5.97	0.00	0.00	5.98	8.11	6.00	0.00	0.00
14th (2)day	11	5.51	7.40	5.54	0.00	0.00	5.57	7.48	5.54	0.00	0.00
14th (3)day	12	9.53	12.74	9.58	0.00	0.00	9.75	13.09	9.70	0.00	0.00
15h (1)day	13	5.41	7.27	5.44	0.00	0.00	5.46	7.35	5.43	0.00	0.00
15th (2)day	14	5.26	7.17	5.44	0.00	0.00	5.48	7.53	5.80	0.00	0.00
15th (3)day	15	4.66	6.27	4.68	0.00	0.00	6.68	8.95	6.64	0.00	0.00

6.2. Experimental results of deflections (mm)

Table 3. Experimental results of deflections in each of the points.

Day, t1st slab	Step	Deflection in point (mm)									
		1	2	3	4	5	6	7	8	9	10
4th (1) day	1	0.09	0.14	0.11	0.00	0.00	0	0	0	0.00	0.00
4th (2)day	2	0.30	0.48	0.27	0.00	0.00	0	0	0	0.00	0.00
7th (1)day	3	0.53	1.01	0.65	0.00	0.00	0.36	0.55	0.20	0.00	0.00
7th (2)day	4	0.99	1.66	1.06	0.00	0.00	0.69	1.06	0.65	0.00	0.00
11th (1)day	5	1.09	1.94	1.29	0.00	0.00	0.79	1.36	0.88	0.00	0.00
11th (2)day	6	1.25	2.06	1.38	0.00	0.00	1.07	1.82	1.17	0.00	0.00
11th (3)day	7	1.83	2.44	1.45	0.00	0.00	1.35	2.47	1.84	0.00	0.00
11th (4)day	8	2.51	3.81	2.75	0.00	0.00	2.22	3.60	2.34	0.00	0.00
11th (5)day	9	3.01	4.46	3.17	0.00	0.00	2.71	4.29	2.95	0.00	0.00
14th (1)day	10	4.17	6.22	4.54	0.00	0.00	3.96	5.84	4.28	0.00	0.00
14th (2)day	11	4.09	6.14	4.31	0.00	0.00	3.71	5.56	3.98	0.00	0.00
14th (3)day	12	8.93	12.42	9.08	0.00	0.00	9.07	12.32	8.97	0.00	0.00
15h (1)day	13	6.81	9.90	7.07	0.00	0.00	6.37	8.70	6.27	0.00	0.00
15th (2)day	14	6.74	9.85	7.06	0.00	0.00	6.45	8.86	6.43	0.00	0.00
15th (3)day	15	6.64	9.67	6.88	0.00	0.00	6.65	9.12	6.58	0.00	0.00

6.3. Cracking in the slabs:

Table 4. Experimental results of cracks in the slabs (mm).

Period	Point									
	1	2	3	4	5	6	7	8	9	10
until Step 11	0.00	0.00	0.00	0.00	0.00	0.00	0.00	0.00	0.00	0.00
after Step 12	0.00	0.10	0.00	0.95	0.80	0.00	0.05	0.00	0.40	0.25

6.4. Graphics of the deflections in combination of points:

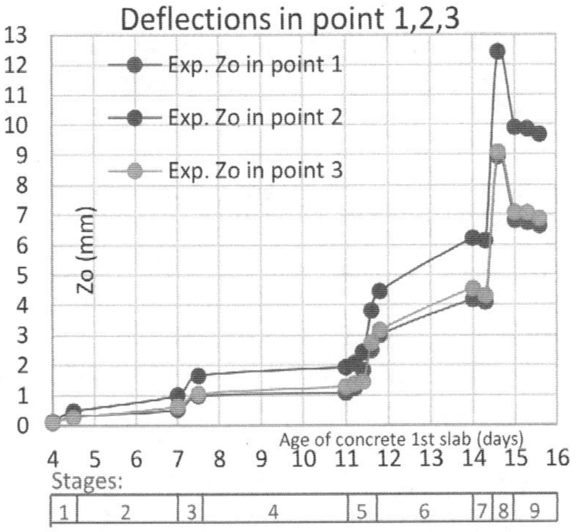

Figure 5. Deflections in point 1,2 and 3.

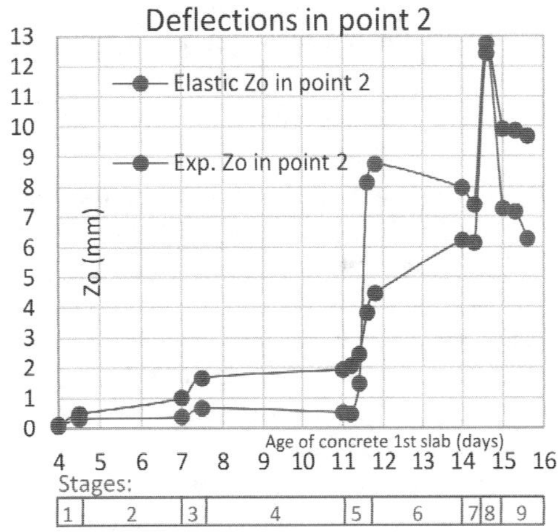

Figure 6. Deflections in point 2.

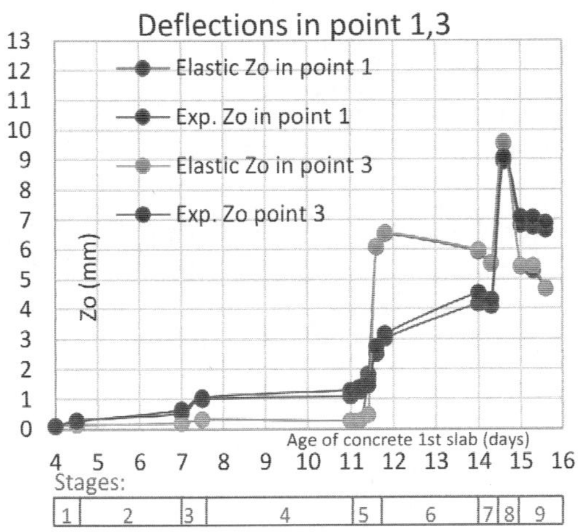

Figure 7. Deflections in point 1 and 3.

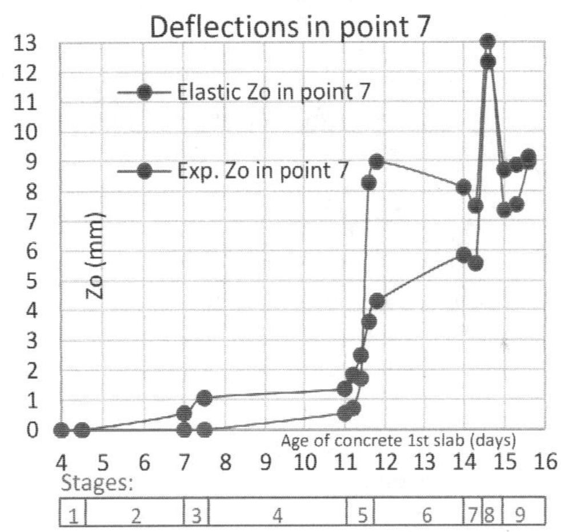

Figure 8. Deflections in point 7.

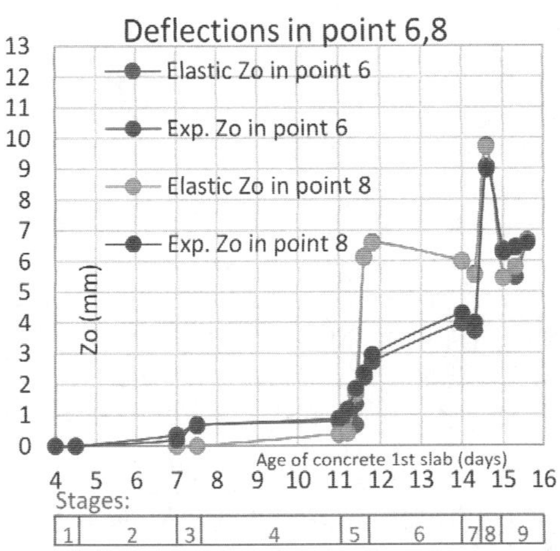

Figure 9. Deflections in point 6 and 8.

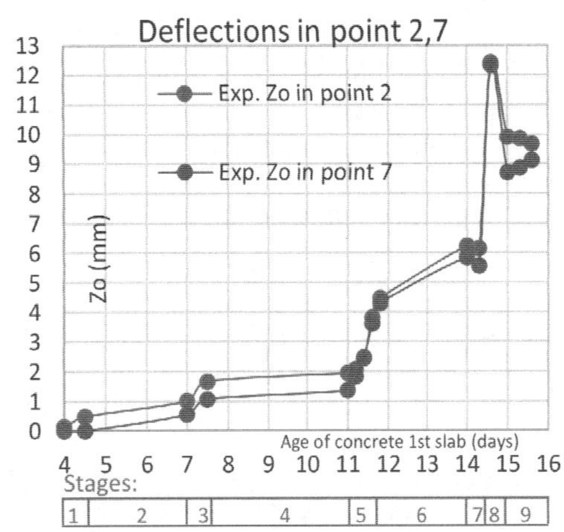

Figure 10. Deflections in point 2 and 7.

6.5. Dependencies between the deflections in the areas of the shores and those in the middle of the slab from the experimental measurements of the first and second slabs

- In stage 2 and stage 4 in Figure 5, there is a gradual increase in the deflections of point 2 from those in points 1 and 3. This means that the eccentrically placed shores of the second slab relative to the first plate affect more deflections between the two propping shores of the first slab.
- In stage 4 and stage 6 in Figure 5, continuing deflections of the slab are reported despite the lack of load change in the stages. It is more clearly expressed in stage 6, due to the higher load and complete striking of the first slab.

6.6. Dependencies between elastic and experimental deflections in the first and second slab

- From stage 1 to stage 5 of Figure 6, the elastic deflections have higher values. This is due to the non-reporting of the mechanism for adjusting the height of the shores in the model. This is the reason for their increased stiffness in the model.
- In stage 6 of Figure 6, where the first plate is completely striked, the elastic deflections have values twice as high. This is due to the modeling of node 5 and 6 as simple joint without bending moment in the node. In real construction, these nodes are not simple joints. The most unfavourable case for the maximum value of the positive moment in the middle of the main field is accepted in the modeling.
- In stage 2 of Figure 8 and Figure 10, deflections in points 6, 7 and 8 of the formwork system are observed for the construction of a second slab, without any concrete still being laid in the formwork form. This is due to the deflections from the first slab on which the formwork for the second slab steps. The deformations in the first slab are obtained as a result of the load of the formwork system for the next level.
- In stage 4 of Figure 10, the graph of point 7 of the second slab is parallel to the graph of point 2 of the second slab, since the second slab is still fully cast and takes on with slightly smaller values, the deformation curve of the first slab.
- From stage 6 to step 7 of Figure 9, the elastic deflections in point 7 are higher due to the fact that again nodes 5, 6, 9 and 10 are modeled as simple joints. In real construction, these nodes are not pure simple joints.
- All these observations are confirmed in Figure 7 and Figure 9, where the comparisons of the elastic and experimentally measured values of the deflections have the same character.

6.7. Dependencies between elastic and experimental deflections in the two slabs in point 2 and 7

- From stage 1 to stage 8 of Figure 10, the relatively parallel movement of the two curves of the graph is visible regardless of the stage of striking and loading of the slabs. This is because the weight is distributed between the two slabs by the shores between them.
- In the middle of stage 9, after the complete striking of the two slabs, a violation of the parallelism of the two curves is observed, because, after the complete striking of the second slab, no more weight is transferred from the shores to the first slab. Each slab begins to work independently.

6.8. Features of the formation of cracks in the slabs

- In stage 6 of Figure 6, after calculation in the numerical model, we should get cracks in the middle of the first slab of the order of 0.24 mm. However, in the real construction cracks do not appear. This could again be due to some inaccuracies in the numerical model in nodes 4 and 5, because again in real conditions the node has more stiffness than that in the model. Thus, the bending moment in the middle of the slab is less than expected with the introduced static scheme in the model.
- In stage 8 of Figure 6 and Figure 9, after calculation in the numerical model, we should get cracks in the middle of the field of the first and second slab of the order of 0.62mm and 0.28mm, respectively. However, in the real construction, we get cracks of 0.10 mm and 0.05 mm. The

difference in the case is that we get serious cracks in nodes 4 and unit 5, respectively 0.95mm and 0.80 mm. In nodes on 9 and 10, we get cracks of 0.45mm and 0.25mm. With the adopted static scheme, we should not have any negative bending moment and cracks in these areas of the slabs. This again clearly shows the appearance of a negative bending moment at slab-column joints.

- During the assessment of cracks, especially at the stage of preparation, at lower strength and deformation characteristics of concrete, it is important to monitor not only the widths but also the crack depths. The non-destructive ultrasonic method is successfully used for this purpose. It should also be borne in mind that in such (unfinished) structures exposed directly to atmospheric conditions, there may be some limitations in the use of the ultrasonic method. Some details as well and experimental data on the subject are given in [12].

6.9. Features regarding the strength characteristics of the two slabs

- We observe a relatively low elastic modulus of concrete. In comparison, the first slab has a cubic compressive strength obtained on day 28 equal to 22.0MPa, and the value of the elastic modulus is 21.8GPa. Data on the development of concrete in the current experimental design are given in [13]. Similar results, typical for the building practice in the region, are also presented in [14] and [15]. This is probably due to the aggregate used for coarse aggregate in the ready-mixed concrete.

7. Conclusions

From the analysis of the results in the study, we come to the following conclusions and recommendations:

- The slabs of the experimental construction work together as the transfer of the loads from the first to the second slab flows through the shores in each of the stages. It follows that each slab to be produced by the Early striking method must be checked for coverage of the bending moments and the shear forces obtained from its self-weight and from the weight of next or more slabs above it, as well as the temporary loads that are likely to occur during construction.
- When inspecting the slabs for unregulated deflections, the fact must be taken into account that each slab propped by shores takes to a large extent the deformations of the slab below it.
- The deformations in the shores directly affect the deformations in the slabs. The more detailed the models of the shores, the more accurate the results will be obtained from the slab checks using the numerical model results will be. Failure to take into account the height adjustment device of the telescopic support in the model would make the shore model a stiffer element in the model than it is. It is also desirable to avoid eccentric arrangement of the telescopic supports supporting the slabs on the floors of the building.
- When performing inspections of flat slabs before applying the Early Striking method, when modeling a column-slab node as a simple joint, it should be borne in mind that the obtained bending moments are in favor of safety in terms of positive moments, but not in favor of safety regarding the negative moments of the end and middle supports of the slab. In addition to checking for cracking and irregular deflection in the main field, an important point is to check in the negative moment areas above the columns (especially the end ones, where the negative moment is usually considered zero). At the discretion of the designer, the node can be modeled as stiffer and attention paid to the values of the negative moment in it at different stages of the checks for striking and loading the slab. This is recommended because in case of possible loading from a higher slab (using shores), placing heavy temporary loads on one of the plates and insufficient reinforcement in the upper zone of the plate at the node (obtained from the calculations at the initially accepted simple joint connection of the node), the cross-section of the slab in the upper zone of the slab and near the column can be compromised by unwanted cracks.
- It is recommended that before starting the construction of buildings, research be done on what elastic modulus the concrete, which will be used for the construction of the structure has.

IOP Conf. Series: Materials Science and Engineering **951** (2020) 012010 doi:10.1088/1757-899X/951/1/012010

Acknowledgments
Acknowledgments to building companies “Tehstroy” Ltd. and “DOKA” Bulgaria, and eng. Radoslav Stafunski for the assistance provided in the construction of the experimental construction.

References
[1] Alvorado Y, Calderon P, Gasch I and Torres B 2016 Study of technical and economical alternatives of a shoring and striking process during the construction of a building with reinforced concrete slab floors *Revista Ingenieria de Construccion* 31 (No 2) 117-126.
[2] Brazas G, Daukšys M, Šadauskienė J, Augonis M and Kelpša Š 2016 Productivity Analysis of Concrete Slab Construction by Using Different Types of Formwork *Journal of Sustainable Architecture and Civil Engineering* **2** (No.15) 38-46.
[3] Alvorado Y, Calderon P, Gasch I and Adam J 2010 Experimental and numerical study during construction of multistory buildings using clearing or partial striking, XXXVII IAHS World Congress on Housing
[4] Gasch I, Alvorado Y, Calderon P and Ivorra S 2013, Construction loads using a shoring– clearing striking process *Structures and Buildings* **167** (Issue SB4) 217-229.
[5] Buitrago M, Alvorado Y, Adam J and Calderon P 2016 Maximum loads on shores during the construction of buildings, *ICE Proceedings Structures and Buildings*
[6] Buitrago M, Adam J, Moragues J and Calderon P 2018 Load transmission between slabs and shores during the construction of RC building structures – A review *Engineering Structures* **173** 951–959.
[7] Pallet P 2019 Backpropping flat slabs, (Pallet Temporary Work), PFP/136 F
[8] Zigmunt O 2010 The analysis of slab resistance during early striking exemplified by Gridflex Peri formwork system, Vilnius, Lithuania The 10th International Conference.
[9] Hrischev L, Tsvetkov S and Nedelchev N 2015 Early striking calculations of formwork for reinforced concrete slabs according to Eurocode 2, VII International scientific conference architecture, civil engineering – modernity Varna Bulgaria.
[10] Buitrago M, Adam J and Calderon P 2018 Structural failures in cast-in-place RC building structures under construction, ICITECH, Universitat Politècnica de València. Camino de Vera s/n, 46022 Valencia, Spain
[11] Kazakov K 2010 Finite Element Method for structural modeling, Prof. Marin Drinov Academic Publishing House, Sofia, Bulgaria
[12] Ivanchev I and Slavchev V 2019 About the Possible Limitations in the Usage of the Non-Destructive Ultrasonic Pulse Velocity Method for Assessment of Cracks in Reinforced Concrete Structures, Subjected to Direct Environmental Exposure *Buildings* **9** (Issue 9) 202
[13] Bakardzhiev S and Lazova O 2019 Research of the strength characteristics of concretes in the quality receiving period, (XIX International scientific conference by construction and architecture VSU’2019)
[14] Ivanchev I, Slavchev V, Dobreva I and Daalov B 2018 Experimental determination of modulus of elasticity and stress-strain diagrams of concrete, XVIII Anniversary international scientific conference by construction and architecture VSU'2018.
[15] Ivanchev I and Slavchev V 2019 Probable compressive strength and modulus of elasticity of concrete, determined by non-destructive ultrasonic pulse velocity method (NDUPVM) in different ways of measuring transducers placement, XXIX International Scientific Symposium 'Metrology and Metrology Assurance; Sozopol, September 6-10, Bulgaria

IOP Conf. Series: Materials Science and Engineering **951** (2020) 012011 doi:10.1088/1757-899X/951/1/012011

Effect of depth of epoxy coated lapped splices on bond efficiency

K B Osifala

Department of Civil Engineering, Yaba College of Technology, Yaba, Lagos, Nigeria.

E-mail: kosifala@yahoo.com

Abstract: This paper gives a report of research investigation carried out to study the effect of depth of epoxy coated lapped spliced bars on bond strength and efficiency. Thirty-six full size beams of varying lengths and sectional dimensions with lap spliced bars in constant moment region were cast and tested in a four-point bending system. Bond strength and efficiency of epoxy coated top reinforcing bars were investigated using beams with 300 mm and 500 mm depths. Three beam sizes (300 × 180 mm, 300 × 200 mm and 300 × 230 mm) and high yield reinforcing bars (16 mm, 20 mm and 28 mm) were coated with epoxy over the lap length and used with each depth. The ultimate moments from the tests were used to determine the stress developed in the reinforcing bars. Coated top reinforcing bars were found to have lower bond strength and efficiency than coated bottom reinforcing bars in the depths investigated. The location factor, α, which is a reduction factor determined using an average of three beams, for epoxy coated top bars were found to be 0.86, 0.84 and 0.89 respectively for 16 mm, 20 mm and 28 mm diameter bars at 300 mm depth and 0.62, 0.80 and 0.86 respectively for 16 mm, 20 mm and 28 mm diameter bars at 500 mm depth. .

1. Introduction

In reinforced concrete structures, it is sometimes necessary to reinforce members at both the top and bottom (doubly reinforced section). The need arises when the depth of beam is limited and the tensile reinforcement is not adequate to take the moment. In continuous beams, top bars are needed at the supports, so also, cantilevered slabs and beams require the use of top reinforcements to take care of the negative moments. In these cases, it has been found that bond resistance between concrete and uncoated reinforcing bars at the top is reduced as a result of the cavitation created by the air trapped under the bars during vibration of the concrete, the air trapped under the bars may further reduce the bond resistance, of coated top bars. Bond strength is known to be reduced for top bars and coating has been severally confirmed to reduce bond resistance. The extent of further reduction in bond strength when epoxy coated bars are used as top reinforcement is expected to be determined by this investigation.

One of the most influential but least precisely defined variable influencing bond is the position of the embedded bar during casting and hence the amount of concrete poured in a single process under the bar. Uncoated reinforcing bars have been used to date in investigating top reinforcing bars, the results showed a reduced bond strength (Ferguson and Thompson [1], Morita et-al [2]). No attempt has yet been made to investigate epoxy coated top reinforcing bars. Investigation of bond efficiency of epoxy coated top reinforcing bars is worthwhile to increase our knowledge on the extent of reduction or otherwise on bond strength.

IOP Conf. Series: Materials Science and Engineering **951** (2020) 012011 doi:10.1088/1757-899X/951/1/012011

2. Design of test beams

Thirty-six (36) beams were cast for the investigation. Twelve (12) and twenty-four (24) beams were cast for the preliminary and confirmatory investigations respectively. Two depths (300 mm and 500 mm) were investigated and the widths were of three sizes, 180 mm, 200 mm and 230 mm. Epoxy coated high yield, 28 mm, 20 mm and 16 mm diameter bars were used.

Contact arrangement of lapped bars was adopted in this investigation because it is the most probable on construction sites. The loading points were 150 mm from the lap ends thereby making the distance between the point loads to be l_s + 300 mm. The ultimate anchorage bond length recommended in BS 8110: Part I 1988 [4] was used.

$$l_s = \frac{0.87 fy\phi}{4\beta\sqrt{f_{cu}}}. \tag{1}$$

Failure mode in flexure is strongly dependent on the shear-span/effective depth ratio, a_v/d. Based on reported experimental studies in Kong and Evans [5] the failure pattern to ensure flexural failure was used. a_v/d was taken as 3, hence a_v = 3d. The depth of all beams was either 300 mm or 500 mm while the length is a function of lap length and shear span, thus length $\ell = \ell_s$ + 300 mm + 2a_v + 200 mm. The lengths varied with the bar diameters as 2.75 m, 2.90 m, 3.20 m, 3.95 m and 4.00 m respectively. The design of the beams is shown in figure 1 and the beam data are shown in table 1.

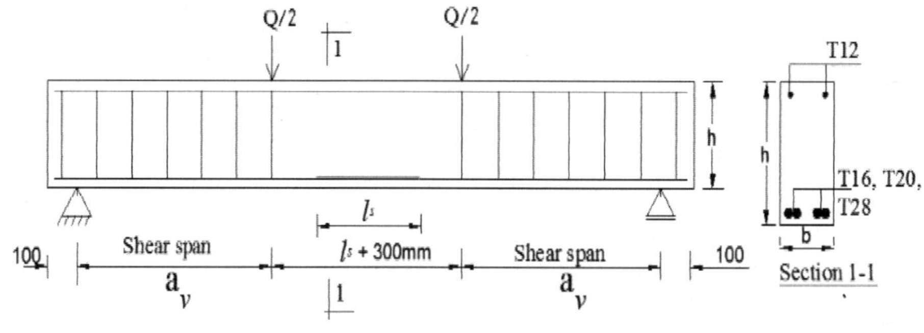

Figure 1. Typical beam without stirrup.

3. Preparation and application of coating materials

The reinforcements were sand blasted to remove rust.

3.1. Data collection

The Epoxy zinc rich is in two parts. Part A, base and Part B, which is the hardener. The parts were measured out by volume in ratio 3:1, as recommended in the manufacturer's manual, and the two parts were thoroughly mixed together. Thinner was added until the mixture was light enough to pass through the nozzle of a conventional spraying gun. The lengths of the reinforcements to be coated were marked out. The mixture was then sprayed, in one coat, on the marked portions of the reinforcements and allowed to dry in the shade without direct exposure to sunshine.

4. Manufacture and test procedure of cubes and beams

The Concrete used for the investigation was Grade 35 with targeted cube strength of 35 N/mm^2. Concrete cubes were cast in batches and cured for 28 days in the laboratory. The strengths achieved at test are in table 2. The reinforcement cages were put in the mould and concrete spacers were placed at interval at various locations; bottom and sides. The ready-mix concrete was poured and vibrated in the mould with poker vibrator. After demoulding, the beams were cured by wetting at intervals for seven days.

Table 1. Beam Data.

Beam Name	*h* (mm)	*b* (mm)	Tensile Reinft.	As (mm^2)	Lap Lgth.+ 300 mm	Shear Span av (mm)	Beam Length (m)
D-T-4A	300	180	2T16	402	940	801	2.75
D-T-4B	300	180	2T16	402	940	801	2.75
D-T-4C	300	180	2T16	402	940	801	2.75
D-T-5A	300	200	2T20	628	1100	795	2.90
D-T-5B	300	200	2T20	628	1100	795	2.90
D-T-5C	300	200	2T20	628	1100	795	2.90
D-T-6A	300	230	2T28	1232	1420	783	3.20
D-T-6B	300	230	2T28	1232	1420	783	3.20
D-T-6C	300	230	2T28	1232	1420	783	3.20
D-T-10A	500	180	2T16	402	940	1400	3.95
D-T-10B	500	180	2T16	402	940	1400	3.95
D-T-10C	500	180	2T16	402	940	1400	3.95
D-T-11A	500	200	2T20	628	1100	1350	4.00
D-T-11B	500	200	2T20	628	1100	1350	4.00
D-T-11C	500	200	2T20	628	1100	1350	4.00
D-T-12A	500	230	2T28	1232	1420	1190	4.00
D-T-12B	500	230	2T28	1232	1420	1190	4.00
D-T-12C	500	230	2T28	1232	1420	1190	4.00
D-B-1A	300	180	2T16	402	940	801	2.75
D-B-1B	300	180	2T16	402	940	801	2.75
D-B-1C	300	180	2T16	402	940	801	2.75
D-B-2A	300	200	2T20	628	1100	795	2.90
D-B-2B	300	200	2T20	628	1100	795	2.90
D-B-2C	300	200	2T20	628	1100	795	2.90
D-B-3A	300	230	2T28	1232	1420	783	3.20
D-B-3B	300	230	2T28	1232	1420	783	3.20
D-B-3C	300	230	2T28	1232	1420	783	3.20
D-B-10A	500	180	2T16	402	940	1400	3.95
D-B-10B	500	180	2T16	402	940	1400	3.95
D-B-10C	500	180	2T16	402	940	1400	3.95
D-B-11A	500	200	2T20	628	1100	1350	4.00
D-B-11B	500	200	2T20	628	1100	1350	4.00
D-B-11C	500	200	2T20	628	1100	1350	4.00
D-B-12A	500	230	2T28	1232	1420	1190	4.00
D-B-12B	500	230	2T28	1232	1420	1190	4.00
D-B-12C	500	230	2T28	1232	1420	1190	4.00

Remarks: D-T - Top cast epoxy coated; D-B - Bottom cast epoxy coated; Beams are without stirrup over lap; Side and bottom covers are 25 mm thick.

4.1. Compressive strength of concrete cubes
The aim of the test was to determine the strength of concrete in the beams and to serve as a control so that the compressive strength reach at least the targeted value of 35 N/mm^2. The cubes were weighed on Avery weighing machine, 50 kg capacity and tested in the Avery Universal testing Machine of 100,000 kg capacity.

4.2. Beam test
The beams were tested with a four-point bending arrangement with an I steel section as the spreader beam in a 100,000 kg capacity Avery Universal Testing Machine to determine the ultimate failure loads, modes and crack patterns. The load was gradually applied until the failure of the beam occurred. The failure load was recorded and analysed as follows

Table 2. Characteristics of Hognestad Et -Al's Stress Block.

S/N	fcu (N/mm^2)	k_1	k_2	Σcu
1.	31.09	0.611	0.457	0.0034
2.	36.98	0.578	0.449	0.0033
3.	35.06	0.586	0.451	0.0033
4.	37.90	0.575	0.450	0.0033
5.	36.98	0.578	0.449	0.0033
6.	37.90	0.575	0.4 5	0.0033
7.	37.90	0.575	0.45	0.0033
8.	36.98	0.578	0.449	0.0033
9.	35.10	0.586	0.452	0.0033
10.	34.44	0.594	0.454	0.0034
11.	37.90	0.575	0.450	0.0033
12.	36.98	0.578	0.449	0.0033
13.	38.70	0.574	0.449	0.0033
14.	33.30	0.596	0.456	0.0 034
15.	30.44	0.614	0.457	0.0034
16.	37.88	0.578	0.451	0.0033
17.	33.12	0.598	0.455	0.0034
18.	30.51	0.616	0.459	0.0034
19.	26.56	0.638	0.463	0.0035
20.	28.66	0.625	0.461	0.0035
21.	32.03	0.603	0.457	0.0034
22.	33.28	0.598	0.453	0.0034

5. Analysis of test results

The steel stress developed in each beam was determined by analysing the section using the general theory for ultimate flexural strength. The ultimate moment for each beam was obtained by applying the principles of statics. The lapped splice was in the constant moment region. The ultimate moment at the point load in figure 2 is:

$$M_u = \frac{Q}{2}(a_v). \quad (2)$$

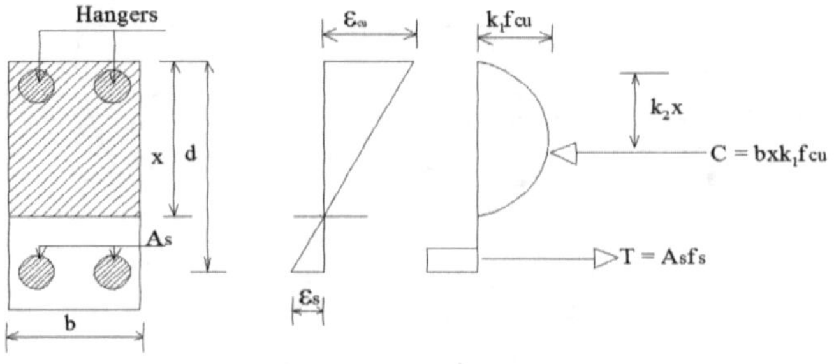

Figure 2. Strain and Stress distribution at failure.

From the equilibrium condition in figure 2

$$k_1 f_{cu} bx = A_s f_s. \quad (3)$$

Taking moment about the compressive force, the ultimate moment of resistance is

$$M_u = A_s f_s(d - k_2 x), \quad (4)$$

from where

$$\frac{M_u}{bd^2} = \rho f_s - \frac{k_2}{k_1}\frac{(\rho f_s)^2}{f_{cu}}. \tag{5}$$

The properties of the concrete stress block are expressed in terms of the characteristic ratio k_1 and k_2. Stress block proposed by Hognestad et-al reported in Kong and Evans [5] and universally accepted was used. Based on the mean cube strength, various values of k_1 and k_2 and Σ_{cu} are tabulated in table 2. The quadratic equation in (5) was solved yielding two values of f_s, one of which was always unreasonable because it is not practicable. The reasonable value was the stress developed in the steel.

5.1. *Test Bond Stress, τ_t*

The bond strength is the average stress along the length of the splice. From equilibrium

$$A_s f_s = \tau_t \pi \phi_e l_s, \tag{6}$$

from where:

$$\tau_t = \frac{f_s \phi_e}{4 l_s}, \tag{7}$$

The steel stress developed in each beam that was determined from equation (5) was substituted in (7) to obtain the τ_t (τ_{test}).

5.2. *Bond stress using the Yield Stress of Tensile Reinforcement, τ_y*

$$\tau_y = \frac{f_s \phi_e}{4 l_s}, \tag{8}$$

Table 3. Tensile Test Results.

Bar Size	Yield		Ultimate		Elongation
(mm)	Load (tons)	Stress (N/mm^2)	Load (tons)	Stress (N/mm^2)	%
T8	3.00	594.27	4.34	851.79	13.00
T8	2.84	554.66	4.24	831.98	13.54
T8	2.94	574.47	4.34	851.79	12.00
AVRG		574.47		845.19	12.83
T10	4.02	510.26	4.90	621.96	12.50
T10	4.16	528.03	5.12	649.88	14.50
T10	4.20	533.11	5.15	653.69	15.00
T10	4.05	514.07	5.00	634.65	14.50
AVRG		521.37		640.045	14.125
T12	5.85	515.38	7.54	660.74	16.50
T12	5.95	524.19	7.55	665.15	15.50
T12	6.00	528.59	7.52	662.51	17.50
T12	5.90	519.78	7.42	653.69	15.50
AVRG		521.985		660.523	16.25
T16	11.10	549.98	13.35	661.46	16.50
T16	10.84	535.11	12.60	624.34	14.00
T16	10.35	512.82	12.70	629.25	14.50
AVRG		532.64		638.34	15.00
T20	16.50	523.25	19.70	624.73	15.00
T20	16.50	523.25	19.90	631.07	16.00
T20	17.20	545.45	20.10	637.42	17.50
AVRG		530.65		631.07	16.17
T28	36.90	599.26	42.00	679.58	15.00
T28	35.70	577.65	43.34	700.62	15.00
T28	35.00	566.32	41.24	666.64	16.00
AVRG		581.08		682.28	15.33

IOP Conf. Series: Materials Science and Engineering **951** (2020) 012011 doi:10.1088/1757-899X/951/1/012011

5.3. Theoretical Bond Stress, τ_{cal}
The theoretical bond stress, τ_{cal}, was determined from a semi-empirical statistical regression equation developed by Orangun *et-al* [6].

For bars with no transverse reinforcement

$$\tau_{\text{cal}} = \left[1.2 + 3\left(\frac{c}{d_b}\right) + 50\left(\frac{d_b}{l_s}\right)\right] f_c^{1(0.5)}, \tag{9}$$

Equations (9) is in imperial unit and concrete cylinder strength. The expressions were converted to SI units and modified for use with concrete cube strength by Cairns and Abdullah [3] to give:

$$\tau_{\text{cal}} = \left[0.09 + 0.24\frac{c}{d_b} + 3.9\frac{d_b}{l_s}\right]\sqrt{f_{cu}}\,. \tag{10}$$

The bond strength equation in (10) proposed by Orangun *et-al* [6] was adopted as the formula for theoretical bond stress in this work. It formed the basis of the bond strength equation in the 1995 and 2002 American Concrete Institute (ACI) codes. Leet, Kenneth [7].

5.4. Bond Stress using BS 8110's Recommendation, τ_{BS8110}
The British Standard, BS 8110's, 1997[8] recommended bond stress was also used to calculate the bond efficiency.

$$\tau_{\text{BS}} = \beta\sqrt{f_{cu}}\,. \tag{11}$$

with the factor β, taken as 0.5 because deformed bars were used in tension.

6. Discussion of test results
Effect of Depth of Epoxy Coated Top Reinforcement on Bond Efficiency

The effect of depth of epoxy coated top reinforcing bars on bond efficiency was investigated using beams with two depths, 300 mm and 500 mm. Three beam sectional sizes (300×180 mm, 300×200 mm, 300×230 mm) and varying high yield reinforcing bars (16 mm, 20 mm and 28 mm diameter) were used for each depth. One beam each was cast in the preliminary investigation. The beam data and the test results are in tables 1 and 4; figure 3.

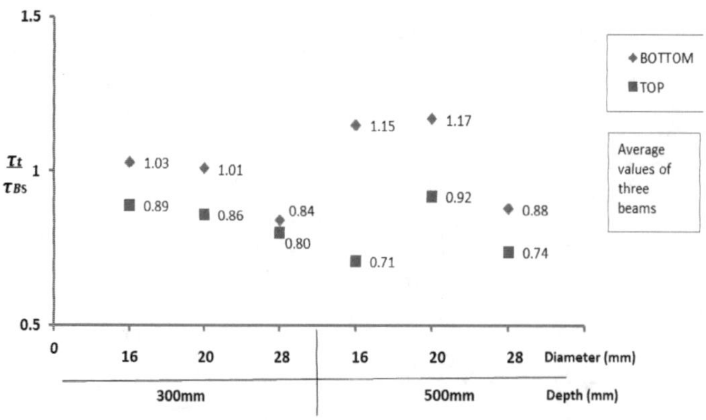

Figure 3. Effect of Depth of Epoxy coated bars on bond efficiency.

Generally, the preliminary tests showed that bond efficiency of coated top reinforcing bars were lower than that of coated bottom reinforcing bars. The decrease in bond efficiency did not follow any regular pattern for the various diameters of bars nor for the depths investigated. Confirmatory tests were carried out with two additional beams each for the bar diameters and depths used in the preliminary beams. The trend observed in the preliminary tests was confirmed in the confirmatory tests. Coated top reinforcing bars were found to have a lower bond strength and efficiency than coated bottom reinforcing bars at the two depths. This is in line with previous research findings on uncoated top reinforcing bars by Tepfers [9], Jirsa [10], Donahey and Darwin [11]. The lower bond efficiency recorded in top coated reinforcing bars may be due to the cavities created underneath the reinforcing bars by the escaping air bubble during the vibration of the concrete.

If the distance from the base of the beam to the top bar is considered, it may be said that the higher the depth the lower the bond strength and efficiency. This analogy is confirmed in beams with 16 mm diameter bars as the difference in the average bond efficiency of top and bottom cast bars is 0.14 for 300 mm deep beams and 0.44 for 500 mm deep beams. This was also the pattern for beams with 20 mm diameter bars, as the difference in the bond efficiency for 300 mm deep beams is lower than that of 500 mm deep beams, by 0.15 and 0.25 respectively. Beams with 28 mm diameter bars had 0.04 as difference for 300 mm deep beams and 0.14 difference for 500 mm deep beams. If the performance of coated 16 mm, 20 mm and 28 mm diameter top cast bars are compared at the two depths, a reduction in average bond efficiency of 20.2%, -6.98% and 7.5% is observed respectively. The negative reduction in bond efficiency for 20 mm was not expected, more tests should be carried out to have a better knowledge. The bottom cast bars showed the reverse, that is, an increase in bond efficiency. Bond efficiency for bottom cast is expected to be higher with the depth of concrete as in figure 3, depending on the cube strength of concrete, level of compaction, lever arm etc.

Table 4. Beam Test Results.

Beam Name	f_{cu} (N/mm^2)	M_u (kN.m)	f_s (N/mm^2)	τ_t (N/mm^2)	τ_{cal} (N/mm^2)	τ_{BS} (N/mm^2)	τ_y (N/mm^2)	$\frac{\tau_t}{\tau_y}$	$\frac{\tau_t}{\tau_{cal}}$	$\frac{\tau_t}{\tau_{BS}}$	$\left(\frac{\tau_t}{\tau_{cal}}\right)_{av}$	$\left(\frac{\tau_t}{\tau_{BS}}\right)_{av}$	$\left(\frac{\tau_t}{\tau_y}\right)_{av}$	Mode of failure
D-T-4A	28.66	24.03	235.86	1.47	3.01	2.68	3.33	0.44	0.49	0.55				Splitting
D-T-4B	30.44	40.05	407.02	2.54	3.10	2.76	3.33	0.76	0.82	0.92	0.79	0.89	0.79	Splitting
D-T-4C	37.88	56.87	589.78	3.69	3.46	3.08	3.33	1.17	1.06	1.20				Splitting
D-T-5A	33.28	58.83	395.74	2.47	2.81	2.88	3.32	0.74	0.88	0.86				Splitting
D-T-5B	33.12	60.42	408.48	2.55	2.81	2.88	3.32	0.77	0.91	0.89	0.88	0.86	0.74	Splitting
D-T-5C	30.51	54.86	369.03	2.31	2.69	2.76	3.32	0.70	0.86	0.84				Splitting
D-T-6A	28.66	75.95	276.62	1.73	2.15	2.68	3.63	0.48	0.80	0.65				Splitting
D-T-6B	33.12	97.29	365.56	2.28	2.31	2.88	3.63	0.63	0.99	0.79	0.99	0.8	0.60	Splitting
D-T-6C	26.56	98.66	393.76	2.46	2.07	2.58	3.63	0.68	1.19	0.96				Splitting
D-T-10A	32.03	54.64	301.32	1.88	3.18	2.83	3.33	0.56	0.59	0.67				Splitting
D-T-10B	33.30	67.20	373.26	2.33	3.25	2.89	3.33	0.70	0.72	0.81	0.63	0.71	0.60	Splitting
D-T-10C	30.44	53.20	293.45	1.83	3.10	2.76	3.33	0.55	0.59	0.66				Splitting
D-T-11A	33.28	105.3	383.16	2.39	2.81	2.88	3.32	0.72	0.85	0.83				Splitting
D-T-11B	37.90	122.85	448.77	2.80	3.00	3.08	3.32	0.84	0.93	0.91	0.94	0.92	0.84	Splitting
D-T-11C	38.70	137.70	506.51	3.17	3.03	3.11	3.32	0.95	1.04	1.02				Splitting
D-T-12A	31.09	164.22	317.28	1.98	2.24	2.79	3.63	0.55	0.89	0.71				Splitting
D-T-12B	37.90	190.4	367.67	2.29	2.47	3.08	3.63	0.63	0.93	0.75	0.92	0.74	0.63	Splitting
D-T-12C	38.70	195.16	377	2.36	2.50	3.11	3.63	0.72	0.94	0.76				Splitting
D-B-1A	31.09	62.64	370.98	2.32	3.14	2.79	3.33	0.70	0.74	0.93				Splitting
D-B-1B	36.98	72.25	471.98	2.94	3.42	3.04	3.33	0.88	0.86	0.97	0.91	1.03	0.91	Splitting
D-B-1C	37.90	85.07	589.58	3.86	3.46	3.08	3.33	1.16	1.12	1.25				Splitting
D-B-2A	31.09	88.40	430.12	2.69	2.72	2.79	3.32	0.81	0.99	0.96				Splitting
D-B-2B	37.90	77.91	539.32	3.37	3.00	3.08	3.32	1.02	1.12	1.10	1.08	1.01	0.89	Splitting
D-B-2C	36.98	92.38	452.16	2.83	2.96	3.04	3.32	0.85	0.95	0.91				Splitting
D-B-3A	31.09	125.44	384.79	3.37	2.46	2.79	3.63	0.93	1.07	0.86				Splitting
D-B-3B	37.90	142.86	452.73	2.83	2.47	3.08	3.63	0.78	1.14	0.92	1.04	0.84	0.78	Splitting
D-B-3C	36.98	122.70	358.61	2.24	2.44	3.04	3.63	0.62	0.92	0.74				Splitting
D-B-10A	37.88	92.40	518.69	3.24	3.46	3.08	3.33	0.97	0.94	1.05				Splitting
D-B-10B	30.44	91	515.84	3.22	3.10	2.76	3.33	0.97	1.04	1.17	1.03	1.15	1.00	Splitting
D-B-10C	33.30	100.1	568.74	3.55	3.25	2.89	3.33	1.07	1.1	1.23				Splitting
D-B-11A	34.44	148.5	554.62	3.47	2.86	2.93	3.32	1.05	1.21	1.18				Splitting
D-B-11B	36.98	124.88	457.29	2.86	2.96	3.04	3.32	0.86	0.96	0.94	1.19	1.17	1.05	Splitting
D-B-11C	35.10	172.13	652.63	4.08	2.89	2.96	3.32	1.23	1.41	1.38				Splitting
D-B-12A	36.98	195.16	378.61	2.37	2.44	3.04	3.63	0.65	0.97	0.79				Splitting
D-B-12B	34.44	209.44	412.68	2.58	2.36	2.93	3.63	0.71	1.09	0.88	1.09	0.88	0.71	Splitting
D-B-12C	35.10	229.67	457.93	2.86	2.38	2.96	3.63	0.78	1.2	0.97				Splitting

From the test results plotted in figure 3, it is generally observed that the bond efficiency of top cast bars was lower for 500 mm deep beam than for 300 mm deep beam except with 20 mm diameter bars. The reverse is the case for bottom cast bars, bond efficiency of bottom cast bars was generally high for 500 mm deep beams than for 300 mm deep beam. The effect of depth appeared to be more obvious with top coated reinforcement.

Cavities under the reinforcing bars, which reduces bond efficiency, could be caused by the escaping air bubbles which settle underneath the top bars, as well as the effect of settlement of the concrete. The reinforcement is not completely rigid in reality and this reduces the effect of the damage due to settlement. Also, the concrete at the top of a member is generally less compacted and also tends to have greater water content, resulting in lower concrete strength and of course lower bond strength.

Location Reduction factor, α

The variation in concrete strength was eliminated by multiplying the test splice strength with

$$\sqrt{35/f_{cu}}\,. \tag{12}$$

From the test results, in figure 3, the bond strength of coated top reinforcements was found to be lower than the bond strength of coated bottom reinforcements. Attempt was made at deducing a factor for the reduction with increasing height using the average value of the splice strength of the three beams for each bar diameter to determine the ratio of the corrected bond strength of coated top bar to the corrected bond strength of coated bottom reinforcement.

$$\alpha = \frac{\tau_t(top)\sqrt{35/f_{cu}}}{\tau_t(bottom)\sqrt{35/f_{cu}}}\,. \tag{13}$$

The average of the location reduction factors, α, for coated top bars were found to be 0.86, 0.84 and 0.89 for 300 mm depth and 0.62, 0.80 and 0.86 respectively for 16 mm, 20 mm and 28 mm diameter bars at 500 mm depth. The higher the depth of concrete, the lower the bond strength and efficiency. The effect of bar diameter was not consistent with the depth of concrete below the bar, table 5.

Table 5. Location Reduction Factor, α

Depth of Concrete (mm)	Diameter of bar (mm)	$\alpha = \frac{\tau_t(top)\sqrt{35/f_{cu}}}{\tau_t(bottom)\sqrt{35/f_{cu}}}$	Average
300	16	0.66	0.86
		0.95	
		0.96	
	20	0.89	0.84
		0.72	
		0.90	
	28	0.53	0.89
		0.86	
		1.29	
500	16	0.63	0.62
		0.69	
		0.54	
	20	0.70	0.80
		0.97	
		0.74	
	28	0.91	0.86
		0.85	
		0.78	

7. Conclusion

Based on the test results and analysis of test data it can be concluded that:

- Coated top reinforcing bars were found to have a reduced bond strength and efficiency when compared to coated bottom reinforcing bars;
- Bond strength and efficiency reduced, the higher the depth of concrete below the coated top bars except for 20 mm diameter bars and
- The reduction factor for 300 mm thick concrete below the reinforcing rods was 0.86, 0.84 and 0.89 for 16 mm, 20 mm and 28 mm diameter bars respectively; and for 500 mm of thick concrete below the reinforcing bars was found to be 0.62, 0.80 and 0.86 for 16 mm, 20 mm and 28 mm diameter bars respectively.
- Comparison made with coated and uncoated reinforcing bar confirmed that bond strength and efficiency of coated bar is lower than that of uncoated bars. (figure 4. see Appendix)

Acknowledgments

Late Emeritus Prof. C. O. Orangun's interest, criticism and useful suggestions are gratefully acknowledged.

Julius Berger Nig. Plc provided the materials, labour used in the construction of the formworks, reinforcement cages, casting, curing and transportation to the Concrete Laboratory, University of Lagos. The assistance is thankfully acknowledged.

Berger Paints Nig. Ltd. supplied the Epoxy coating used for the investigation. Vinyl Chloride and other coatings were supplied by International Paints for West Africa (IPWA). The assistance is appreciated and acknowledged. Sand blasting and coating of the steel rods were carried out free of charge by Messrs John Alapo Nig. Ltd. The assistance is thankfully acknowledged. The technical staff of the Concrete Laboratory, University of Lagos assisted in testing the beams, cubes and reinforcements, the assistance is acknowledged.

Appendices

Appendix A.

Notations

A_s	area of tension reinforcement	l	anchorage length
α_v	shear span	l_s	lap length
b	width of beam	l_d	development length
c	minimum cover	x	depth of neutral axis
c_b	thickness of bottom concrete cover	Q	applied load
c_s	thickness of side concrete cover	M_u	ultimate test bending moment
d	effective depth of beam	BS	British Standard
d_b	diameter of bar	β	bond coefficient
f_{bu}	ultimate anchorage bond stress	ε_{cu}	strain in concrete in compression
f^1_c	cylinder strength of concrete	ε_s	strain in tension reinforcement
f_{cu}	characteristics cube strength	ρ	steel ratio, (A_s/bd)
f_y	characteristics strength of reinforcement	τ_{cal}	theoretical bond stress
k	ratio of steel stress	τ_t	test bond stress
k_1	ratio of the average compressive stress to the characteristic cube strength f_{cu}	τ_{BS}	ultimate bond stress recommended in BS 8110: 1997
k_2	ratio of the depth of the centroid of the stress block to the neutral axis depth	ϕ_e	effective bar size or nominal diameter of bar

Appendix B.

Table 6. Beam Data for Uncoated Bars.

Beam Name	h (mm)	b (mm)	Tensile Reinft.	As (mm^2)	Lap Lgth.+300mm	Shear Span av (mm)	Beam Length (m)
D-UBR-10A	300	180	2T16	402	940	801	2.75
D-UBR-10B	300	180	2T16	402	940	801	2.75
D-UBR-10C	300	180	2T16	402	940	801	2.75
D-UBR-11A	300	200	2T20	628	1100	795	2.90
D-UBR-11B	300	200	2T20	628	1100	795	2.90
D-UBR-11C	300	200	2T20	628	1100	795	2.90
D-UBR-12A	300	230	2T28	1232	1420	783	3.20
D-UBR-12B	300	230	2T28	1232	1420	783	3.20
D-UBR-12C	300	230	2T28	1232	1420	783	3.20
D-UTR-4	300	180	2T16	402	940	801	2.75
D-UTR-5	300	200	2T20	628	1100	795	2.90
D-UTR-6	300	230	2T28	1232	1420	783	3.20

Remarks: UBR – Uncoated Bottom Reinforcement, UTR – Uncoated Top Reinforcement

Appendix C.

Table 7. Beam Test Results for Uncoated Bars.

Beam Name	f_{cu} (N/mm^2)	M_u (kN.m)	f_s (N/mm^2)	τ_t (N/mm^2)	τ_{cal} (N/mm^2)	τ_{BS} (N/mm^2)	τ_y (N/mm^2)	$\frac{\tau_t}{\tau_y}$	$\frac{\tau_t}{\tau_{cal}}$	$\frac{\tau_t}{\tau_{BS}}$	$\left(\frac{\tau_t}{\tau_{cal}}\right)_{av}$	$\left(\frac{\tau_t}{\tau_{BS}}\right)_{av}$	$\left(\frac{\tau_t}{\tau_y}\right)_{av}$	Mode of failure
D-UBR-10A	34.67	54.07	561.53	3.51	3.31	2.94	3.33	1.05	1.06	1.19			1.22	Splitting
D-UBR-10B	36.47	55.27	573.31	3.58	3.40	3.02	3.33	1.08	1.05	1.19	1.01	1.02		Splitting
D-UBR-10C	34.44	57.06	576.65	3.00	3.30	2.93	3.33	0.90	1.09	1.28				Splitting
D-UBR-11A	34.67	61.17	473.92	2.96	2.87	2.94	3.32	0.89	1.03	1.01				Splitting
D-UBR-11B	36.47	82.28	555.30	3.47	2.94	3.02	3.32	1.05	1.17	1.20	1.15	1.14	1.13	Splitting
D-UBR-11C	34.44	79.10	556.86	3.48	2.86	2.93	3.32	1.05	1.22	1.19				Splitting
D-UBR-12A	34.67	94.44	371.23	2.32	2.37	2.94	3.63	0.64	0.98	0.78				Splitting
D-UBR-12B	36.46	121.32	476.37	2.98	2.43	3.02	3.63	0.82	1.23	0.99	0.73	1.12	0.90	Splitting
D-UBR-12C	34.44	111.19	492.99	2.69	2.36	2.93	3.63	0.74	1.14	0.92				Splitting
D-UTR-4	33.22	45.0	457.7	2.86	3.24	2.88	3.33	0.58	0.88	0.99				Splitting
D-UTR-5	33.22	72.8	438.14	2.74	2.81	2.88	3.32	0.38	0.98	0.95				Splitting
D-UTR-6	33.22	99.8	379.13	2.36	2.32	2.88	3.63	0.64	1.02	0.82				Splitting

Remarks: UBR – Uncoated Bottom Reinforcement, UTR – Uncoated Top Reinforcement

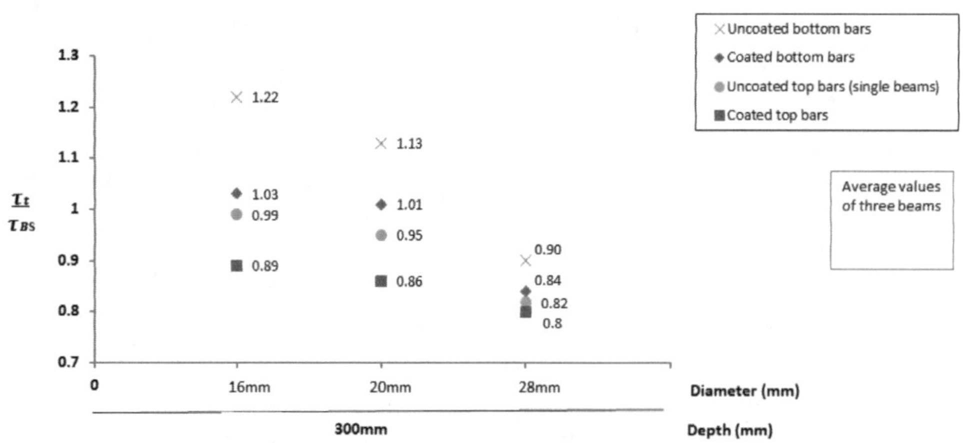

Figure 4. Comparison of Coated and Uncoated reinforcing bars cast at the top and bottom with 300 mm deep beam.

IOP Conf. Series: Materials Science and Engineering **951** (2020) 012011 doi:10.1088/1757-899X/951/1/012011

References

[1] Ferguson P M and Thompson J N 1962 Development Length of High Strength Reinforcing Bars in Bond *ACI Journal* **59** (7) 887-922
[2] Morita S and Kaku T 1977 Splitting Bond Failures of Large Deformed Reinforcing Bars Symposium Paper at Interaction Between Steel and Concrete *ACI Convention 1977* 93-110
[3] Cairns J and Abdullah R 1994 Fundamental Test on the Effect of an Epoxy Coating on Bond Strength *ACI Materials Journal* **91** 331-337
[4] British Standards Institution 1997 BS 8110: Part I: Structural Use of Concrete. Code of Practice for Design and Construction.
[5] Kong F K and Evans R H 1987 *Reinforced and Prestressed Concrete* 3rd Ed (London: Routledge) p 508
[6] Orangun C O, Jirsa J O and Breen J E 1977 A Re- evaluation of Test Data on Development Length and Splices *ACI Journal, Proceeding* **74** (3) 114-122
[7] Leet M K 1997 *Reinforced Concrete Design* 3rd Edition (McGraw Hill Companies) chapter 6 202-247
[8] British Standards Institution 1985 BS 8110: Part I: Structural Use of Concrete. Code of Practice for Design and Construction.
[9] Tepfers R 1976 *A Theory of Bond Applied to Overlapped Tensile Reinforcement Splices for Deformed Bars* Publication No. 73: 2 (Goteborg: Division of Concrete Structures, Chalmers University of Technology) p 328
[10] James O J, Breen J E, Luke J J and Hamad B S 1982 „Effect of Casting Position on Bond" Proc. Int Conference on Bond in Concrete (Scotland: Applied Science Publishers) 300-307.
[11] Donahey R C and Darwin D 1985 Bond of Top-Cast Bars in Bridge Decks *ACI Journal* **82** (1) 57-66

IOP Conf. Series: Materials Science and Engineering **951** (2020) 012012 doi:10.1088/1757-899X/951/1/012012

Application of natural fiber composite materials in the strengthening of reinforced concrete structures

I Ivanova[1]*, J Assih[2], D Dontchev[1]
[1]Department of Applied mechanic, University of Chemical Technology and Metallurgy, Kliment Ohridski Boulevard, 8 Sofia, 1756
[2.] Civil Engineering Department Sustainable Construction, ITHEMM EA 7548, University of Reims Champagne-Ardenne,

*E-mail: i.ivanova@uctm.edu

Abstract. This paper presents an experimental investigation of strengthened reinforced concrete structure by bonded natural fiber composite materials. Nonetheless, the composite plates are externally constituted of a matrix based on epoxy adhesive in which the aligned continuous natural fibers are embedded. In recent years, the natural fibers are of great interest as a result of policies to reduce the negative impact on the environment. They are a renewable source, helping to capture carbon dioxide during their cultivation. Numerous technical, economic, and environmental advantages such as low cost, high strength-to-weight ratio, low density, and non-corrosive properties make them very attractive to use. The objective of this research is to assess the strengthened capacity of natural fibers. This research mainly used hemp and flax natural fibers. So, these natural fibers are an alternative to synthetic reinforcing fibers like carbon fiber or glass fiber, which are widely used in civil engineering. This article briefly presents the results of the tests carried out on concrete prismatic specimens strengthened by external bonding different composite plates based on natural fibers. The tests were presented by examining the influence of reinforcement on the ultimate capacity and the element of rigidity. The results show a significant increase in the ultimate load of the strengthened concrete prismatic specimens compared to the reference specimen. The cracking and failure modes are presented. The results show that the natural fibers are a good alternative to glass and carbon fibers.

1. Introduction

Over the last decade increases scientific research of the natural plant fibers owing to environmental issues on global warming and in the term of environment protection. They are a renewable source, helping to capture carbon dioxide during their cultivation and their using allows for reducing environmental impact. Biomaterials are mainly used in many sectors of human activities such as in construction sector (building, flooring, sound insulation bricks...), and in the cars industry sector, which requires light materials with best performance. These materials have a potential for recycling and natural degradation, which contributes to the minimum impact on the environment. The natural composite materials are constituted of a matrix based on epoxy adhesive and aligned continuous plant fiber reinforcements having a structural function presents interesting mechanical properties. Numerous technical, economic, and environmental advantages such as low cost, high strength-to-weight ratio, low density, and non-corrosive properties make them very attractive to use [1-2].

The widely used composite carbon fiber plates in civil engineering, has become a real alternative in the field of rehabilitation of structures, but they have drawbacks such as high costs during the manufacturing and end-life services, cause adverse effects to human health and less environmental-friendly.

The technique of the strengthening of concrete structure has begun with bonding steel plates [3], then the appearance of synthetic Fiber Reinforced Polymer (FRP) composites [4-5] consists of bonding a composite plate on the beam underside bending reinforcement [6], or the side areas flexural and shear reinforcement [8-12], or prestressing reinforcement [7]. This technique enables us to make structures safe and suitable in service when the environment changed to improve the durability of the structures. The advantages of this strengthening technique are: better distribution of stresses in the adhesive layer compared with tightening or welding technique, easy and speed handling, the work can be carried out during construction sites.

Scientific research is being carried out to replace the most common composites (e.g. carbon and glass fibers) by materials less harmful to the environment, as natural fibers in the field of civil structures. The natural composite materials by plant fibers such as jute, flax, sisal, cotton, bamboo, hemp, coconut fibers are studied [13-17]. Therefore, these fibers have interesting physical and mechanical properties. This technique of reinforcement allows to increase remarkably the bending and shear stresses of the structures and in addition to new stresses.

The objective of this study is to investigate the possibility of reinforcing structures using natural biobased fibers. Hemp and flax natural fibers were used in this case. These natural fibers are an alternative to synthetic reinforcing fibers like carbon fibers or glass fibers, which are widely used in civil engineering. The results of the tests carried out on prismatic concrete specimens strengthened by external bonding different composite plates based on natural fibers are presented. The results show that the natural fibers are a good alternative to glass fibers.

2. Materials

The natural reinforced fiber composites are attracting increasing attention because of their low cost, low density, biodegradability and availability, ease of processing, high specific modulus, and their capacity to be recycled, minimum impact on the environment, and a reduction in their cost. The natural fibers are hydrophilic because they are composed of lignocellulose, which contains hydroxyl groups. They are therefore incompatible with hydrophobic thermoplastics such as polyolefins and have a low resistance to moisture. This weak point has improved a matrix of polymer type in order to highlight their mechanical properties (Table 1).

Table 1. Properties of used materials

Material	Young's modulus [GPa]	Strength [MPa]	Poisson ratio
Concrete	30	35	0.25
Adhesive	11	24	0.40
Flax FCP	62	22	
Hemp2,5x2,5 FCP	60	21	
Hemp3x3 FCP	63	24	
Glass FCP	70	31	
UCFCP*	90	1055	0.45

*UCFCP: Unidirectional Carbon Fiber Composite Plate.

The normal concrete was used for all specimens. The cement:sand:gravel proportions in the concrete mix were 1:1,7:2,9 by weight. The water/cement ratio was 0.5 and cement type II Portland was used. The maximum size of the aggregate was 16 mm. Four 160 mm x 320 mm concrete cylinders were cast and tested to determine the mechanical proprieties of the concrete. The average compressive strength was 35 MPa and the flexural tensile strength was 4.5 MPa. The concrete had an average elastic modulus of 30 GPa. A summary of the mechanical properties of the "Sikadur 30" adhesive is given as Young's Modulus of 11000 MPa, the tensile stress of 24 MPa, and Poisson ration of 0.40. Fiber fabrics of hemp, flax, glass, and carbon have been tested. Figure 1 shows the natural fiber fabrics used with different meshes. One bidirectional flax fiber fabric (Fig.1a) and two bidirectional hemp fiber fabrics (Fig.1b and 1c) with a similar quality were tested to show the difference between them. The results showed a good penetration of resin in composite plate manufacture. Figure 2 shows the synthetic reinforcing fiber fabrics (bidirectional glass fiber fabric (Fig.2a) and unidirectional carbon fiber fabric (Fig.2b)).

a) Flax Fabric of mesh of 2x2 b) Hemp Fabric of mesh of 2,5x2,5 c) Hemp Fabric of mesh of 3x3

Figure 1. Natural reinforcing plant fiber fabrics used in the experimental tests

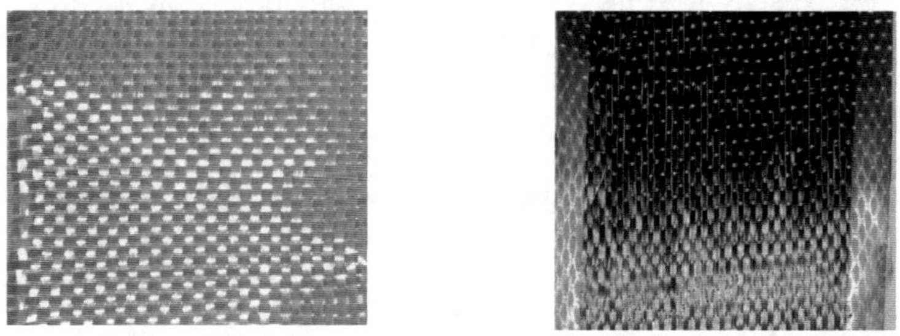

a) Glass Fabric b) Carbon Fibers Fabric

Figure 2. Synthetic reinforcing fiber fabrics used in the experimental tests

3. Experimental setup

3.1. Influence of different types of fiber fabrics

Five types of materials of Figures 1 and 2 were tested. For each type of fiber, four to eight specimens are stressed in tension. First the fabrics were tested and secondly, the composite plates (adhesive and fiber) were made and tested as well.

3.2. Influence of different composite plates

The composite plates were prepared manually with one layer of fabric and bonded together with an epoxy matrix, then they were pressed. The lay-up normally cures at room temperature for 24 hours. Pull off tests were carried out, with a tensile machine at constant speed of 0.25 mm/s, until sample collapses (Fig.3a). The extensometer technique based on electrical strain gauges was used to study the local behavior. This technique contributes to measure the strain of composite fabrics (Fig.3a) and concrete in specific point of the structure. In fact, the strains are measured using strain gauges bonded to the composite fabric sheets in the middle of the specimens where the bending moment is maximum.

The bonded area of both the composite plate, need a special preparation to carries out to remove all weak material, oil, grease, etc. The concrete was sand-blasted down to the level of the aggregate and was cleaned by using a water jet to ensure a good bond between the epoxy glue and the concrete surface. The resin and hardened were just hand-mixed and hand-applied on the concrete surface. The glue has hardened for 24 hours. The adhesive thickness was 1 mm. The composite fibers were adhered to the lower area of the reinforced concrete structures by using an epoxy resin (Fig.3c). It allows for increasing the bending and shear stresses of the structures. Of course, this approach also contributes to the durability of civil engineering structures and at the same time it using a strengthening system with environment respect.

a) Experimental device for tensile test set-up

b) Flexural four points test set-up

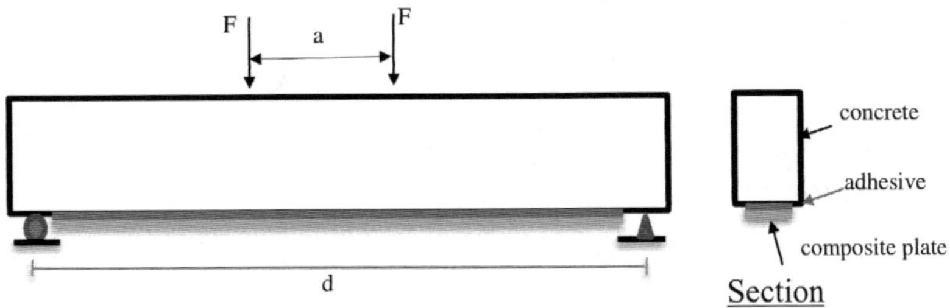

c) Strengthened concrete specimen

Figure 3. Experimental tests set-up and strengthening concrete specimen

In this series, five prismatic concrete specimens (100 mm by 100 mm by 400 mm) were strengthened with epoxy-bonded fiber composite plates, and one without strengthening - control specimen. The specimens were instrumented with electrical gauges and the displacement captor LVDT, Figure 3b. All concrete specimens were tested under a four-point load. So, this last device allows one domain where shear strengths are null. For each test, mid deflection, concrete strain distribution, strain of the composite plate, and the cracking state were noted. The evolution of the neutral axis position was determined by the deformation of strain gauges on the central section of the specimen. During the tests, load strain and deflection were recorded by an automatic data acquisition system.

4. Results and discussion

Figure 4 shows the experimental results obtained for composite materials. We usefully compare the result of fiber fabrics and composite plates.

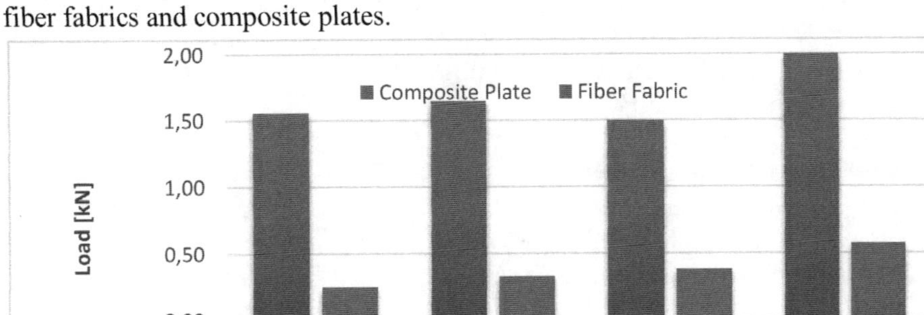

	Hemp2,5x2,5 FCP	Hemp3x3 FCP	Flax FCP	Glass FCP
Composite Plate	1,56	1,65	1,50	2,01
Fiber Fabric	0,25	0,33	0,38	0,57

Figure 4. Ultimate load of fiber fabrics and composite plates

Figure 5 shows the comparison between natural fiber composites plates and Glass FCP in tensile test. Composites reinforced with natural fibers have a tensile behavior similar to that reinforced with glass fibers. The composite materials plates have a linear elastic behavior up to failure.

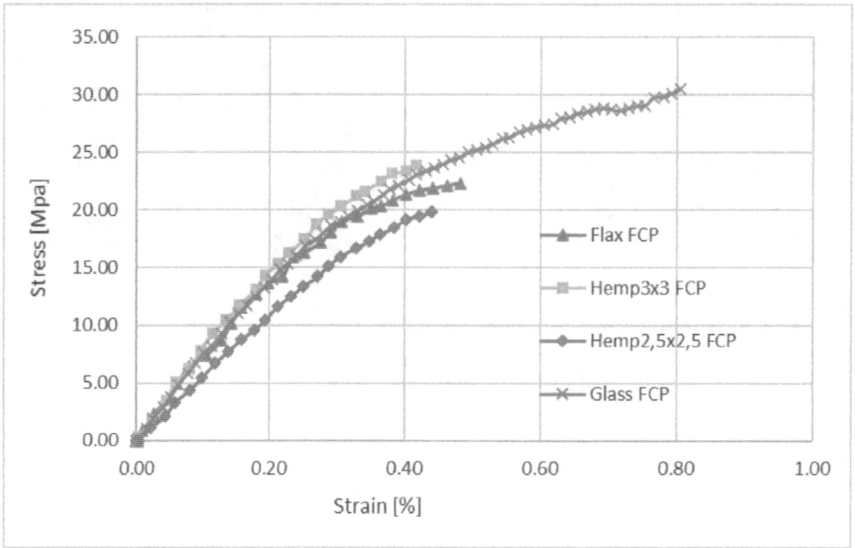

Figure 5. Stress versus strain of fiber fabrics and adhesive in tensile tests

The results in Table 2 show the effect of strengthening. The results show that the flexural properties of flax FCP for strengthening was similar to Glass FCP. Contrariwise the hemp flexural properties were lower.

Table 2. Results of strengthened concrete

	Control specimen	Hemp 3x3 FCP	Flax FCP	Carbon UCFCP	Glass FCP
Ultimate load F_u [kN]	12,09	13,02	15,74	23,50	15,72
F_u/F_0 [%]	0	8	30	94	30
Displacement [mm]	5,23	8,84	4,83	7	8,76

5. Conclusion and perspectives

In this paper, the mechanical properties of reinforced natural plant fiber epoxy composites have been described. The tensile, flexural and hardness properties of reinforced natural fiber polymer composites show a significant improving (positive effect) on strengthened concrete specimens. The results present:

- The bio-based materials such as hemp and flax fibers can contribute to a reinforcement of a concrete beam from 1% to 30%.
- In comparison to bidirectional glass fiber fabric, the natural fiber fabrics (flax and hemp) are a good alternative to glass fibers.
- The tests carried out concern a single layer of hemp and flax fibers, in fact, it would be interesting to carry out other studies in order to determine the optimum thickness for effective reinforcement on reinforced concrete beam.

Acknowledgements

This research is fully supported and was financed by University Agency Francophonie, (Franch: Agence Universitaires de la Francophonie -AUF), Fund for Scientific Research (FSR) in Bulgaria through the Programs For Bilateral Cooperation 2018 - University Agency of Francophonie, within the framework

of the project Study of biomaterials for strengthening reinforced concrete structures, BioRenfort, Ref. КП-06-франкофония/4, which makes this important and effective research.
Thanks also to the support provided by the Project Ref. 12046 with Scientific Investigation Sector of and University of Chemical Technology and Metallurgy (UCTM).

References

[1] Baley C 2013 Fibre naturelles de renfort pour matériaux composites *Technique de l'ingénieur*

[2] Tong F S, Siew Choo Chin S C, Doh S I and Gimbun J 2017 Natural Fiber Composites as Potential External Strengthening Material – A Review, *Indian Journal of Science and Technology,* Vol **10** 2

[3] Raithby K D 1980 External strengthening of concrete bridges with bonded steel plates, TRRL Supp. Report 612, crowthorne, Transport & Road Research Laboratory

[4] Bresson 1971 Nouvelles recherches et application concernant l'utilisation des collages dans les structures - béton plaqué, Ann ITBTP, série béton et béton armé 278 fev

[5] Saadatmanesh H. and Eshani M.,1989, Application of fiber-composites in Civil Engineering, *Proceeding of the sections related to structural materials at structures congress 89 ASCE*

[6] Kaiser H 1986 Strengthening of Reinforced Concrete with Epoxy Bonded Carbon Fiber Plastics, Doctorat thesis ETH (in Germany)

[7] Traintafillou T and Deskovic N 1991 Innovative Prestressing with FRP sheets: Mechanics of short term Behavior, *Journal of Engineering Mechanics*, ASCE **117** 7 pp 1652-1672

[8] Assih J. 1998, "Contribution à l'étude du Renforcement et de la Réparation de Poutre en Béton Armé par Collage de Plaques Composites en Fibres de Carbone", Thèse de doctorat, UFR Sciences de Reims

[9] BarrosJ.AO, Dias S J E and Lima J L T 2007 Efficacy of CFRP-based techniques for the flexural and shear strengthening of concrete beams, *Cement&Concrete Composites,* **29** pp 203-217

[10] Ivanova I 2013 Comportement mécanique de console courte en béton armé renforcée ou réparée par collage des matériaux composites, Thèse de doctorat 2013, Université de Reims Champagne-Ardenne, Reims France

[11] Ivanova I, Assih J, Li A, Dontchev D and Delmas Y 2015 Experimental investigation into strengthened short reinforced concrete corbels by bonding carbon fiber fabrics, *Journal of Adhesion Science and Technology*, **29** 20 pp 2176–2189

[12] Ivanova I, Assih J, Dontchev D 2018 Influence of anchorage length of composite fabrics and bonded surface on the strengthened short reinforced concrete corbel by bonding CFRF, *European Journal of Environmental and Civil Engineering, ISSN:1964-8189* https://doi.org/10.1080/19648189.2018.1498395

[13] Abot A 2010 Caractérisation des fibres longues de chanvre (Cannabis sativa) en vue de leurs utilisations dans des matériaux composites, Thèse de doctorat, l'Université de Poitiers.

[14] Do Thi Vi Vi 2011 Matériaux composites fibres naturelles/polymère biodégradables ou non, Thèse de doctorat 2011, Université de Grenoble France

[15] Misnon M I , Islam M M, Epaarachchi A J and Lau K T 2015 Analyses of woven hemp fabric characteristics for composite reinforcement", *Materials and Dasign,* **66** pp 82-92

[16] Benzarti K, Chlela R, Zombre W, Quiertant M and Curtil L 2018 Durability of flax/bio-based epoxy composites intended for structural strengthening, *MATEC Web of Conferences*, **199** 07014 ICCRRR 2018

[17] Velmurugan G, Sekar S and Suresh K S 2018 Woven Hemp and Glass Fiber Hybrid Composite - A Comparative Study of Tensile and Shear Properties with and without NaOH Treatment, *IAETSD Journal For Advanced Reserch In Applied Sciences*, **5** 4 pp 994-998

Conformity assessment of two orthotropic steel deck bridges in Bulgaria according to present requirements

L D Georgiev[1] and S D Ivanov[2]*

[1] Roads and Transport Structural Facilities Department, Faculty of Transportation Engineering, University of Architecture Civil Engineering and Geodesy, Sofia, Republic of Bulgaria
[2] Department of Metal, Timber and Plastic Structures, Faculty of Civil Engineering, University of Architecture Civil Engineering and Geodesy, Sofia, Republic of Bulgaria

* E-mail: sdenkov_fce@uacg.bg

Abstract. Two orthotropic steel deck (OSD) bridges along "Hemus" highway in Republic of Bulgaria are discussed in the paper. They are built at 1985 and are in exploitation over 30 years up to present times. The first of the bridge superstructures is with closed section trapezoidal longitudinal ribs of the OSD and steel plated main girders with 72,45 m span. The second bridge superstructure is with open type section longitudinal ribs of the OSD and steel box main girder with 162m span. Visual inspection and some measurements are performed for evaluation of the overall condition of these bridges. It must be mention that the traffic has considerably increased. Some checks for fatigue for the second bridge are performed in accordance with Eurocode. The condition in respect to fatigue behaviour in accordance with the used structural details in the OSD and the prescriptions of the present design standards is analysed and discussed.

1. Introduction

In the focus of this article fall two steel road viaducts built in 1985 in Bulgaria. They are situated at km 35+144 (Viaduct 1) and km 35+760 (Viaduct 2) of the "Hemus" highway, which is envisaged to connect Sofia with Varna town, the second biggest Bulgarian harbour on the Black sea.

The superstructure of Viaduct 1 is continuous beam over six spans 60m+4x72,45m+60m. It consists of two steel plate girders with an orthotropic steel deck plate on top (see figure 1). In this section of the highway the alignment falls into horizontal curve with a constant radius of 850m and longitudinal slope of 4,61%.

The superstructure of the Viaduct 2 is continuous steel box girder beam over three spans 100m+162m+100m (see figure 2). It still holds the span record in Bulgaria for road bridge. The road alignment is again falling into horizontal curve with a constant radius of 1467,25m and longitudinal slope of 4,61%.

Both bridges have orthotropic steel deck (OSD) plates and are in exploitation for 35 years. In this period the traffic intensity as well as the traffic volume on the highway has increased considerably. These facts and the reported defects of similar bridges with OSD [1] abroad, trigger research project in the University of Architecture, Civil Engineering and Geodesy, Sofia. Under this project visual inspections of the two viaducts were performed. The analysis of the findings as well as the conformability assessment of the OSD to Eurocode is presented in this article.

IOP Conf. Series: Materials Science and Engineering **951** (2020) 012013 doi:10.1088/1757-899X/951/1/012013

Figure 1. General view of Viaduct 1.

Figure 2. General view of Viaduct 2.

2. Visual inspection and conformity analysis

Both Viaducts have orthotropic steel deck plates (see figure 3). The difference is that in Viaduct 1 the longitudinal stiffeners are closed type trapezoidal, while in Viaduct 2 they are open type single plate stiffeners.

(a) (b)

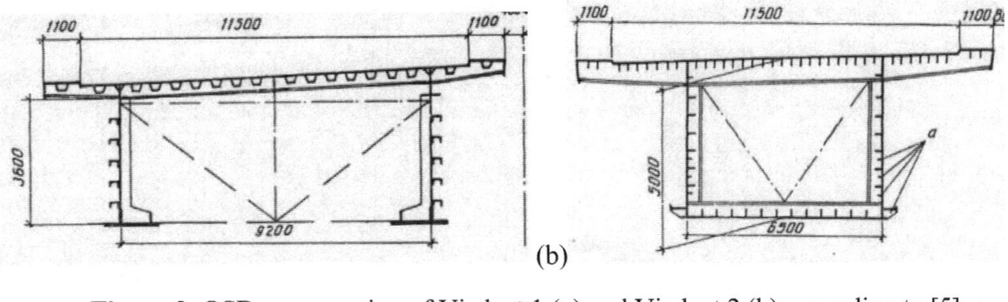

Figure 3. OSD cross section of Viaduct 1 (a) and Viaduct 2 (b), according to [5].

IOP Conf. Series: Materials Science and Engineering **951** (2020) 012013 doi:10.1088/1757-899X/951/1/012013

The main findings during the visual inspections of the viaducts are as follows:

Misalignment between the adjacent assembly elements of the steel structure in Viaduct 1. In order to compensate it the connecting splices were produced with considerable initial deformations. Usually, prescribed misalignment tolerances for welded details, in the standards, are in the range of a few millimetres. In this case they are obviously larger.

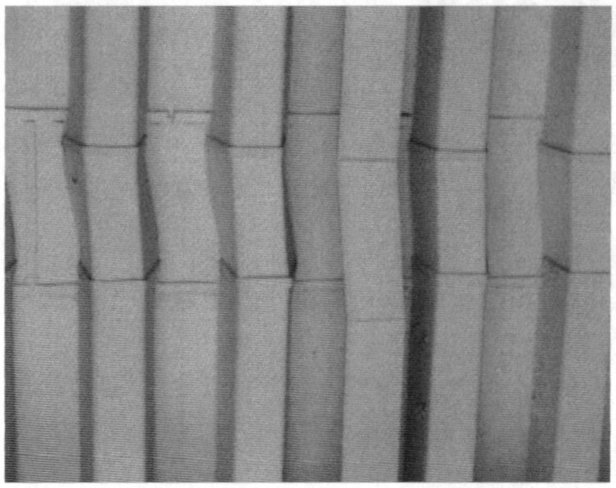

Figure 4. Misalignment between adjacent assembly elements in the connection of the stiffeners of the OSD of Viaduct 1.

Fatigue crack is found into the zone of the connection of the longitudinal stiffeners close to the backing strip (see figure 5) in Viaduct 1. In [2] this detail is prescribed with backing strip from the inside of the trapezoidal stiffener and tack weld falling into the place of the future butt weld. Moreover, the sequence of the welding in this detail is very important. It should be such that the residual stresses are as small as possible and resulting in compression at the bottom flange of the stiffener (see figure 5 (c)).

(a) (b) (c)

Figure 5. Crack noticed into the zone of the backing strip of the longitudinal stiffener (a) and detail of the connection according to [2] (b) and (c).

The connection detail between the trapezoidal longitudinal stiffeners of Viaduct 1 and the web of the cross beams is realized without holes in the web. In [2] it is recommended that stiffeners are continuous, passing through cut outs in the webs of the cross beams. Discontinuous stiffeners are allowed only for light weight traffic (pedestrian bridges and under footpaths) and for distances between cross girders smaller than 2,75m. As far as the cope hole at the bottom is concerned, both options are allowed in [2], with and without it (see figure 6 (b)). In any case, one should bear in mind, that the reduced plasticity in cold formed trapezoidal stiffener in the bended zone, combined with high tension residual stresses from welding leads to low fatigue strength detail –category 36, according to [3].

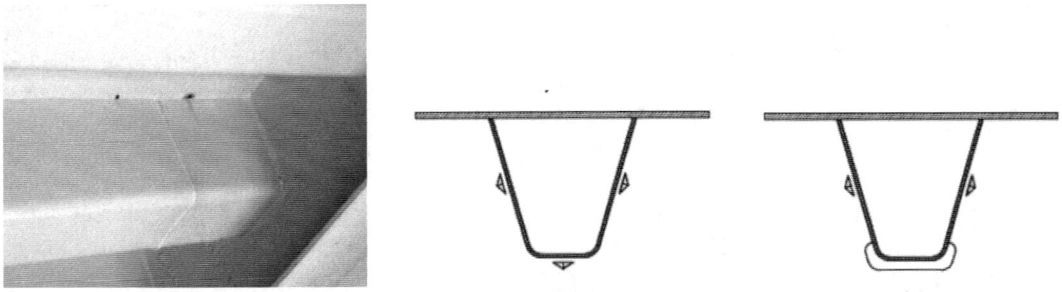

(a) (b) (c)

Figure 6. Connection between longitudinal stiffener and cross girder (a) and recommended detail of the connection according to [2] (b) and (c).

Fatigue crack is also found at welded connection between closed trapezoidal longitudinal stiffener and 12mm thick deck plate in the zone of the first intermediate support of the main girders. In order to prevent such cracks, following prescriptions are given in [2] - minimum plate thickness 14mm(16mm) and minimum stiffness of the longitudinal stiffeners should be provided as well. In addition, the weld could be either automatic or manual, according to figure 7 (b and c), but its thickness should not be smaller than the thickness of the stiffener. In this case, as seen from the picture, the weld quality is poor, and the weld thickness is smaller at the location of the crack.

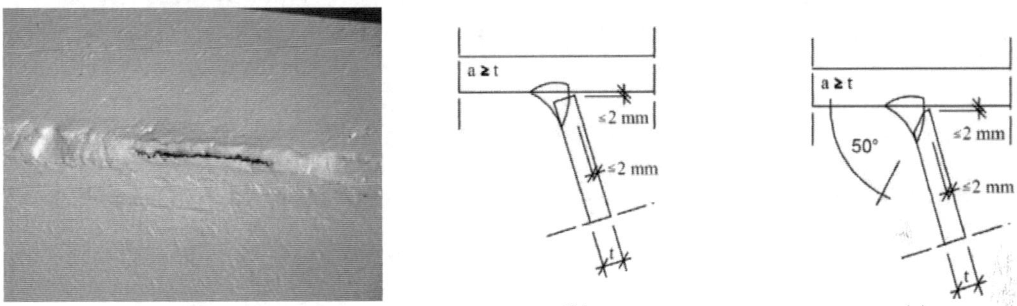

(a) (b) (c)

Figure 7. Fatigue crack found in the connection weld between longitudinal stiffener and deck plate (a) and recommended detail of the welding according to [2] (b) and (c).

Open flat stiffeners in Viaduct 2 are passing through the web of the cross beams, welded only from one side. In this case the load is transferred eccentrically. According to [2] open flat stiffeners should be welded to the web of the cross beam from both sides. Single sided welding is not allowed.

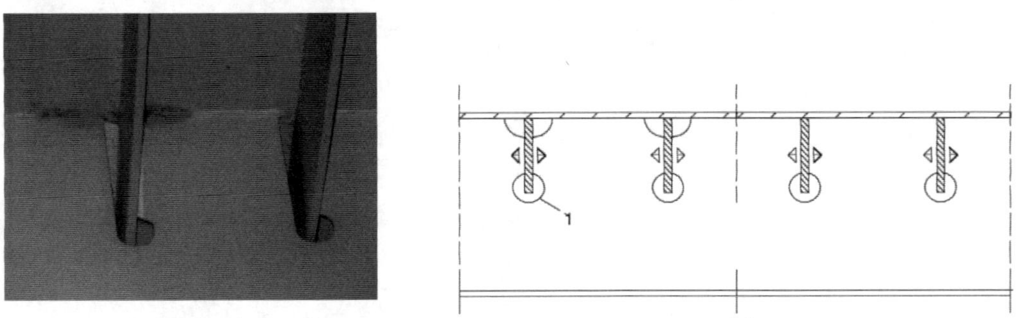

(a) (b)

Figure 8. Connection detail between longitudinal open flat stiffener and web of the cross beams (a) and recommended detail according to [2] (b).

3. Fatigue check of specific OSD details according to EN 1993-1-9

In addition to the conformity assessment, fatigue checks of some of the details of the OSD plates are performed, according to [3]. The selected method for the verification, according to БДС EN 1993-2:2007/NA:2011, is the Safe Life Concept. According to this method an acceptable reliability should be provided during the whole design life of the structure without the need for regular inspections and maintenance. The fatigue checks are performed according to the damage equivalent factor concept, using fatigue load model FLM3, according to [4]. The partial safety factor for fatigue strength of the OSD elements is chosen as 1,15. The internal forces and moments are determined from elastic analysis of FE model of the OSD plate segment of Viaduct 2. In longitudinal direction the segment is 14m (7 spans between cross beams) and in transversal 6,50m (the distance between the webs of the box girder). More information on the FE model is given in [6]. The additional stresses from global bending of the box girder are not considered.

Four details are verified, summary of the results is presented below.

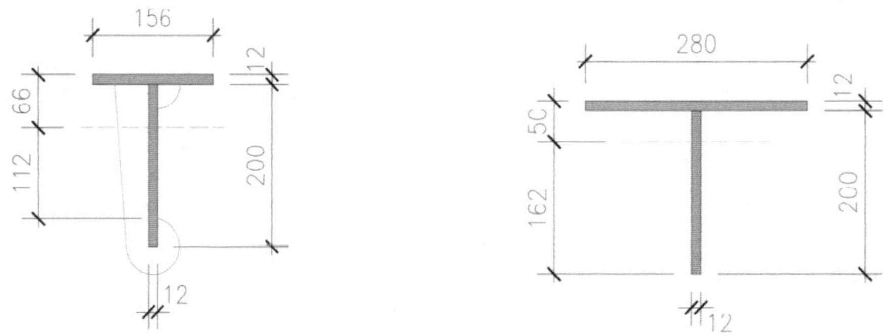

Figure 9. Longitudinal stiffener effective cross sections, over support (a), mid-span (b).

Deck plate connection with cross beam web – Detail category 71.

$$\Delta\sigma_{\mathrm{FLM3}} = \frac{\max M_{\mathrm{FLM3}} - \min M_{\mathrm{FLM3}}}{W_y} \tag{1}$$

$$\Delta\sigma_{\mathrm{FLM3}} = \frac{[1.0-(-8.8)]100}{300.6} = 3.3kN/cm^2 = 33MPa \tag{2}$$

$$\Delta\sigma_{E,2} = \lambda_1\lambda_2\lambda_3\lambda_4\Delta\sigma_{\mathrm{FLM3}} = 1.77\Delta\sigma_{\mathrm{FLM3}} = 58MPa \tag{3}$$

$$\gamma_{Ff}\Delta\sigma_{E,2} = 58 < \frac{\Delta\sigma_c}{\gamma_{Mf}} = \frac{71}{1.15} \approx 62 \tag{4}$$

Deck plate connection with longitudinal stiffener – Detail category 100.

$$\Delta\sigma_{\mathrm{FLM3}} = \frac{[10.5-(-2.5)]100}{540.2} = 2.4kN/cm^2 = 24MPa \tag{5}$$

$$\Delta\sigma_{E,2} = \lambda_1\lambda_2\lambda_3\lambda_4\Delta\sigma_{\mathrm{FLM3}} = 2.19\Delta\sigma_{\mathrm{FLM3}} = 53MPa \tag{6}$$

$$\gamma_{Ff}\Delta\sigma_{E,2} = 53 < \frac{\Delta\sigma_c}{\gamma_{Mf}} = \frac{100}{1.15} \approx 87 \tag{7}$$

Stiffener connection to the web of the cross beam – Detail category 80.

$$\Delta\sigma_{\text{FLM3}} = \frac{[1.0-(-8.8)]100}{177.1} = 5.5kN/cm^2 = 55MPa \tag{8}$$

$$\Delta\sigma_{E,2} = \lambda_1\lambda_2\lambda_3\lambda_4\Delta\sigma_{\text{FLM3}} = 1.77\Delta\sigma_{\text{FLM3}} = 97MPa \tag{9}$$

$$\gamma_{Ff}\Delta\sigma_{E,2} = 97 > \frac{\Delta\sigma_c}{\gamma_{Mf}} = \frac{80}{1.15} \approx 70 \tag{10}$$

Web of the cross beam – Detail category 112.

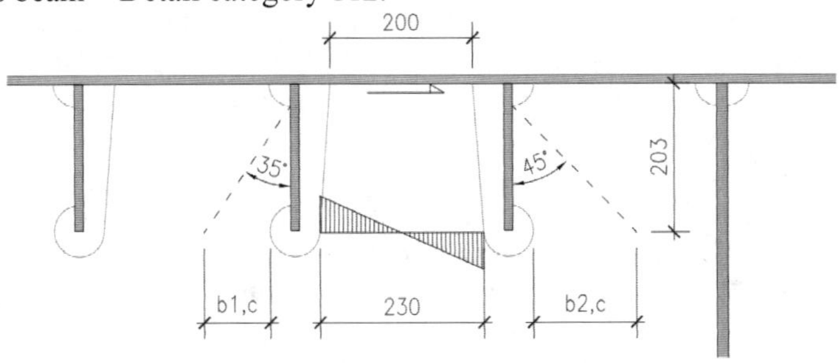

Figure 10. Stresses in the web of the cross beam.

$$M = 0.20v_L 0.203 = 10kNm \tag{11}$$

$$\sigma = \frac{M}{W} = 8.1kN/cm^2 = 81MPa \tag{12}$$

$$\Delta\sigma_{\text{FLM3}} = \sigma - 0 = 81MPa \tag{13}$$

$$\Delta\sigma_{E,2} = \lambda_1\lambda_2\lambda_3\lambda_4\Delta\sigma_{\text{FLM3}} = 2.18\Delta\sigma_{\text{FLM3}} = 177MPa \tag{14}$$

$$\gamma_{Ff}\Delta\sigma_{E,2} = 177 > \frac{\Delta\sigma_c}{\gamma_{Mf}} = \frac{112}{1.15} \approx 97 \tag{15}$$

4. Summary and conclusions

Two steel road bridges with OSD on "Hemus" highway in Bulgaria have been inspected after 35 years of exploitation. Their general condition, based on visual inspections, looks satisfactory.

Problems with fatigue cracks in steel bridges with OSD plates, having the same age, have been reported worldwide. On European level this has led to specific recommendations for their structural detailing, published as Annex C in [2]. The two bridges under consideration were designed and detailed according to the best bridge practice and standards in force at the time of their construction. Nevertheless, conformity analysis with recent recommendations has shown that some of their details are inappropriate in terms of fatigue.

Fatigue verification of the OSD plate of Viaduct 2, in accordance with Eurocode, were performed. Damage equivalent factor concept and FLM3 were used. The results show that the check is not fulfilled for two of the four details. It is important to state that the stresses from global bending of the steel box girder are not considered into these verifications. They should be added in cumulative way to the local stresses: , and are of importance to all details subjected to longitudinal normal stresses. One of the critical details is the one-sided connection detail of open flat stiffener, which could be the reason for removing this option from the contemporary standards. For Viaduct 1 verifications were not performed, but fatigue cracks were detected during the visual inspection.

Inappropriate detailing, together with the increased traffic on the highways since the time of their construction as well as the execution quality 35 years ago, are all factors that might lead to crack initiations at different locations of the OSD plates. In order to prevent future accidents more thorough investigation of the two viaducts is recommended. The relevant managing authorities should take the appropriate measures such as regular inspections in order to control and prevent any crack propagation. Since these viaducts have passed only one third of their design life, different advanced strengthening methods could also be an option for prolonging their life.

Acknowledgments
Authors would like to thank the "Research, Consultancy and Design Centre (RCDC)" in UACEG, Sofia, Republic of Bulgaria for the financial support of the research project "Investigation of road steel bridge superstructures with orthotropic deck in Republic of Bulgaria" in accordance with Contract BN-220/2019 as well as to the "Avtomagistrali" Ltd for providing access to the superstructure of Viaduct 2.

References

[1] Jong F B P August 25-27, 2004 Orthotropic Bridge Conference, Sacramento, California, USA
[2] European Committee for Standardization - CEN (2006). EN 1993-2:2006. Eurocode 3: Design of steel structures - Part 2: Steel bridges. Brussels, Belgium
[3] European Committee for Standardization - CEN (2005). EN 1993-1-9:2005. Eurocode 3: Design of steel structures - Part 1-9: Fatigue. Brussels, Belgium
[4] European Committee for Standardization - CEN (2006). EN 1991-2:2006. Eurocode 1: Actions on structures - Part 2: Traffic loads on bridges. Brussels, Belgium
[5] Staykov P 1985 Steel bridge constructions along "Hemus" highway - "Viaduct 1" and "Viaduct 2" Construction Journal 12/1985 (in Bulgarian)
[6] Georgiev L et al. 2019 Investigation of road bridge superstructures with orthotropic steel deck in republic of Bulgaria research project under contract БН-220/19, RCDC in the UACEG, Sofia (in Bulgarian)

Progressive Collapse of flat slab structures – key elements and connections

G Sokolova

Department "Reinforced Concrete Structures", University of Architecture, Civil Engineering and Geodesy, 1 Hristo Smirnenski Blvd., Sofia 1046, Bulgaria

E-mail: galq.sokolova@gmail.com

Abstract. The loss of even one vertical support member can lead to catastrophic consequences and the progressive collapse of big portion or even the entire structure of the building. This report presents a computational study of a single-story, non-scaled 3D fragment of a high-rise concrete flat slab structure building with peripheral beams along the facade in case of dynamic failure of a column. The main objective is to examine alternative load paths for the transfer of stress from the peripheral beams to the columns, across the slab and its punching areas. In case of accidental failure of an outer column, the facade beams serve as prevention which reduces the redistribution of loads to the interior, ie. to the columns with critical slab punching areas. Research on the performance of peripheral beams and the key role of corner columns is the focus of this report.

1. Introduction

Accidental actions are extremely short-lived and unlikely to occur, but their consequences could be catastrophic. For example, a progressive collapse with a high rate of destruction. Examples for accidental situations are explosions, vehicle impacts, floods, landslides.

In accidental design situations, limited damage to structural elements is possible and allowed. In these cases, instead of the criterion of ultimate limit state, the requirement for *robustness* is introduced: the ability of a structure to withstand events like fire, explosions, impact or the consequences of human error, without being damaged to an extent disproportionate to the original cause [1].

Dozens of examples of disproportionate (or even total) collapse of buildings result from terrorist attacks or unintentional human errors, lead to comprehensive experimental and computational research in recent years worldwide.

The approach to survey the robustness of structures in case of accidental actions through scenarios of local failure of a single element and the search of alternative load paths for transferring the stress is considered to be an universal approach, and is applied in experimental research and in applied problems solving computational studies [3].

Critical for flat slab structures is the "fragile" punching shear phenomenon in the areas of the slab-column connections. Punching shear resistance has been studied in detail under static load and is established with sufficient certainty by modern norms and standards. The experience with accidents involving these punching areas in cases of seismic impacts, especially in irregular structures with significant rotation, is one of the reasons for the construction of beams in the peripheral of the building, especially when the slabs are without cantilever parts.

In any case, flat slab structures are most sensitive to accidental actions and the control over horizontal and vertical progressive collapse process is most challenging. (see figure 1). In this regard, the study of the above-described modified flat slab structures, set in a situation of a dynamic failure scenario of a penultimate-external column, caused (for example) by external explosive impact, can be assessed as relevant.

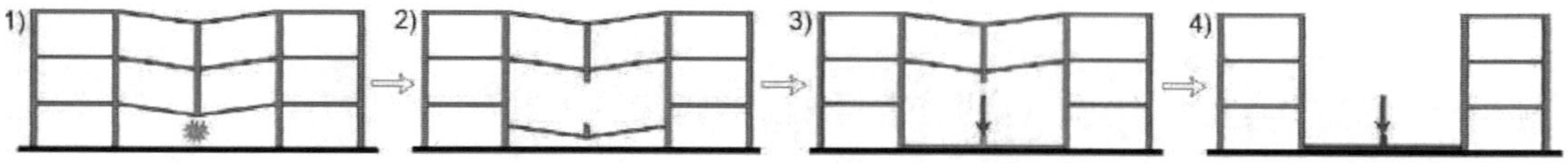

Figure 1. The stages of the "pancake"-type progressive collapse

Modern research on critical dynamic punching shear areas of flat slabs is extremely insufficient, and also any kind of experimental or computational survey on parts of flat slab structures is lacking. This report presents a computational study of a single-story, non-scaled 3D fragment of a high-rise concrete flat slab structure building with peripheral beams along the facade after a dynamic failure of a column. The main objective is to examine alternative load paths for stress transfer from the peripheral beams to the columns, across the slab and its punching areas.

2. Experimental Program and FEM analysis

This research is a part of a broader program of computational analyses examining the behaviour of flat slab structures in case of dynamic failure of a column. The research is analytical and is necessary due to the extremely complex nature of the phenomenon of progressive collapse. This requires an explicit nonlinear dynamic analysis of the intended tasks - finite element modelling is applied using the explicit reinforcement contact model in ANSYS Autodyn [4].

Input values for concrete properties were those of the Riedel-Hiermaier-Thoma (RHT) concrete model with compressive strength class is C30/37. The failure mode of the concrete model was selected as a tensile failure to simulate the behaviour of a real concrete specimen. The tensile strength was assumed 10% of the concrete compressive strength [7]. The material model of structural steel is elastoplastic with linear hardening and strength class B500C [2].

The considered problems are tested in the conditions of modelled earth gravity. This acceptance aims to take into account simultaneously both the distributed permanent loads and the normalized quasi-constant part of the other variable actions.

The speed of implementation of the external force and displacement in the segment of local damage (failure of reinforced concrete column) is consistent to the fragile nature of elements under pressure.

3. Progressive collapse analysis - objective and models

3.1. Objective

In case of accidental failure of an outer column, the facade beams serve as prevention that reduces the redistribution of loads to the interior, i.e. to the columns with critical slab punching areas. The focus of this experimental program is to examine the behavior of peripheral beams and the key role of corner columns in increasing the robustness of the whole structure.

3.2. Problems

The model is a vertical part of reinforced concrete structure of an 8-story administrative building with flat slabs. The spans are regular, 6.0 m in both directions and the floor height is 3.60 m (see figure 2). The slabs are without cantilever parts and the beams are distributed along the facade.

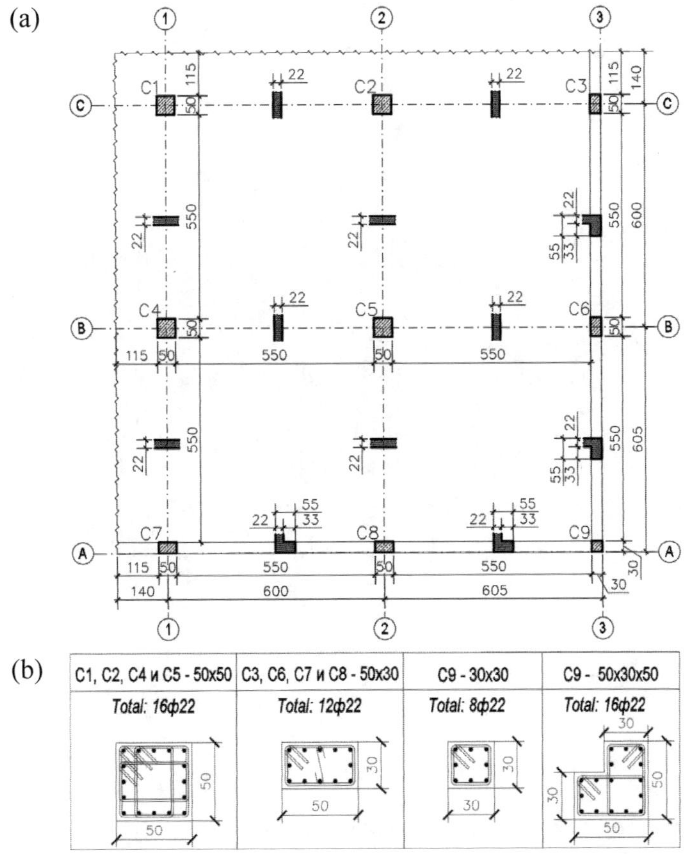

Figure 2. Details of model frame (unit: cm). (a) Plan of the model; (b) Column sections

The model reviews the full height of a vertical fragment, with two intermediate axes from the corner of the structure, mentally separated from it by conservative cantilever parts with a length of 1.40 m (see figure 2a).

The design of the structure takes under consideration the requirements of the Eurocode standards system. Without having to participate in the seismic protection, the frame elements respond to structural requirements of EN 1998-1.

The columns above the first floor, except for those that are subject to a dynamic failure, are loaded with the calculated values of the vertical forces, determined for accidental design situation of 7 floor levels. The aim is for the columns, which are expected to be subjected to redistribution of forces, to already have objective values of the pressure forces. The failure of the column is modeled by increasing dynamic force from 0 to 500 kN, with speed of dynamic failure about 110 kN/s (see figure 3).

Model 1-C: Simulates failure of the corner column C9.

Model 1-0: Simulates failure of column C8 on the first floor. In this model, the corner column C9 has a minimum admissible cross-sectional size of 30/30 cm.

Model 1: The main model, in which the datum from Model 1-0 has led to the experiment performed with an L-shaped corner column C9 (see figure 2b).

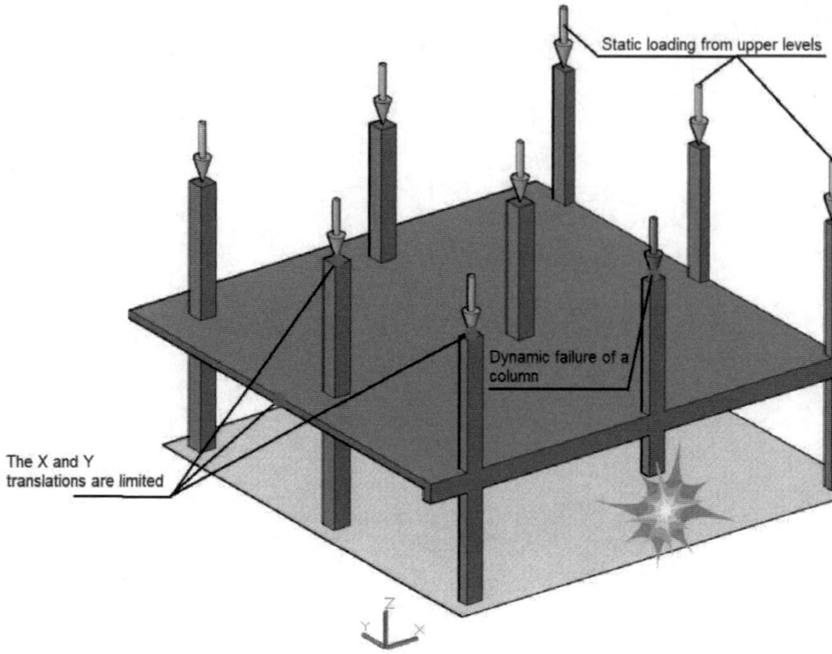

Figure 3. Analysis model

4. Results

4.1. Model 1-C: failure of the corner column C9

The goal is to analyze the magnitude of this accidental action in comparison to the failure of the adjacent penultimate-external column C8.

The simulation has a vertical dynamic action increasing from zero, (see point 3.2), applied in C9.

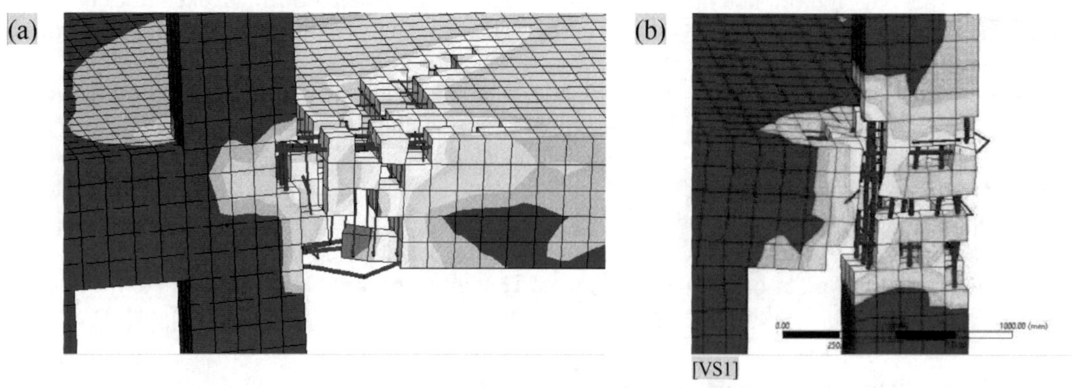

Figure 4. (a) Potential destruction at the node beam - column C8 at 3.94s; (b) Condition of the node with C9, where the force is applied at 3.94s

Critical for the structure is the moment 3.94s right before the failure of longitudinal reinforcement in the facade beam. The load is 458 kN with a displacement of 455 mm in the spot of the removed column C9. The deformed state of the beam and the connections with the columns (see fig.4) define the shape of the next failure. The critical areas are the columns C8 and C6 (they are symmetrical), which support the cantilever part, formed after the failure of column C9.

The total collapse of the model's angle at moment 4.5 s is with relatively very large value of the dynamic force, simulated in the corner (see figure 5).

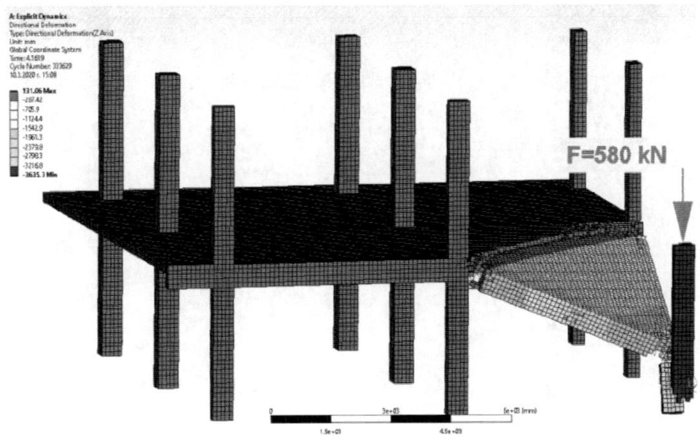

Figure 5. Final condition (Model 1-C)

4.2. Model 1-0: failure of the penultimate-external column C8 with a cross-sectional size of column C9 -30/30 cm

The objective is to establish the effectiveness of the constraints in the angle of the structure. The corner column has to ensure the alternative load paths in the peripheral beams.

The deformation of the beam and the connections with the columns (see figure 7) define the manner of the next collapse. The catenary action in large deformations of the beam "retracts" column C9 to the inside. (see figure 6a). This bending, accompanied by high pressure from the upper levels, destroys the corner column before the failure of longitudinal reinforcement in the facade beam (see figure 6b). Such behaviour can be defined as fragile. The scenario of dynamic failure of column C8 and the subsequent collapse of column C9 is a typical case of chain breakdown (or progressive collapse).

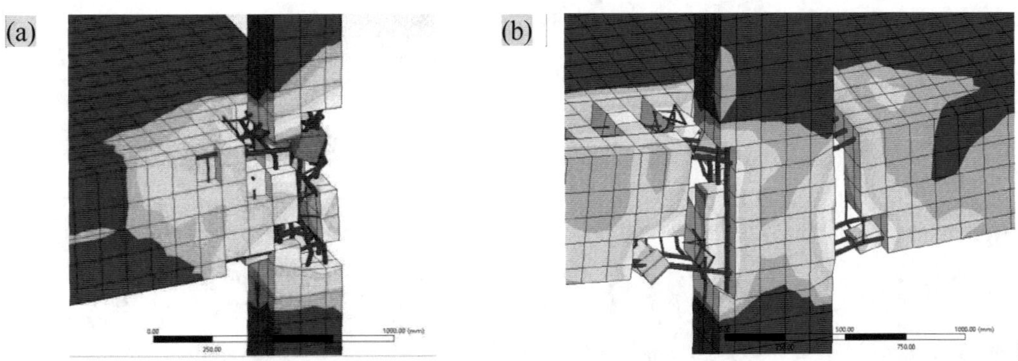

[VS2]

Figure 6. Column C9 failed before (a) the longitudinal reinforcement in the beam reaches ultimate tensile stress at the connection with C8 (b).

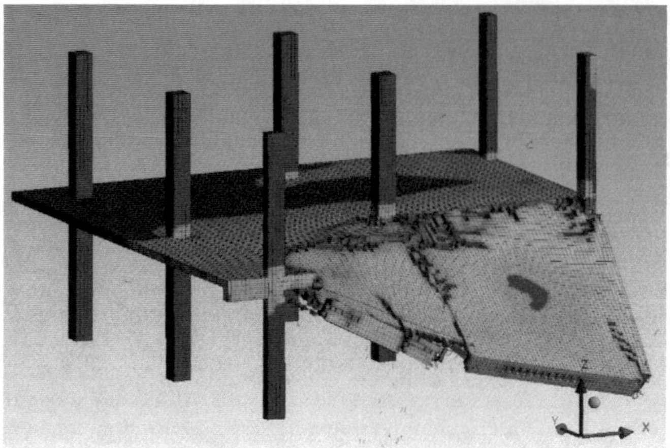

Figure 7. Final condition (Model 1-0)

4.3. Model 1: failure of the penultimate-external column C8 with a L-cross section of column C9 - 30/50/30cm

Critical for the structure is the moment 4.1 s, right before the failure of longitudinal reinforcement in the facade beam. The load is 400 kN with a displacement of 235 mm in the spot of the removed column C8.

The failure of longitudinal reinforcement is at 4,19 s (see figure 8) with vertical deformations 320 mm.

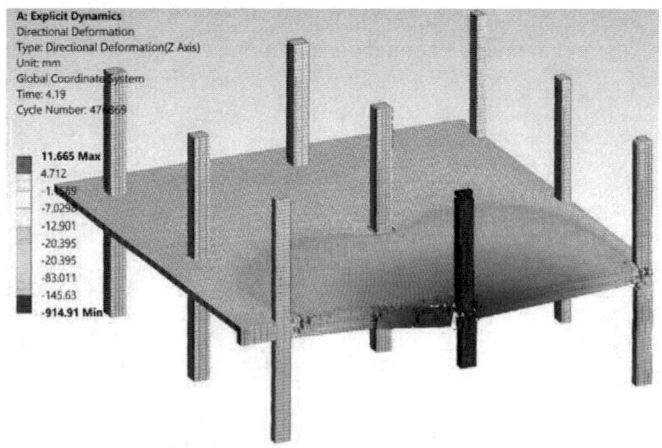

Figure 8. Longitudinal reinforcement of the beam reaches ultimate tensile stress at 4,20 s.

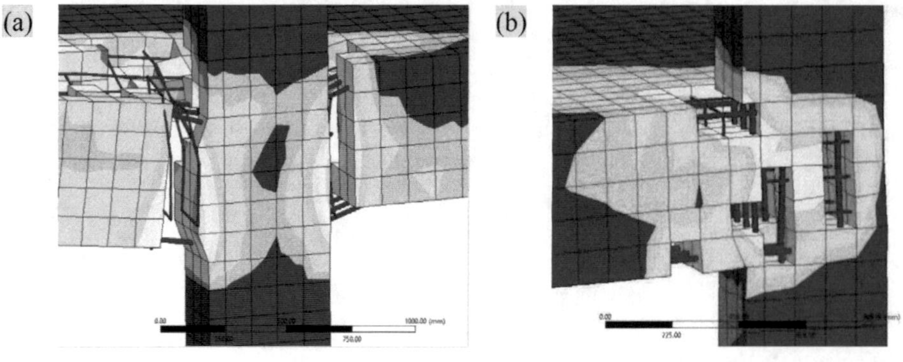

[VS3]

Figure 9. Beam - column details at 4.20 s: (a) column C8; (b) column C9.

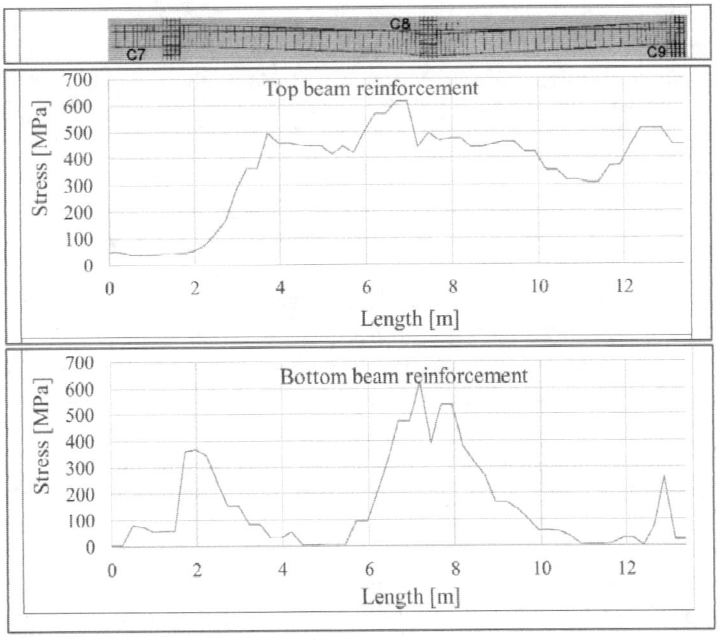

Figure 10. Tensile stress in longitudinal reinforcement of the beam.

The stress distribution in the longitudinal reinforcement (see figure 10) in the beam demonstrates its combined action - vierendeel action in the beam-column connections in combination with the catenary action of the frame beam [6].

Plastic hinges are formed in the middle near column C8 and at the end of the beam near column C9.

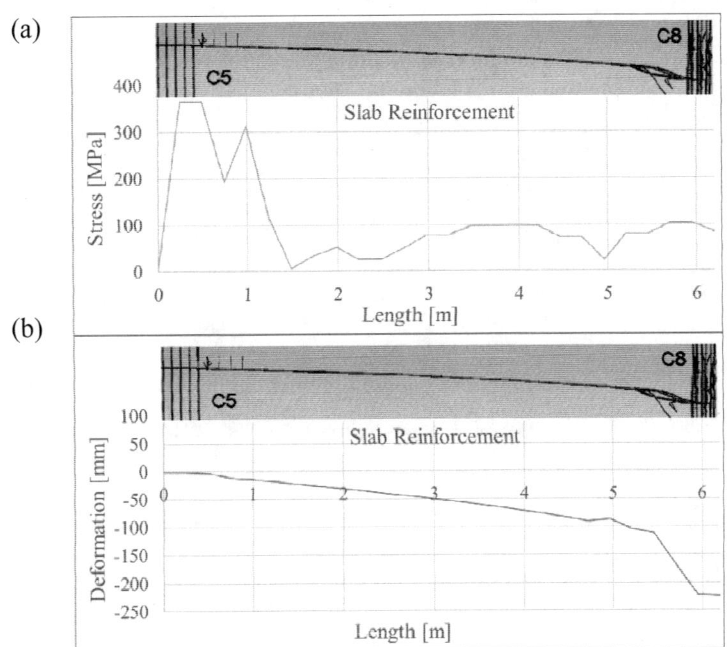

Figure 11. Longitudinal bottom reinforcement in the slab (a) tensile stress; (b) deformation.

The deformed scheme of the bottom longitudinal reinforcement of the slab between columns C5 and C8 shows larger displacements at the connection with the peripheral beam, but they are local. The stresses in this area are limited (see figure 11). Near the column, the vertical displacements of the slab and the reinforcement have induced significant tension stress in the longitudinal reinforcement, typical for a failure of transverse punching shear reinforcement. This is a display of alternative load path behaviour of the longitudinal reinforcement [5].

5. Conclusions

The manner of destruction shown in Model 1 can be assessed as ductile. The catenary action in the beam and the vierendeel action in the beam-column connections are established and evident. Those alternative load paths ensure the robustness of the structure in case of dynamic failure of a column and are an effective prevention against the redistribution of the large forces to the punching shear areas in the flat slabs.

The analyzes show that the corner column is a key element that guarantees the alternative load paths in the beams, and the L-shaped column is one of the most proper solutions.

Acknowledgments

The author would like to thank prof. PhD eng. Atanas Georgiev for his overall support.

This work is supported by Karoll Knowledge Foundation and the author is sincerely grateful for the support and the confidence.

References

[1] EN 1991-1-7: Eurocode 1: Actions on structures - Part 1-7.

[2] Sokolova G 2019 *Survey of reinforced concrete beams behavior after ultimate bending limit state,* IX-th International scientific conference on architecture and civil engineering ArCivE- 2019, Varna, Bulgaria

[3] GSA Alternate Path Analysis and Design Guidelines for Progressive Collapse Resistance, General Services Administration (GSA), 2013

[4] Heckotter Ch., *et al.* 2017 *Comparison of the RHT Concrete Material Model in LS-DYNA and ANSYS AUTODYN*

[5] Georgiev A 2011 *Punching Shear*

[6] Department for Communities and Local Government 2011 *Review of international research on structural robustness and disproportionate collapse*

[7] Kim H 2016 *Computational simulation ofprogressive collapse of reinforced concrete rigid frame with reinforcement contact model,* p.982

Response of a reinforced concrete cantilever beam subject to impulse impact loads

I Stoynova* and T Christov

USEA "Lyuben Karavelov" Bulgaria, Faculty of Construction, Department "Mechanics and Mathematics", 1373 Sofia, 175 Suhodolska str.

* E-mail: stoynova@vsu.bg

Abstract. To study the response of a RC cantilever beam subjected to impulse excitations is based computational model. Static and dynamic analysis of the computational model was performed using a software product based on the Finite Element Method. Two numerical models have been builded for static analysis. In the first model, the concrete modelled with a linear - elastic material model and the reinforcement as deformable bars. In the second model, the concrete modelled using a nonlinear model of concrete - the material model of Willam and Warnke, and the reinforcement as deformable bars. For dynamic analysis, concrete modelled as a linearly elastic material and reinforcement as deformed bars. Under impulse loading, the applied force acts for a very short but limited time interval. This article presents five cases of impulse duration. The dynamic time-history analysis used to determine the dynamic response of the RC cantilever beam under the action of time variables. This type of analysis used to determine time-varying displacements, stresses, and internal forces caused by a combination of static and dynamic (pulse and harmonic) loads. A dynamic time-history analysis used to study the dynamic response to time-varying loads. The time-determined displacements, deformations, stresses and internal forces give the behaviour and reaction of the structure at any time in the studied interval.

1. Introduction

Transient or time-history dynamic analysis is a technique used to determine the dynamic response of a structure under the action of time-dependent excitations. Can use this type of analysis to determine the time-varying displacements, strains, stresses, and forces in a structure as it responds to any combination of static, transient, impulse and harmonic loads [1].

The basic equation in Finite Element Method of motion solved by a transient dynamic analysis is (1):

$$[M]\{\ddot{u}\} + [C]\{\dot{u}\} + [K]\{u\} = \{F(t)\} \quad (1)$$

where:

[M] – Mass matrix;
[C] – Damping matrix;
[K] – Stiffness matrix;
$\{\ddot{u}\}$ – Nodal acceleration vector;
$\{\dot{u}\}$ – Nodal velocity vector;
{u} – Nodal displacement vector;
{F(t)} – Load vector.

2. Response to impulse excitation – unit impulse

Where very large force acts for a very short time but with a time integral that is finite is called impulsive force. As shown in Figure 1 – force $F(t)=\frac{1}{\Delta t}$, with time duration Δt starting at time t_1. Where value of Δt approaches to zero, force becomes infinite, however, the magnitude of impulse, defined by the time integral of $F(t)$ remains equal to unity. Such a force in the limit case $\Delta t \to 0$ is unit impulse [3].

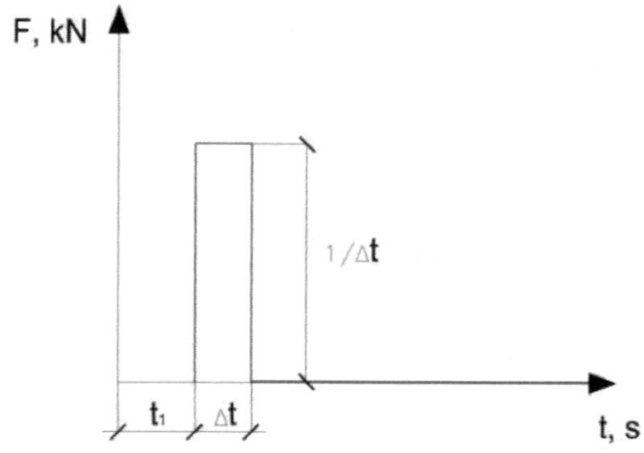

Figure 1. Impulse excitation – unit impulse

According to Newton`s Second law of motion, if a force F acts on a body with mass m, the rate of change of momentum of the body is equal to applied force (2):

$$\frac{d}{dt}(m\dot{u}) = F \tag{2}$$

If mass is constant: $F = m\ddot{u}$.

Integrating both sides with respect to t and gives (3):

$$\int_{t_1}^{t_2} pdt = m(\dot{u}_2 - \dot{u}_1) = m\Delta\dot{u} \tag{3}$$

The integral at the left side is magnitude of impulse, and product of mass and velocity is momentum [3].

3. Numerical models

The concrete and reinforcement material properties are according to Eurocode 2. [4]

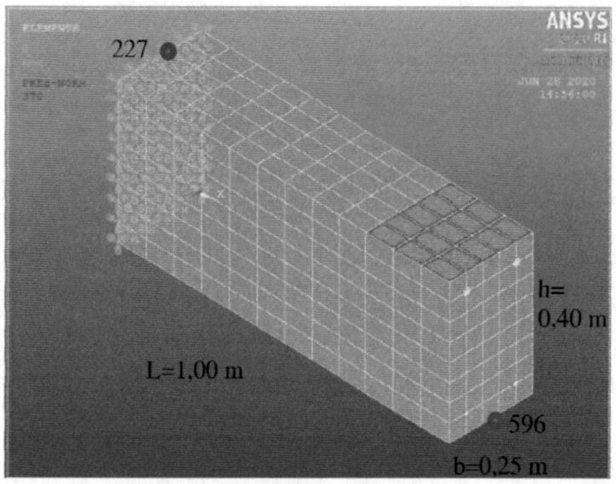

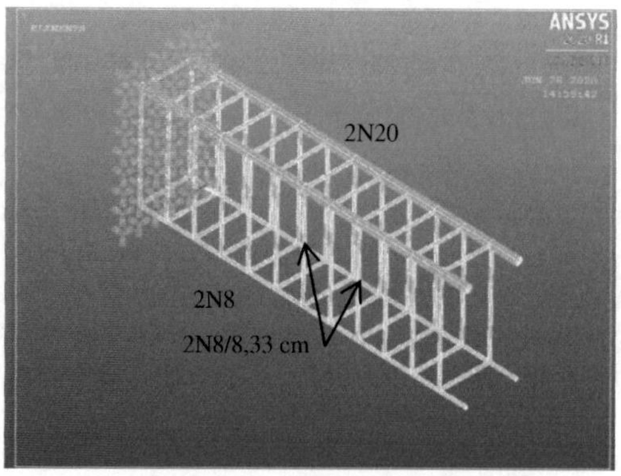

Figure 2. a) Scheme and load area b) Position of reinforced bars

The concrete properties are: C20/25, f_{cm}=28 MPa, E_{cm}=30 GPa, Poison`s ratio is 0,2.
Reinforcement diameters and position is shown on Figure 2b).
Cross section of the cantilever beam – b=0,25 m, h=0,40 m, L=1,00 m

3.1. Static analysis. Linear elastic concrete model.

First numerical model (M1) is modeling by ANSYS (Figure 4). The load stamp size is 0,25/0,25 m. The static load of 370 kN/m^2 is applied on the load stamp. The cantilever beam is modelled by SOLID 65 (CONCRET 65) finite elements as linear elastic concrete material model – Table1. Reinforcement bars is modelled with BEAM188 finite elements – Table 2.

Table 1. Concrete properties

Concrete material model for C20/25
Linear isotropic
Young's modulus E_c=3x10^7 kN/ m^2
Poison`s ratio ν=0,2

Table 2. Reinforcement properties

Reinforcement steel B500 model
Linear isotropic
Young's modulus E_s=2x10^8 kN/ m^2
Poison's ratio ν=0,3

3.2. Static analysis. Nonlinear inelastic concrete model.

Second model (M2) is modelling with material properties according to Table 3, are presented. The boundary conditions adopted are restraints applied on the left side of the cantilever beam - displacements and rotation by axes X, Y and Z. The load stamp size are 0,25/0,25 m. The applied force start at 160 kN/m^2 by step of 160 kN/m^2. The beam is modelled by SOLID 65 (CONCRET 65) finite elements nonlinear plasticity material model. Different constants for application of the Willam and Warnke's material model [5] in ANSYS are require (Table 3). Reinforcement bars is modeled with BEAM188 finite elements – Table 2. In Figure 4 is shown the distributtion of the normal stresses in the support zone and displacements in the free end of the cantilever beam. In Figure 5 is shown formed cracks at model M2.

Table 3. Concrete material nonlinear inelastic model in tension zone

1. Shear transfer coefficient for an open crack(β_t)	0,2
2. Shear transfer coefficient for an close crack (β_c)	0,9
3. Uniaxial cracking stress (f_r)	2800 kN/m^2
4. Uniaxial crushing stress (f_c)	-1

Tensile strength of bending concrete - the average tensile strength in bending of reinforced concrete elements depends on the average tensile strength and the height of the cross section. [8]

3.3. Dynamic analysis. Impulse load.

The most important point in the study of building structures of dynamic impacts is the choice of a suitable dynamic model. [6]

The numerical model of cantilever beam is modeling by ANSYS. Applied load at the end of the beam on the load stamp (Figure 2a). The load stamp size is 0,25/0,25 m. The stamp load case is presented. The impulse force with magnitude of 370 kN/m^2 is applied. The concrete is modelled by SOLID 65 (CONCRET 65) finite element as linear elastic concrete material model – Table 1. Reinforcement bars is modelled with BEAM188 finite elements – Table 2.
Five load cases are presented – Table 5.

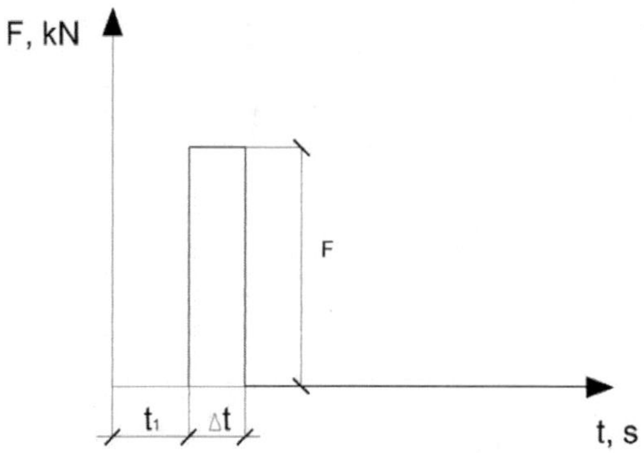

Figure 3. Impulse excitation, applied on the load stamp

Table 4. Impulse load cases

№	Load magnitude, kN/m²	Duration Δt, s	Duration t, s
1	q=370	0,005s	2,00 s
2	q=370	0,025s	2,00 s
3	q=370	0,05s	2,00 s
4	q=370	0,10s	2,00 s
5	q=370	0,20s	t=2,00 s

4. Numerical results and discussion

4.1. Static analysis

In Figure 4b) displacement u_y is presented and in Figure 4a) presented distribution of σ_x at load of 370 kN/m². The results from static analysis at node 596 are presented in Table 5.

4.1.1. Linear elastic material models

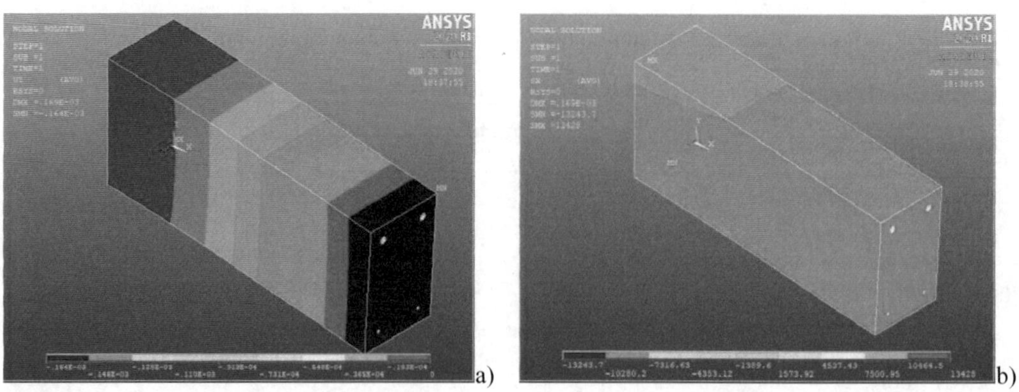

Figure 4. a) Displasement u_y, b) Distribution of normal stress σ_x at the cantilever beam, q=370 kN/m²

4.1.2. Nonlinear material model

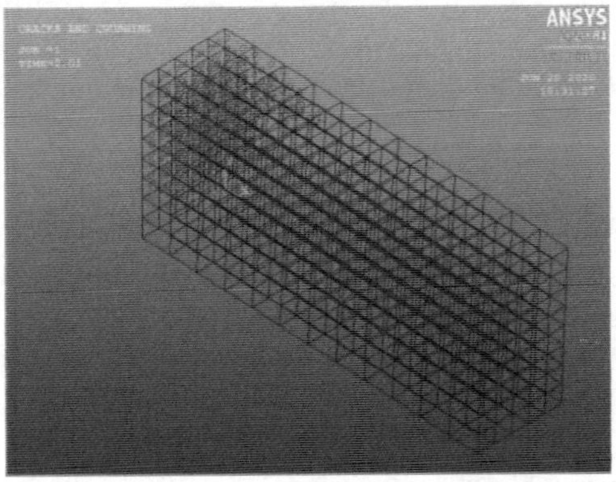

Figure 5. Nonlinear concrete material model. Crack initiation at applied force q=370 kN/m^2

Table 5. Results from static analysis at node 596 and node 227, q=370 kN/m^2.

Parameter	Linear elastic material model (M1)	Plasticity material model (M2)
$\|u_y\|$ - node 596	0,000161757 m	0,00102674 m
σ_x – node 227	2911,42 kN/m^2	2517,98 kN/m^2

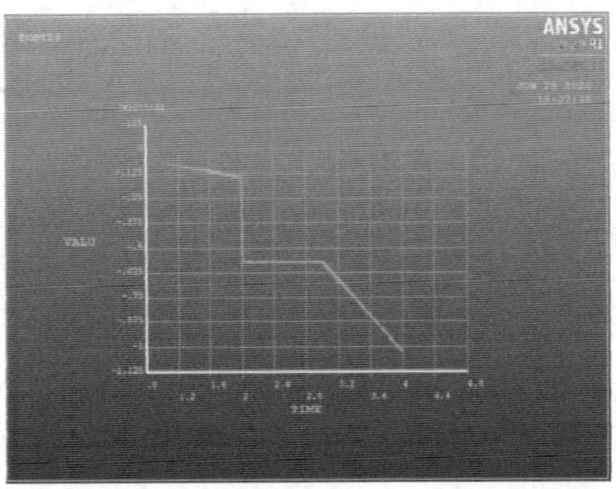

Figure 6. Displacement - Load relation in node 596 by nonlinear static analysis.

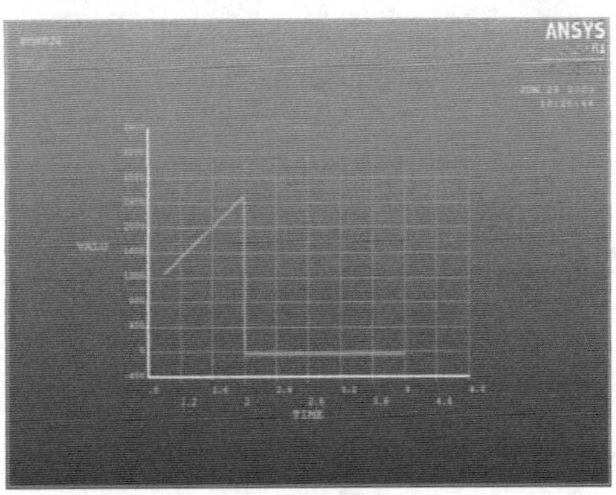

Figure 7. Normal stress σ_x - Load relation in node 227 by nonlinear static analysis.

4.2. Modal analysis

In the Table 6 natural frequencies from modal analysis are presented.

Table 6. Results from modal analysis

Mode	Frequency, s^{-1}
1	42,994
2	65,236
3	148,393

4.3. Dynamic analysis

In the Figure 8 is shown variations of displacement u_y in the node 596 at time duration of impulse load: Δt=0,005s, Δt=0,025 s, Δt=0,05 s, Δt=0,10 s and Δt=0,20 s for first numerical model with material properties from Table 1 and Table 2 (M3).

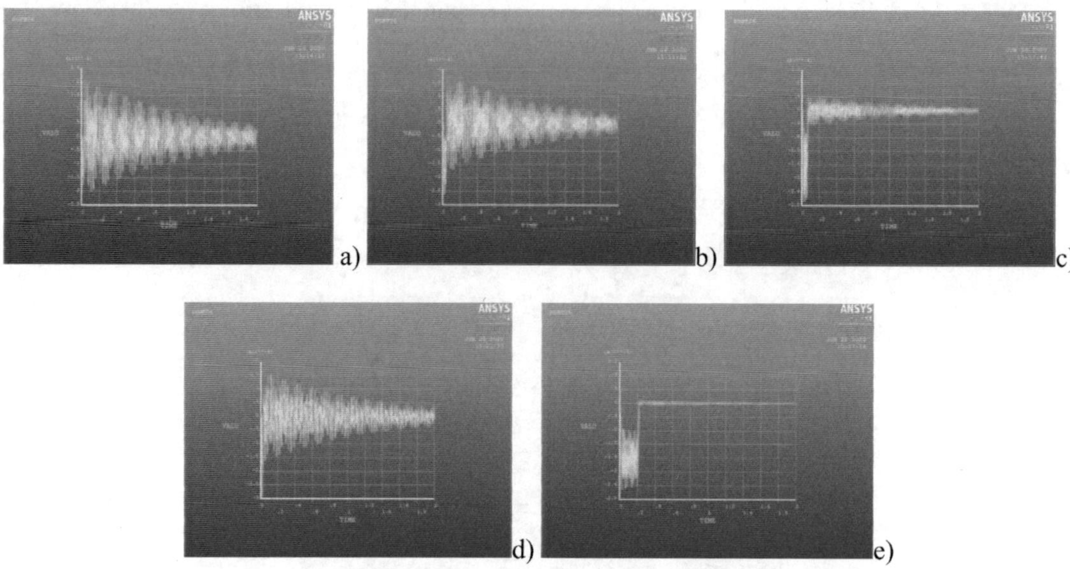

Figure 8. Displacement u_y – time t relation at node 596, q=370 kN/m^2; a) Δt=0,005s, t=2,00s; b) Δt=0,025 s, t=2,00s; c) Δt=0,05 s, t=2,00s d) Δt=0,10 s, t=2,00s; d)Δt=0,20 s, t=2,00s

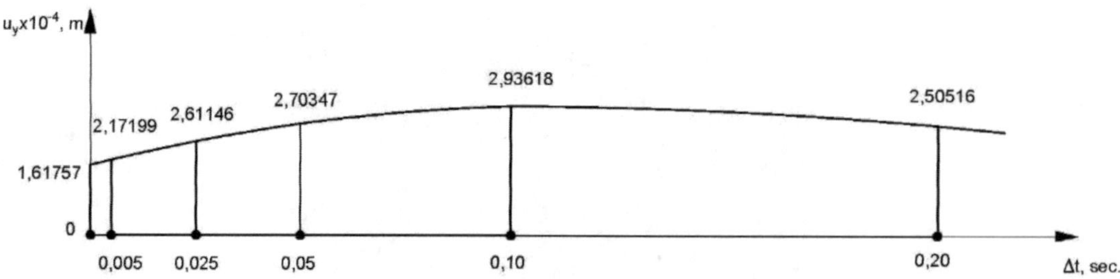

Figure 9. Response of the impulse load excitations. Relation " u_y - Δt " in the node 596.

Table 7. Results for displacements at node 596 and normal stress at node 227: static vs. dynamic analysis

Parameter	Static analysis Linear elastic material model	Dynamic analysis				
		Δt=0,005s	Δt=0,025s	Δt=0,05s	Δt=0,10s	Δt=0,20s
max. $\lvert u_z \rvert x10^{-4}$, m node 596	1,61757	2,17199	2,61146	2,70347	2,93618	2,50516
Difference from static analysis	-	34,27%	61,44%	67,13%	81,52%	54,87%
max σ_y, kN/m^2 node 227	2911,42	3931,40	4894,65	5076,84	5426,91	4616,71
Difference from static analysis	-	35,03%	68,11%	74,38%	86,40%	58,57%

5. Conclusions

Normal stresses at magnitude load considerably exceed those applied static load the same size – from 68 to 87% - Table 7. The stress state near the load stamp characterized by stresses increasing.

With the applied impulse action, the displacements in a node from the free end of the cantilever beam increase smoothly, as their maximum value is reached at a duration of action of the impulse force of Δt = 0,1 s (Figure 9). Then, with a decrease in the time interval of action of the impulse force, a gradual decrease in displacements is observed. The maximum values of displacements increase by about 82% compared to the value of displacements from static load. It follows that multiplying the values of displacements / respectively stresses and strains / coefficient equal to 1,9 get the values of displacements from the impulse action.

The FEM also provides opportunities for the study of stress and strain state of the constructive members subjected to local concentrated static load and dynamic excitations, taking into account the material and geometric nonlinearity. The results obtained by these calculation models are much closer to the actual behavior of the studied systems and constructive members. The conclusion is that by multiplying the values of displacements coefficient, we will obtain the values of displacements from impulse effects.

Acknowledgements
This paper is financially supported by USEA (VSU) “Lyuben Karavelov”, Sofia, project 08/2019.

References

[1] Bankov B., “Theory of elasticity, stability and dynamic of elastic systems”, UACEG, Sofia, 2004;
[2] ANSYS Academic 2020 R1 Reference Manual, www.ansys.com;
[3] Chopra A., “Dynamic of structures”, Prentice Hall Inc. 1995, ISBN 0-13-855214-2;
[4] European standard Eurocode 2 – BDS EN 1992-1-1;
[5] Willam, K. J. and Warnke, E. P. (1975), “Constitutive models for the triaxial behavior of concrete”. *Proc. Int. Assoc. for Bridge and Structural Engineering*, vol **19**, pp. 1- 30;
[6] Handruleva An., Matuski V., Kazakov K., Bankov B.,“Practical examples on Theory of elasticity and dynamics of structures”, VSU Publ. house, Sofia, 2011, ISBN 978-954-331-033-3, p.121
[7] Kazakov K., Yanakieva An.,Handruleva An., “Comparison of plain problems modeled by different FE formulation”, *In Proc.: X International Scientific Conference VSU2010*, ISSN 1314-071X, pp. I-1-I-6
[8] Cvetkov St., Georgiev G., “Methodological guide for development of course project reinforced concrete according to Eurocode 2”, VSU Publishing house, Sofia, 2012, ISBN : 978-954-331-037-1, p.45.

Basic recommendations for strengthening of RC beams with Ultra High Performance Fiber Reinforced Concrete (UHPFRC) – conclusions from experimental investigation

E Abdulahad and M Traykova*

Department of reinforced concrete structures
University of Architecture, Civil Engineering and Geodesy
1046 Sofia, Bulgaria

*E-mail: marina5261@abv.bg

Abstract. In modern engineering practice, it is very often necessary to look for alternative solutions for increasing the load-bearing capacity and stiffness of various structural elements of buildings, due to unsatisfactory requirements of the current regulations, increasing the load, occurring damages and defects. The modern approach, which implies that construction contributes to the sustainable development of society, requires the use of high-tech materials with significantly higher strength characteristics than traditional ones. One such material is the Ultra High Performance Fiber Reinforced Concrete (UHPFRC), which differs significantly from conventional concrete because of its specific structure and composition. In this aspect, there are a number of challenges that need to be overcome in order to provide the practical application of UHPFRC in Bulgaria – optimal mixture, technological problems, national regulations, economic indicators, etc. All this necessitates a thorough study of the actual behaviour of structures strengthen by UHPFRC. This paper presents an experimental study of the stress and strain conditions of a reinforced concrete beam with rectangular cross section strengthen with a 5 cm layer of UHPFRC concrete in the compressive zone. The analysis of the results from the experimental tests provides a number of important practical conclusions and general recommendations for the application of UHPFRC for strengthening of reinforced concrete beams in real structures.

1. Introduction

The contemporary structural solutions for rehabilitation and strengthening and the philosophy of the sustainable design are related to the application of new generation construction materials. A lot of challenges shall overcome during rehabilitation and strengthening of structures, which imposes high performance materials to be applied – usually based on fiber reinforced composites. In our country the challenges are related additionally to the requirements for resistance to seismic actions, and thus the use of conventional materials and processes is very often unsatisfactory.

During the last decade, UHPFRC has been used in structural rehabilitation and strengthening works. The concept was born in 1999 to enhance conventional reinforced concrete. UHPFRC is a smart, light and durable material. The main objective is making existing structures durable and strong enough for future needs and to update their performance and increase their structural capacity (stiffness, ultimate resistance, fatigue resistance). The strengthening method using a layer of strain hardening UHPFRC is an effective method in terms of technical performance. The application of one single layer can fulfil several functions: 1) increase resistance in bending and shear, 2) increase stiffness, 3) reduce fatigue stresses, 4) act as a waterproofing layer if necessary.

The UHPFRC technology to improve the resistance and the durability of existing concrete structures is cost-effective at execution. Actually, lower intervention costs (when compared to traditional methods) are the main reason for the increasing number of applications of this technology.

The very first scientific studies on this subject in Bulgaria and the possible steps to the implementation of UHPC (Ultra High Performance Concrete) as structural material of new generation are obtained within the project „Ultra High Performance Concrete – alternatives for construction of high-rise buildings and transport equipment in Republic of Bulgaria", which has been funded (2010-2014) by the National Research Fund within Ministry of Education and Science in Bulgaria [1].

In [2], [3] is made further analysis of the challenges and the opportunities in the application of UHPFRC for rehabilitation and strengthening of existing reinforced concrete structures.

Up to 2018 the research for the application of UHPFRC in Bulgaria was mainly focused on the properties of the material and recipes. In 2018 a research project started in the University of Architecture, Civil Engineering and Geodesy (UACEG), Sofia. One part of the investigation was dedicated to the application of UHPFRC for strengthening of reinforced concrete beams [4]. The main results concerning the beam specimens are discussed in [4] and [5].

This paper is a continuation of [5] and presents the experimental test of one of three specimens of reinforced concrete beams. The goal of this investigation is to explore the potential of the strengthening in the local Bulgarian conditions.

2. Theoretical determination of the capacity of the unstrengthen and strengthen beam

For the analysis, the main characteristics for the unstrengthen beam are presented:

2.1. *Moment of resistance of the beam before strengthening*

M_{Rd}=33,26 kN.m → F=47,51 kN (available bars 3Ø12)

2.2. *Design shear resistance of the beam before strengthening*

$V_{Rd,max}$=72,9 kN

$V_{Rd,s}$=36,92 kN (available stirrups Ø6/15)

2.3. *Theoretical calculation of deflections*

For the calculation of the deflections the following geometrical characteristics of the cross section are found:

2.3.1. *Geometrical characteristics of the non- cracked and cracked section*

$I_I = 288380935\ mm^4$ – second moment of area of the non-cracked section;

$\frac{1}{r_I} = 0{,}0000129\ \frac{1}{mm}$ – curvature of the non-cracked section;

$I_{II} = 180955426\ mm^2$ – second moment of area of the cracked section;

$\frac{1}{r_2} = 0{,}00002063\ \frac{1}{mm}$ – curvature of the cracked section

2.3.2. *Average stiffness:*

$W_{red,1} = 2323217\ mm^3 \rightarrow M_{crc} = f_{ctm} W_{red,1} = 4{,}41\ kN.m$

W- first moment of area;

$$\psi_s = 1 - \beta\left(\frac{M_{crc}}{M}\right)^2 = 0{,}975$$

$$\frac{1}{r}=\psi_s\frac{1}{r_2}+(1-\psi_s)\frac{1}{r_1}=0,00002044\ \frac{1}{mm} \rightarrow \text{The theoretical deflection is: } f=s\frac{1}{r}L^2=13,34\text{ mm}$$

3. Experimental programme and results from the experiments

The research project in UACEG includes the analysis of the behaviour of strengthen with UHPFRC orthotropic steel decks and RC beams [4].

The three RC beam specimens have length of 3000 mm each and cross section 250/150 mm. The layout of the reinforcement is presented in [4], [5]. The characteristics of the materials used for the test are:

3.1. Materials for test specimens

- Concrete C20/25 with compressive strength f_{ck}=20 MPa and modulus of elasticity E_{cm}= 30000 MPa;
- Ultra-high performance fiber reinforced concrete (UHPFRC) with cube strength $f_{frc,cube}$=141,3 MPa and cylinder strength $f_{frc,k}$=0,8.141,3=113 MPa;
- Reinforcing steel B500 for longitudinal reinforcement with yield strength f_{yk}=500 MPa and modulus of elasticity E_s= 200000 MPa;
- Reinforcing steel B235 for transverse reinforcement with yield strength f_{yk}=235 MPa and modulus of elasticity E_s= 200000 MPa.

Figure 1. Laboratory test of strengthen beam

3.2. Experimental set up and experimental programme

The test of the specimens was carried out in two stages: before strengthening and after strengthening. The strengthening consists in the application of UHPFRC with 50 mm thickness in the compressive zone. Initially the top surface of the beam is cleaned and a layer of special additive ADIPLAST for good bond between existing and new concrete is placed. Later the layer of UHPFRC is applied. The production of the UHPFRC was done by Hidrobeton according to their original recipe. The main components include cement 52,5R, micro silica fume, metakaolin, quartz powder (sand), superplasticizer, steel fibers and water. Three different mixes were made. The details for the production of UHPFRC are given in [4]. The general goal was to obtain an optimal mix of UHPFRC on site, not only in laboratory.

All specimens were tested as "simply supported" beams. The load was applied as concentrated in the middle of the beams, statically. Figure 1 and 2 show the experimental test set up of the strengthen beam. For the relative deformations of the reinforcement, strain gauges were used (X_i). The relative deformations of concrete were measured by Hugenberger's strain gauges (T_i). The deflections of the beam were registered by deflection indicators (U_i).

IOP Conf. Series: Materials Science and Engineering **951** (2020) 012016 doi:10.1088/1757-899X/951/1/012016

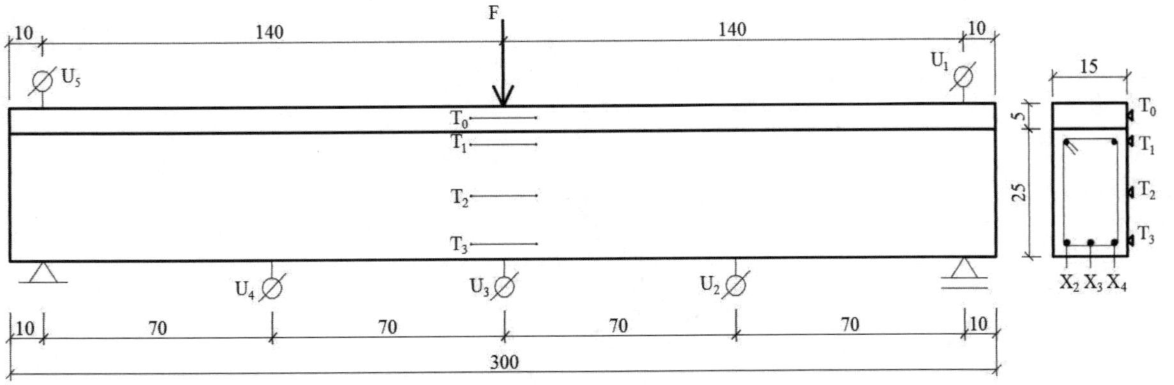

Figure 2. Experimental test set up

Before strengthening RC beams (specimens 1 and 2), the experimental moment of resistance (bearing capacity) was found after the loading of 47,7 kN, when the width of the cracks was 0,3 mm (taken as a limit value according to [6]):

M_{Rd}=33,41 kN.m → F=47,7 kN

The third beam (specimen 3), presented in this paper, was subjected to loading of 49 kN before strengthening. The opening of cracks was less than 0,3 mm.

After strengthening, the experimental moment of resistance (bearing capacity) was found at loading of 66,41 kN:

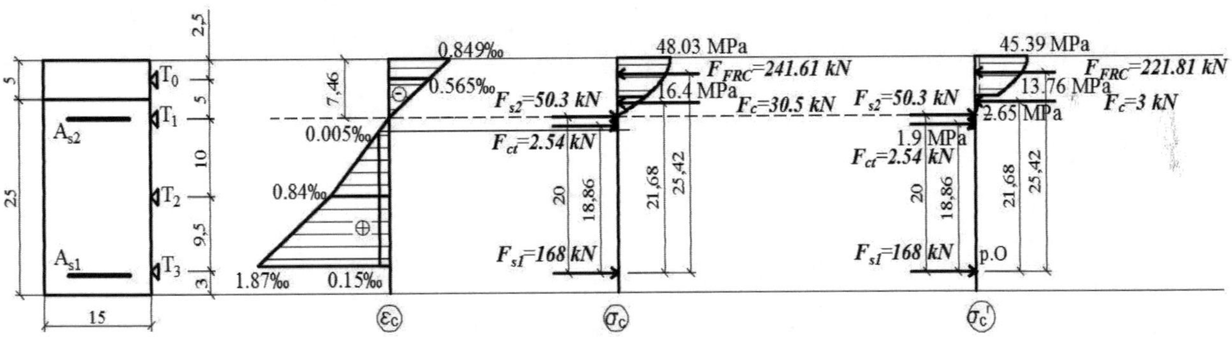

Figure 3. Stress and strain distribution in the strengthen beam

$$\sigma_{ci}=f_{frc,k}\left(1-\left(1-\frac{\varepsilon_{frci}}{\varepsilon_{frc2}}\right)^{2}\right)=113\left(1-\left(1-\frac{0,849}{2,6}\right)^{1,4}\right)=48,03\text{ MPa} \qquad (1)$$

M_{Rd}=46,49 kN.m → F=66,41 kN

When the load reached 75 kN, the opening of the cracks was 0,45 mm and the yielding of the reinforcement was reached.

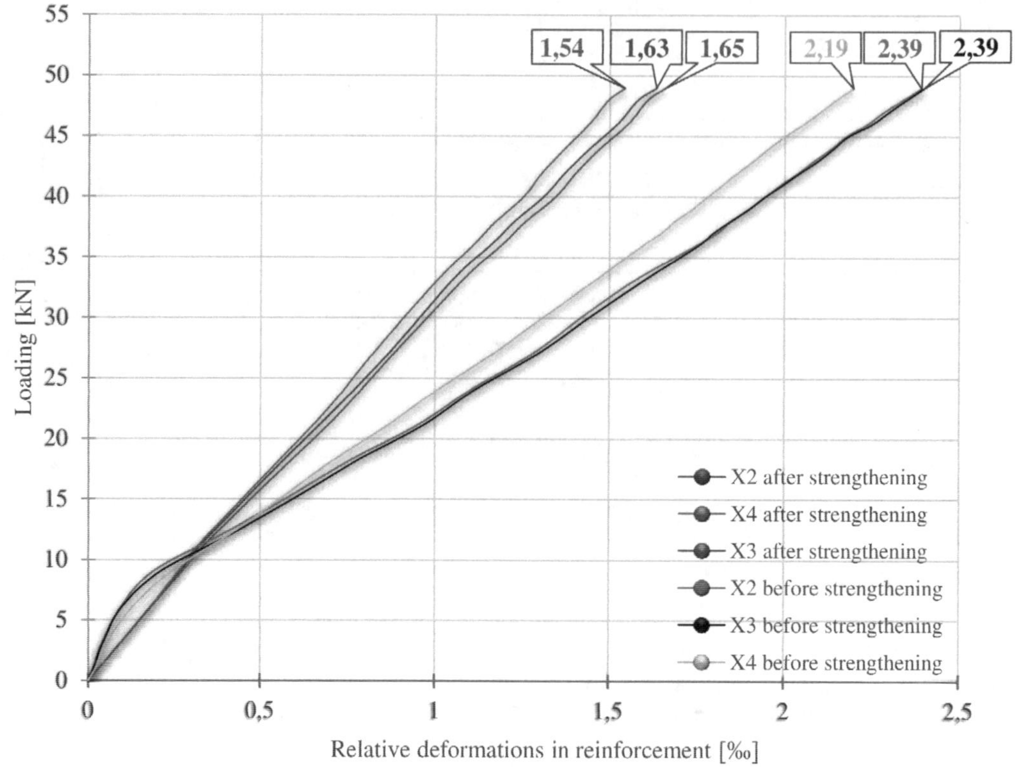

Figure 4. Diagramme „Relative deformation in reinforcement - Loading" before and after strengthening

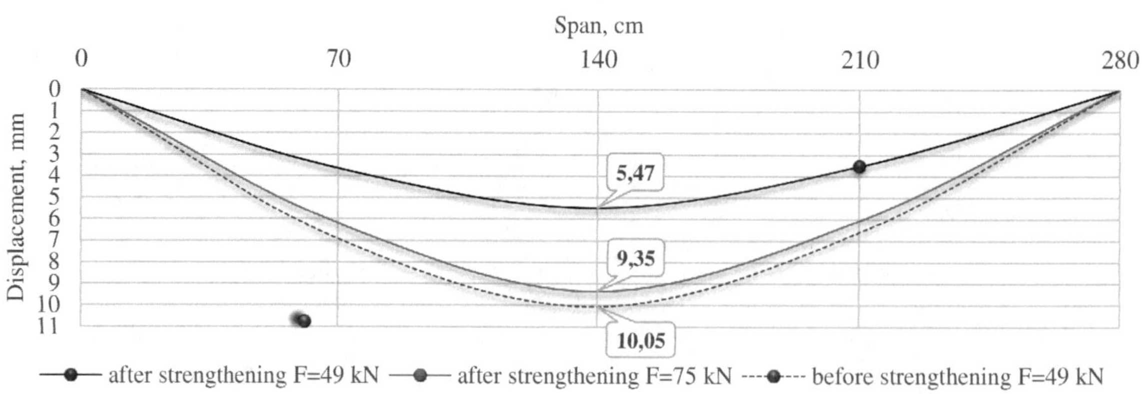

Figure 5. Beam's displacements before and after strengthening

3.3. Evidence and results of the laboratory work

Based on the test and the calculations for the stresses as in equation (1), the paper presents the following results:

- Stress and strain distribution for concrete in the strengthen beam – figure 3;
- Diagramme „Relative deformation in reinforcement - Loading" before and after strengthening- figure 4
- Beam's displacements before and after strengthening -figure 5.

The analysis of the diagrammes could be summarized in some very important conclusions:

- The stress diagramme in concrete is reduced, because of the big difference between the strength of the original concrete and the UHPFRC layer and the additional stresses in the interface-figure 3. There is a good agreement between the theoretical calculation and the experimental results.
- The crack 0,3 mm in the strengthen beam appeared when the load is F=66 kN. For the beam before strengthening the crack 0,3 mm is fixed at loading F=49 kN.
- The relative deformations in the tensile zone (T3) in the strengthen beam decrease with 40% at loading F=49 kN in comparison with the unstrengthen beam.
- The deformations of concrete in the compressive zone (T1) for the strengthen beam at loading F=46 kN, decrease significantly in comparison with the same results in the unstrengthen beam.
- At loading 49 kN the deformations in the reinforcement of the strengthen beam decrease with 42-46%.
- The bearing capacity of the strengthen beam at loading F= 49 kN increase with 40%.
- At level of loading F= 49 kN, the deflections after strengthening decrease twice.

4. Summary and recommendations

Analysing the results of the experimental test, the following summary and recommendations could be drawn:

1) The application of UHPFRC is one effective contemporary solution for rehabilitation and strengthening. UHPFRC can considerably enhance the performance of existing reinforced concrete structures. The properties and the characteristics of this material are excellent for improvement of the maximum strength and the stiffness of reinforced concrete beams. The results obtained in the laboratory test in UACEG and some other experiments [7] show that the technique using UHPFRC gives significant results in strengthening of reinforced concrete beams with a slight change in the own weight. Future application for reinforced concrete structures, using thin layers 4-8 cm of UHPFRC, is very good opportunity to upgrade their behaviour without additional increasing of the self-weight. A possible application of UHPFRC in real RC structures are available in [8], [9].
2) The bond between the old concrete and the new layer of UHPFRC is decisive for the behaviour of the strengthen element. Usually the bond between UHPFRC and concrete is obtained by preparing the concrete substrate surface by high pressure water jetting or sand blasting. The concrete substrate has to be wetted and needs to be wet when the layer of UHPFRC is cast. This surface preparation provides a full bond between the UHPFRC and the concrete substrate. In the presented laboratory tests the application of a special additive on the interface, showed very good results.
3) In terms of life cycle cost, the application of UHPFRC is related to serious reduction of maintenance and environmental impact added value for owner and society.
4) The cost for high fiber content is largely outbalanced by improved mechanical properties. The use of fiber content as high as possible to obtain tensile strain hardening is recommended. The content of fibers should be kept around 2,5% -3% to avoid cracking and provide durability.
5) Next generation of UHPFRC is related to the reduction of energy and carbon footprint while maintaining high packing density and mechanical properties of UHPFRC. The trend is to replace clinker by limestone filler, replace steel fibers by high modulus PE-fibers.
6) The aim of all future investigation is to overcome the challenges related to the composition and technology of preparation of UHPFRC and its in-situ application.

7) It is necessary to develop a local national regulatory framework /recommendations for design and/or to adapt existing recommendations [10] to the specifics of UHPFRC in Bulgarian conditions.
8) Due to the higher cost of UHPFRC, the use for the complete construction of structures of buildings and facilities in Bulgaria will probably be difficult to access for investors at the actual moment, but in the activities of rehabilitation and strengthening, as well as for special purposes, the application of UHPFRC is one sustainable and effective solution.
9) It is necessary to make serious efforts to conduct long-term monitoring of the behaviour of the already strengthen elements.

Acknowledgements
The authors would like to express their acknowledgements to: "Hidrobeton" and especially to Mr. Rosen Nikolov, for his great support and contribution in this research project and the Research, Consultancy and Design Centre of UACEG, Sofia, Bulgaria for the financing grant agreement N БН-206/2018.

References
[1] Zaharieva R and Nazarski D 2014 *Ultra high performance concrete and the opportunities to its application in Bulgaria* First Scientific-Applied Conference with International Participation "Project Management in Construction" (Sofia) pp 333-340.
[2] Traykova M and Chardakova-Nackova T 2016 *Application of UHPFRC for rehabilitation and strengthening of exiting RC structures - challenges and opportunities* DCB 2016 International Conference in Civil Engineering Design and Construction (Science and Practice) (Varna) pp 506- 511.
[3] Zaharieva R 2015 *Ultra high performance concrete as a material for rehabilitation and strengthening of buildings and facilities structures* First Scientific-Applied Conference with International Participation "Rehabilitation and Strengthening of Buildings and Facilities Structures" (Sofia) pp 178-186.
[4] Traykova M, Abdulahad E, Georgiev L and Ivanov S 2019 *Research in the application of UHPFRC for strengthening existing structures in Bulgaria* IABSE Congress (New York).
[5] Traykova M and Abdulahad E 2020 *Opportunities for application of Ultra High Performance Fiber Reinforced Concrete (UHPFRC) for strengthening concrete beams - results of an experimental research* Annual of the University of Architecture, Civil Engineering and Geodesy **53** Issue 3 (Sofia) (in print).
[6] *BDS EN 1992-1-1:2004 Eurocode 2: Design of concrete structures- Part 1-1: General rules and rules for buildings*.
[7] Lampropoulos A, Paschalis S, Tsioulou O and Dritsos S 2015 *Strenghthening of existing reinforced concrete beams using UHPFRC* ICCRRR IV (Germany)
[8] Traykov A 2017 “Rehabilitation of the Complex Reinforced Concrete Shell Roof Structure of an Industrial Building,” in E. Pelke and E. Bruhwiler, *Engineering history and heritage structures - Viewpoints and approaches*, Structural Engineering Document 15, ISBN 978-3-85748-154-3, IABSE (Zurich) pp 67-70.
[9] Traykova M and Boiadjieva R 2014 *Assessment and rehabilitation of the structure of existing reinforced concrete tanks*, 37 Symposium of IABSE “Engineering for progress, nature and people” (Madrid).
[10] *Ultra High Performance Fibre-Reinforced Concretes – Recommendations* 2013 AFGC (France).

IOP Conf. Series: Materials Science and Engineering **951** (2020) 012017 doi:10.1088/1757-899X/951/1/012017

Seismic behaviour of brickwork chimneys in buildings

R I Ivanov

Section "Seismic Engineering", Department "Seismology and Seismic Engineering", National Institute of Geophysics, Geodesy and Geography – Bulgarian Academy of Sciences, Acad. G. Bonchev St., bl. 3, Sofia 1113, Bulgaria

E-mail: rivanov@geophys.bas.bg

Abstract. The construction of chimneys of solid bricks in buildings with sloped roofs was commonplace in Bulgaria for almost a century. The collapse of a chimney during an earthquake could potentially lead to damages greatly exceeding the loss of the chimney itself, e.g. partial damage to the roof tiling and leaks, as well as material damage, injury or loss of life due to debris fall. A FEM model was created, in which the storeys of the building are represented in a generalised way, while the chimney is modelled explicitly as a cantilever supported at roof level. The internal forces in chimneys with heights ranging from 0.5 m to 2.0 m, belonging to buildings with height ranging from two to seven storeys were computed. Acceleration records from real earthquakes acting at the base of the building with varying peak ground acceleration and predominant period were used for input loading. The maximum tensile stresses at the bed joints were computed and were compared to the typical tensile strength of the mortars used for chimney construction, to assess the possibility of collapse. A simple, low-tech method for upgrading of existing chimneys by applying a coat of cement-based plaster with embedded fiberglass mesh is proposed.

1. Introduction

The widespread construction of chimneys made by solid brick units and lime or lime-cement mortar in buildings with sloped roofs began after Bulgaria became an independent country in 1878, and continued until industrialized construction became commonplace. During the 2012 earthquake with magnitude M = 5.8 and epicentre in the vicinity of the city of Pernik, a large number of masonry chimneys in the city collapsed, figure 1. Debris of masonry chimneys were also observed on the pavements in the capital city, Sofia, although virtually no structural damage occurred there. Such behaviour could potentially lead to damages greatly exceeding the loss of the chimney itself, e.g. partial damage to the roof tiling and leaks, as well as material damage, injury or loss of life due to debris falling down from the roof. The problem is further aggravated by the location of buildings with masonry chimneys within a city. Given their relatively old age, they are concentrated in or close to the city centres, where streets are generally narrower and the number of pedestrians and parked vehicles larger, thus increasing the probability of secondary damage.

Research on the seismic behaviour of chimneys is focused mostly on industrial ones, [1], [2], [3] and reference therein, which pose a greater threat, are structurally more challenging and possess some heritage value, which is rarely the case for chimneys of residential or office buildings. No past research on the seismic response of 'small' chimneys is known to the author. The objective of this study is to clarify their behaviour when subjected to ground shaking, and to assess the likelihood of failure for different levels of ground motion and mortar tensile strength.

(a) a house with collapsed chimneys (b) a block with heavy damage to chimneys

Figure 1. Damage to chimneys of residential buildings in the city of Pernik.

2. Analysis model

This study only considers a scenario whereby a chimney collapses before significant damage has occurred to the building it belongs to. Indeed, if the building collapses or suffers heavy damage, a damage to the chimney will be of little or no consequence. In other words, cases where collapse of the chimney is caused by damage or collapse of the building are not considered. This assumption allows us to use elastic material behaviour, considering dissipative mechanisms through increased damping at larger ground motion levels, instead of explicitly specifying nonlinear material behaviour. Also, with tensile failure at the bed joints being the reason for the onset of collapse, elastic analysis is justified as far as the behaviour of the chimney is concerned, considering the low, sometimes negligible strength of mortars used for chimney construction. The modelling and the analyses were carried out by the computer program SeismoStruct (https://seismosoft.com).

2.1. Model parameters

The analysis model is shown in figure 2. It is a finite element model, in which the storeys of the building are represented in a generalised way by their intrinsic stiffness and mass, while the chimney is modelled explicitly as a cantilever supported at roof level. The lumped mass value of 258 t at each floor level is calculated assuming it represents a floor in a building with floor area of 200 m^2, 0.25 m thick external walls, 0.12 m thick internal walls and standard finishes. A single frame element is specified between two floors. Its bending stiffness is adjusted so as a seven story building has a natural period of 0.7 sec, considering that as rule of thumb, buildings like the ones being studied have a natural period of 0.1 sec per story. The rotational degrees of freedom at each floor level are restrained, and the base node at which the ground motion is input is fully restrained. The resulting model is thus of a symmetrical shear building. The range of building heights investigated is between 2 stories and 7 stories. Eigenvalue analysis produced fundamental periods of 0.237, 0.334, 0.423, 0.512, 0.605 sec for the 2, 3, 4, 5 and 6 story buildings respectively, which agrees well with the 0.1 sec per story assumption. Single-story buildings were excluded from the study for two reasons: a) they are very rare (at least in Bulgaria), and b) their natural periods are so short that they are not expected to significantly amplify the base ground motion. On the other hand, buildings with brick chimneys higher than 7 stories are extremely rare. A separate model was created for each of the six individual building heights.

A chimney is fixed to the top floor node and is modelled by four frame elements of equal length. The Young modulus was specified as 10,000 MPa, a typical value for brick masonry, while the specific weight was specified as 20 kN/m^3, a little higher than the usual value of 18 kN/m^3 to allow for additional loads such as plaster and concrete crowns on top of the chimney. Also due to limitations of the program, the wall between the two flues of the stack is not included in the model, so the increased specific weight caters for this too. For the chimney the mass is automatically lumped at the nodes by the program. The range of chimney heights investigated is between 0.5 m and 2.00 m at 0.25 m increments, a total of seven chimney heights. This range was decided upon by visual inspection of the skyline of the city of Sofia and other Bulgarian cities. In order to reduce the number of runs all seven chimneys were included in a model of a particular building height. This would not alter the response significantly, as the total weight of the seven chimneys is 3 t compared to the floor mass of 258 t. The cross section of the chimney used in the analyses represents a double-flue stack. This choice was done for two reasons: a) single-flue stacks which would undoubtedly produce higher stresses are very rare; b) the contribution of the bending moment about the strong axis to the total normal stress is small enough as to assume that the behaviour of double-flue stack is not overly conservative, and may be used to represent the behaviour of stacks with more flues.

The resulting model is deemed simple, yet sufficient for the purpose of this study. The building is modelled so that its most important modes of vibration can be activated, and the ground motion filtered and modified accordingly while it reaches the top of the building and becomes input motion for the chimney.

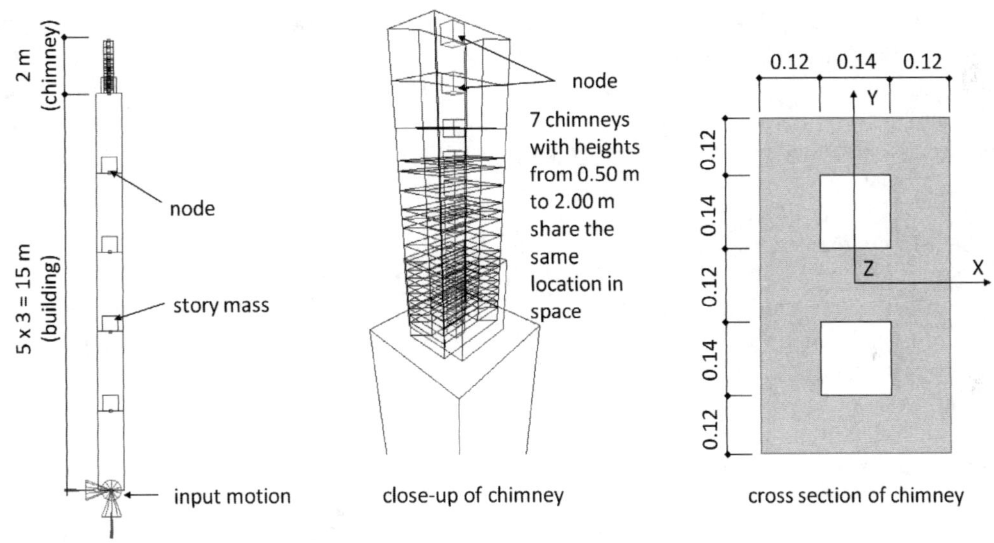

Figure 2. Model of a 5-story building.

2.2. *Input ground motion and damping*

The 2012 Pernik earthquake was recorded at three locations in Sofia, with the ground motions described and analysed in detail in [4]. The shapes of these three ground motions were used as reference, while the peak ground acceleration (PGA) of the stronger horizontal component was scaled to levels of 0.05 g, 0.10 g, 0.15 g and 0.20 g in order to investigate the behaviour with increasing strength of the ground shaking. The buildings studied here are of different structural typology given the time span of almost a century during which they were built. They can be masonry buildings with deformable or RC floors, with walls unframed or framed by RC elements. In terms of seismic design, they can be pre-code, low-code or sometimes moderate-code buildings. The upper bound of 0.20 g was chosen because according to the results of vulnerability studies presented in [5], [6] and [7], the investigated building types are likely to develop significant damage for higher PGA, and thus may collapse together with the chimneys. To account for the different degree of energy dissipation from all sources within the framework of linear elastic analysis, a different damping ratio was adopted for each PGA level listed above, namely 2%, 3%, 4% and 5% of the critical respectively. The damping was specified as Rayleigh type, whereby for the fundamental period of the building it has the values given above, according to the PGA, and for the

IOP Conf. Series: Materials Science and Engineering **951** (2020) 012017 doi:10.1088/1757-899X/951/1/012017

fundamental period of the 2.00 m high chimney it is a constant of 2%, considering the low levels of stress the chimney is exposed to. Note that the use of varying damping ratios precludes the simple scaling of response values which could be done if a constant damping were adopted throughout the analyses. All three components of the ground motion were applied simultaneously at the base node of the model and time-history analysis carried out with duration of 25 sec., sufficient to include the strongest part of each record.

3. Results and discussion

Three tri-axial ground motion records, four levels of PGA, seven chimney heights and six building heights were used in the analyses to produce a total of 504 data sets. For each of them the maximum tensile stress perpendicular to the bed joint at the node where the chimney stack is attached to the building is computed by the formula,

$$\sigma_z = \frac{|M_y|}{W_y} + \frac{|M_x|}{W_x} - \frac{|N|}{A} \tag{1}$$

where M_y, M_x, and N are the bending moments about the principle axes of the cross section and the axial force, while W_y, W_x, and A are the section moduli about the principle axes of the cross section and the cross-sectional area. The resulting stresses are compared to the tensile strength of the bed-joint mortar. If the stress is higher than the strength, the stack is assumed to have failed in tension, and therefore in high risk of collapsing. Whether collapse would actually occur cannot be established with the analysis implemented here, and is beyond the scope of this study.

3.1. *Common features of the response*

A number of important trends were observed for all data sets as follows:

The bending moments about the principle axes of the chimney stack obtained from the analyses have the same values as the theoretical bending moments in a cantilever subjected to an uniformly distributed load $q = ma$, where m is the mass per unit length of the stack, and a is the acceleration at the top of the building/base of stack.

The influence of the fluctuation of the axial force due to the vertical vibrations is negligible when obtaining the maximum of the time history of stress values computed by equation (1). Practically the same maximum is obtained when the contribution of the axial force is computed using the constant self-weight.

The reason for the above is that the chimney stack itself is practically rigid, having a fundamental period of 0.027 sec. Thus no dynamic amplification occurs within the chimney stack. Therefore, for practical purposes, the bending moments in a chimney stack can be obtained quickly and reliably by knowing the acceleration at their base in the direction of the two principle axes. It is not necessary to include the chimney stack explicitly in the computational model.

It was confirmed that the chimney stack can be regarded as an independent (secondary) structure fixed to the building it belongs to instead of to the ground. Therefore, the way the building amplifies the ground motion is crucial to the response of the chimney stack. Indeed, it was observed that due to matching of the response spectrum peaks of the ground motion and the fundamental period of the model, the response of a particular stack height to a particular ground motion strongly depends on the height of the building (in terms of stories) it is attached to.

3.2. *Vulnerability of the chimneys*

The tensile strength f_t of lime or lime-cement mortars used for masonry varies, with values ranging from 0.04 MPa to above 0.20 MPa been reported by technical recommendations of professional organizations [8], or experimental studies [9]. Given the old age and low maintenance of the buildings under consideration, it is assumed that 0.20 is the highest tensile strength that can be realistically expected. A number of lower strength levels were also used in the vulnerability study, namely 0.10 MPa,

0.05 MPa and 0.005 MPa. The last value was meant to represent extreme cases, e.g. very poor maintenance, substandard quality of the mortar, poor workmanship or the use of mud instead of mortar.

The results of the vulnerability study are summarized in figure 3. Chimney heights missing from the figures either have a failure rate of 0% (no failure) or 100% (total failure) for the whole range of PGAs. Many valuable insights can be gained from the results, which may be used to shape disaster mitigation strategies. If a chimney is constructed by good quality mortar (f_t = 0.20 MPa), it is likely to fail only if its height is above 1.50 m. At the same time, even chimneys with height 2.00 m would most likely survive a shaking with PGA of 0.10 g.

At the other end, with poor tensile strength (f_t = 0.005 MPa), only chimneys with height of up to 0.75 m may be considered safe, with this being only for very low levels of ground shaking – PGA of up to 0.50 g. It is interesting to note, that for PGA = 0.05 g, the stress at the base of the 0.50 m chimney remains compressive, thus precluding collapse. For PGA = 0.20 g, chimneys of all heights would fail. If we assume that low tensile strength is prevalent, these results are in good agreement with field observations from the 2012 Pernik earthquake. In the city of Pernik, which experienced PGA of around 0.20 g, the majority of the chimneys in the block of houses show in figure 1, collapsed. These houses are very old pre-war buildings, showing no signs of maintenance. In the city of Sofia, which experienced PGA between 0.05g in the city centre and 0.10 g at locations closer to the epicentre [4], brick debris possibly from chimneys, were observed on the pavements.

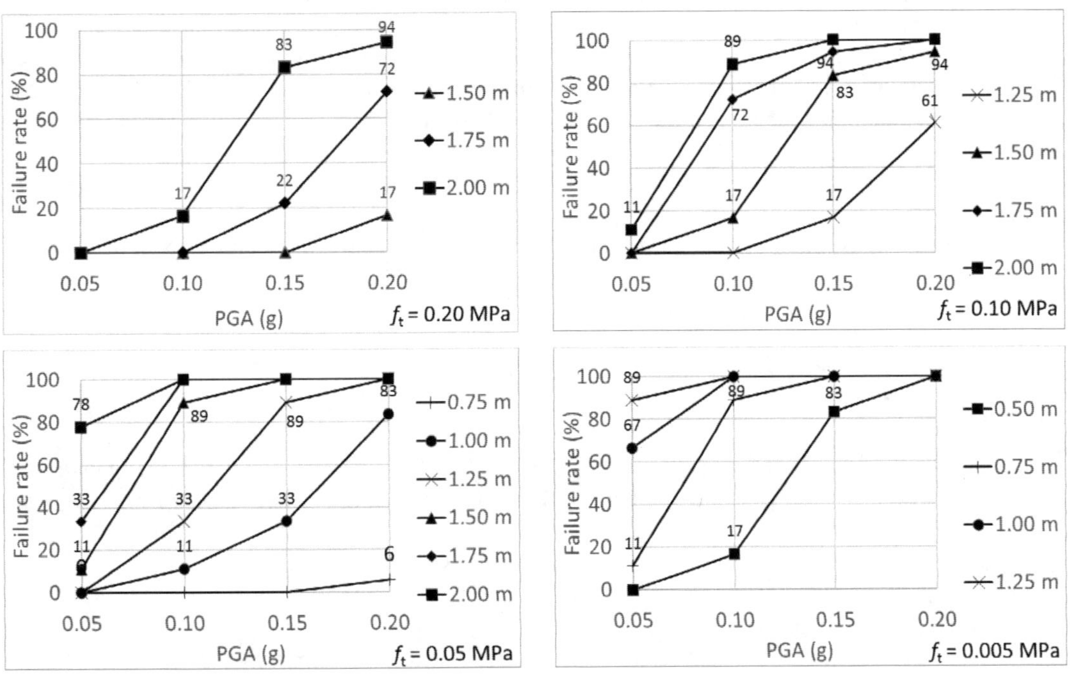

Figure 3. Fragility curves of chimneys for different tensile strengths of mortar.

4. Upgrading method for collapse prevention

Considering the demonstrated high vulnerability of brick chimneys to relatively low levels of ground shaking it is desirable to propose a strengthening method which at best should be cost effective and low-tech to make it implementable by an average bricklayer or plasterer. A design methodology for improving the flexural capacity of masonry walls using CFRP sheets is proposed in [10]. Using FRP technology however, requires somewhat specialized skills, and the sheets are not cheap either. In the following it will be demonstrated that adequate seismic retrofit of the chimneys may be achieved by using ordinary glass fibre mesh for plastering.

Although glass fibre mesh is not meant for structural use, its strength is occasionally quoted by manufacturers, e.g. 1.88 kN/5cm strip is reported at https://terazid.com/shop/produkti-za-toploizolaciq/stuklofiburna-mreja-terazid/, which is equivalent to 37.6 kN/m. Next, we compute the flexural capacity about the weak axis of the cross section shown in figure 2, assuming it is cracked. The mesh will provide a tensile force N_t = 0.64 x 37.6 = 24.1 kN. Assuming very conservatively the compression zone to spread over the width of the whole brick, 0.12 m, will result in lever arm of 0.32 m, and capacity of the section M = 24.1 x 0.32 = 7.71 kNm. This is well over the largest bending moment obtained from the analyses $M_{y,max}$ = 6.85 kNm for PGA = 0.20 g and chimney height 2.00 m. Hence a single layer of glass fibre mesh embedded in suitable plaster will be sufficient to prevent failure of the chimneys considered in this study.

5. Summary and conclusions

The seismic response of chimneys in buildings was investigated. It was shown that for a given ground motion, the internal forces in the chimneys depend entirely on the dynamic properties of the building.

Fragility curves for chimneys of heights ranging from 0.50 m to 2.00 m and tensile strengths of mortar ranging from 0.005 MPa to 0.20 MPa were developed. The high strength fragility curves indicate that if good quality materials and workmanship are applied, most chimneys will be safe against weak to medium ground shaking. On the other hand, low strength fragility curves which agree well with field observations from the 2012 Pernik earthquake, indicated that most chimneys will be damaged even for weak ground shaking.

A simple low-tech retrofit method using glass fibre mesh for plaster was proposed and shown to be adequate for the range of structural and loading parameters of the current study. It is recommended that this method is used whenever large scale roof repairs are done. During such repairs chimneys are often plastered, so embedding a glass fibre mesh in the process can be done at very little extra cost.

References

[1] López-Patiño G, Adam J, Gimeno P and Milani G 2017 Causes of damage to industrial brick masonry chimneys *Engineering Failure Analysis* **74** 188–201

[2] Lopes V 2009 Ambient vibration testing and seismic analysis of a masonry chimney *J. Building Appraisal* **5** 2 101–21

[3] Bru D, Ivorra S and Baeza F 2017 Seismic behavior of a masonry chimney retrofitted with composite materials: a preliminary aproach *Int. J. of Safety and Security Engineering* **7** 4 486–97

[4] Hadjiyski K, Simeonov S and Botev E 2012 Seismic effects of the 22.05.2012 earthquake for the city of Sofia – analysis and assessment based on strong-motion records *Engineering Forum* **7** 28-35 (in Bulgarian)

[5] Kostov M, Kaneva A, Vaseva E, Stefanov D, Varbanov G and Koleva N 2007 An advanced aproach to earthquake risk scenarios of Sofia *Proc. 8th Pacific Conf. on Earthquake Engineering* Singapore 5-7 December 215

[6] Chieffoa N, Clementi F, Formisano A and Lenci S 2019 Comparative fragility methods for seismic assessment of masonry buildings located in Muccia (Italy) *J. Building Engineering* **25** 100813

[7] Halder L, Dutta S and Sharma R 2020 Damage study and seismic vulnerability assessment of existing masonry buildings in Northeast India *J. Building Engineering* **29** 101190

[8] Brick Industry Association 1992 Technical notes 3A – brick masonry material properties *Technical Notes on Brick Construction* Reston USA

[9] Hetherington S 2015 A comparative study into the tensile bond strength of the brick mortar interface of Naturally Hydraulic lime and Portland cement mortars *Masonry Int.* **28** 2 67-74

[10] Dizhur D, Griffith M and Ingham J 2014 Out-of-plane strengthening of unreinforced masonry walls using near surface mounted fibre reinforced polymer strips *Engineering Structures* **59** 330–43

Seismic behaviour before and after strengthening of solid confined masonry walls with glass fiber reinforced polymers - analysis of hysteresis curves, obtained by shaking table tests

E Abdulahad and E Mahmud*

Department Reinforced Concrete Structures, Faculty of Structural Engineering, University of Architecture, Civil Engineering and Geodesy (UACEG), 1 Hristo Smirnenski Blvd., Sofia 1046, Bulgaria

* E-mail: dipl.ing.eminmahmud@gmail.com

Abstract. Masonry is one of the oldest building construction methods used by man thousands of years ago, and one of those used today. The masonry buildings look stable and give a sense of security and comfort, leaving a lasting impression of strength. Shaking table tests are planned in order to deepen the understanding of the behaviour of confined masonry structures. This paper presents the observations from a series of shaking table tests done on a 1:1 scale confined masonry wall model without openings. The specimen was tested under constant gravity load and incrementally increasing in-plane loading seismic up to the occurrence of damage. The horizontal load applied is in-plane of the specimen. Then, the seismically damaged specimen was strengthened, using glass fiber reinforced polymers (GFRPs), and tested again. The specimen was strengthened on the surface of two sides. Hysteresis curves before and after with GFRPs wall strengthening are analysed. The tests results indicate that the GFRP system restores and increases the lateral load capacity of strengthened walls. Moreover, GFRP helps to increase the area of hysteresis curves, leading to increased seismic energy dissipation.

1. Introduction

Masonry systems have a wide variety of forms and have been used as structural material for thousands of years. Some very old stone and brick masonry buildings still exist, proving that masonry successfully resists loads and impacts of environment. For their importance and value, many of those building have been classified into mankind's historical and cultural heritage of highest category. Considering that many of these buildings were built in the past, most of them do not meet the requirements of the Eurocode 8 recommendations and should be properly reinforced. Furthermore, from past earthquakes, it has been observed that the behaviour of the masonry systems under seismic impacts is substantially influenced by the presence and the location of the openings as well.

The main purpose of the research is to determine the behaviour of the masonry walls with different types of openings under seismic/dynamic action and to evaluate innovative retrofitting technique for strengthening existing structures with vulnerable confined masonry structural systems in order to increase their seismic resistance. The experimental program was fully realized in March 2019, within the scientific cooperation between University of Architecture, Civil Engineering and Geodesy

(UACEG) – Sofia, Republic of Bulgaria and University "Ss. Cyril and Methodius" Institute of Earthquake Engineering and Engineering Seismology, Skopje, Republic of North Macedonia [1].

In order to fulfil the purposes of the project three confined masonry walls with the same geometrical and material characteristics but with different configuration of the openings were tested. All specimens were designed and detailed by UACEG, Sofia. The first specimen was designed without opening (CMF), the configuration of the second specimen was with window opening (CMW) whereas the third specimen configuration was with door opening (CMD). The models were built in scale 1:1, in the laboratory of the Institute of Earthquake Engineering and Engineering Seismology. Moreover, the scheme of instrumentation for the models was conceived in a way to get as many as significant and valid experimental results, defining the types of measurement instruments, optimal location and number of measuring points, according to the available capacity of instruments and systems acquisition in the laboratory of IZIIS. The specimens were examined separately on the shaking table, with uniaxial, in-plane excitation following a defined testing methodology consisting of 2 phases - definition of the dynamic characteristics of the specimens and definition of the dynamic behaviour of the specimens under earthquake/dynamic excitation. After the testing, the entire surface of the damaged walls was coated with three-component thixotropic epoxy mortar, whereas the larger cracks were further strengthened with externally glued GFRP (Glass Fiber Reinforced Plastic) plate. The strengthened specimens were tested following the defined testing procedure for the original models, respectively.

This paper presents the observations from a series of shaking table tests done on a 1:1 scale confined masonry wall model without openings. The resulting hysteresis curves from the specimen tests before and after strengthening will be analysed. The remaining results are in the scope of future publications.

2. Description of the specimens and instrumentation set-up

2.1. Description of the specimens

Specimen CMFuS is a masonry wall, scale 1:1, confined with reinforced concrete tie-column, reinforced concrete tie-beam and reinforced concrete foundation. The wall was designed without an opening. The masonry wall is 2.60 m height, with length of 2.80 m and thickness of 25 cm, whereas the dimensions of the confined wall are 2.85 m, 3.30 m and 25 cm, respectively. Geometrical characteristics of specimens CMFuS and CMFS are presented in figure 1. The size of reinforced concrete tie-column and tie-beam are 25x25 cm and the foundation is 400x95x40 cm. The foundation provides 8 holes with a diameter of 50 mm for fixing the specimens to the shake table. The wall was built with materials typical of the Republic of Bulgaria and the region. Detailed information on the specimens can be found in [2]. The strengthened specimen (CMFS) presents the same geometry of the unstrengthened one (CMFuS) – see figure 2. After testing of specimen CMFuS, the entire surface of the damaged walls was coated with a three-component thixotropic epoxy mortar, whereas the larger cracks were further strengthened with the externally glued GFRP (Glass Fiber Reinforced Plastic) plates. The retrofitted specimens were tested following the defined testing procedure for the original models, respectively.

The sequence followed while building the wall was: building the foundation first, then the masonry, and finally the reinforced concrete tie-column and tie-beam. Eurocode 1998 [3] requires that in order to obtain an effective connection between the tie-elements and the masonry, the concrete in the tie-elements must be poured after the masonry has been built. Good bonding between RC tie-columns and a masonry wall can be achieved and by toothing or horizontal reinforcement anchored into tie-columns [4]-[7]. In the implementation of CMWuS, the connection between the RC tie-elements and the masonry is by toothing, and the concrete in the restraining elements is poured after the masonry has been built.

IOP Conf. Series: Materials Science and Engineering **951** (2020) 012018 doi:10.1088/1757-899X/951/1/012018

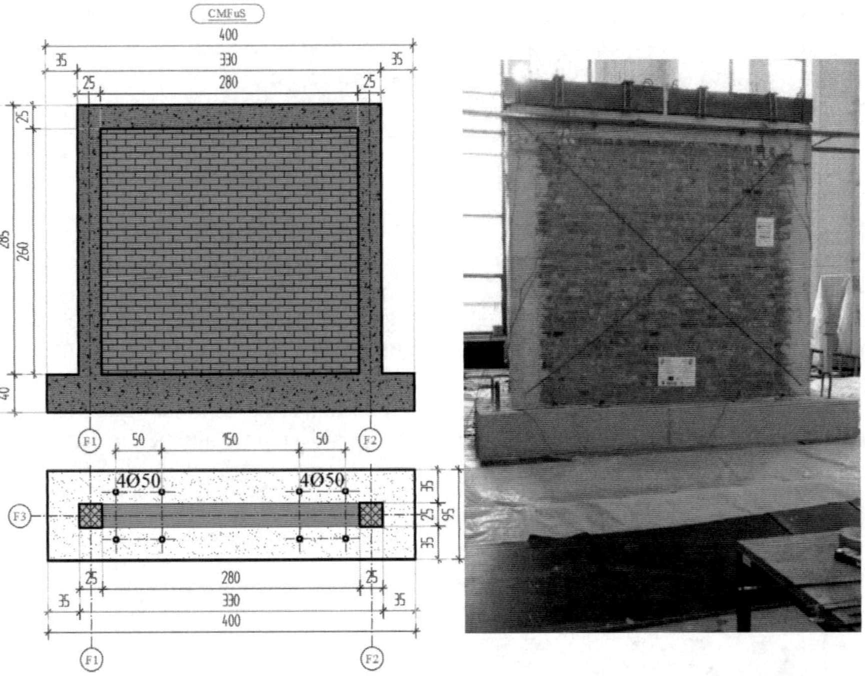

Figure 1. Geometrical characteristics of specimens CMFuS (wall before strengthening).

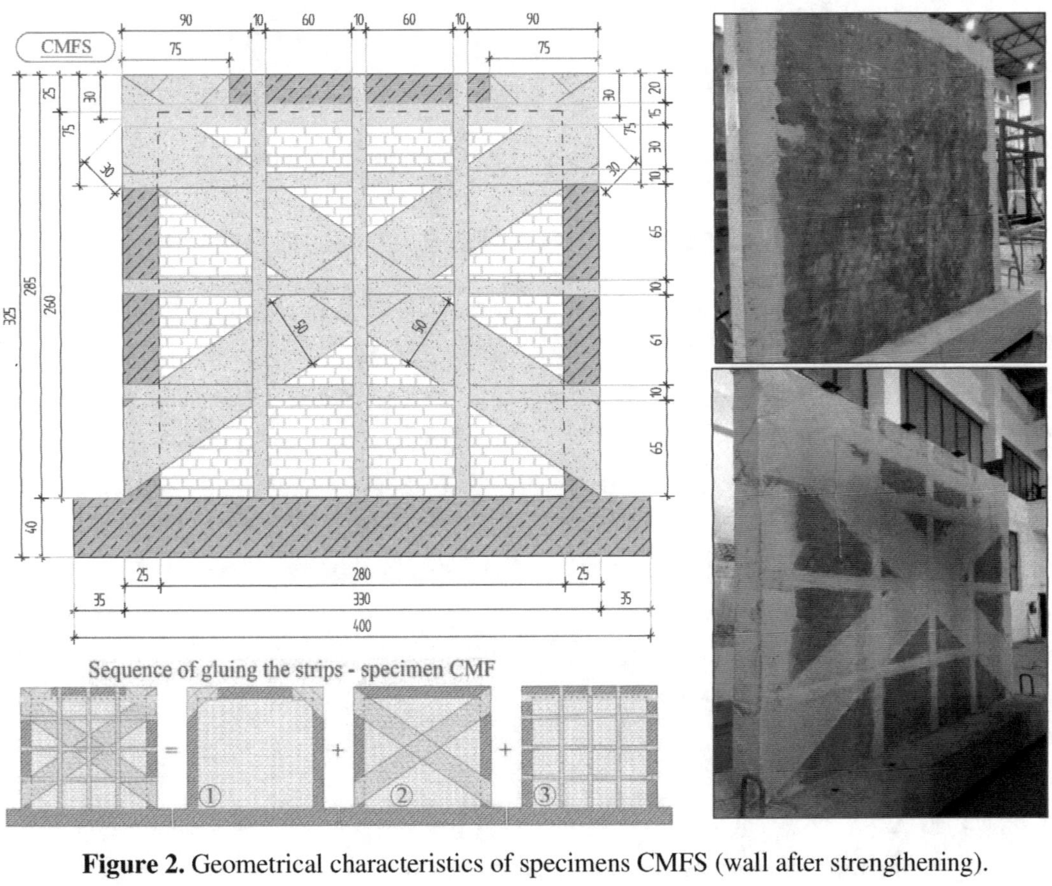

Figure 2. Geometrical characteristics of specimens CMFS (wall after strengthening).

The implementation of the samples complies with the requirements of Eurocode 6 [8] and Eurocode 8 [3]. The vertical joints are completely filled with mortar. The horizontal and perpendicular joints made of general-purpose masonry mortar are between 6 mm and 15 mm thick.

According to Eurocode 8 [3] the longitudinal reinforcement of confining elements may not have a cross sectional area less than 300 mm^2 or 1,00 % of the cross-sectional area of the confining element. The longitudinal reinforcement in the confining element is 4Ø14. Lap splice is 60 bar diameters in length. Stirrups are 8 mm in diameter spaced in 100 and 150 mm are provided around the longitudinal reinforcement. Reinforcing steel is of Class B in accordance with [9].

2.2. Instrumentation Set-up

The specimens' response was monitored by a high-speed data acquisition system consisting of accelerometers (ACC), linear variable differential transformers (LVDT), linear potentiometers (LP) and strain gages (SG), providing information about accelerations at different levels and points, relative displacements and deformations and strains at selected points. Also, input parameters are obtained from the fixed instruments under the shaking table. The position and the general information for the instruments are given in figure 3 and figure 4. For specimens CMFuS and CMFS, detailed information on the instrumentations can be found in [2].

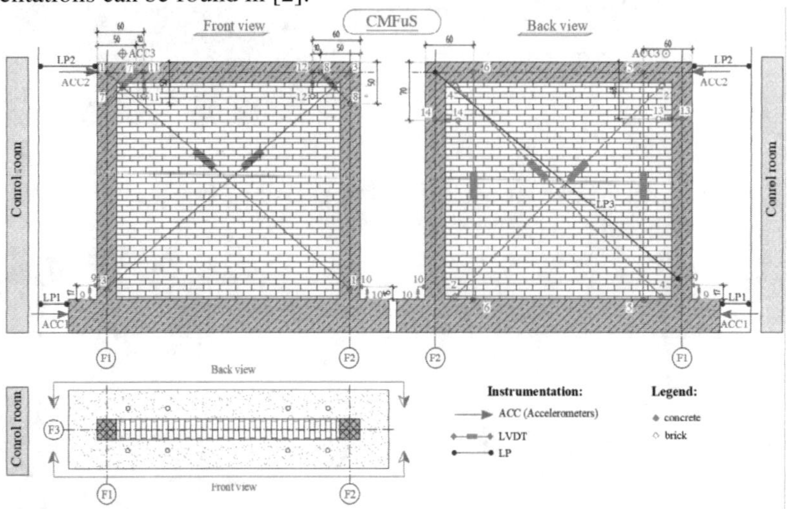

Figure 3. Instrumentation scheme of specimens CMFuS – (front and back view).

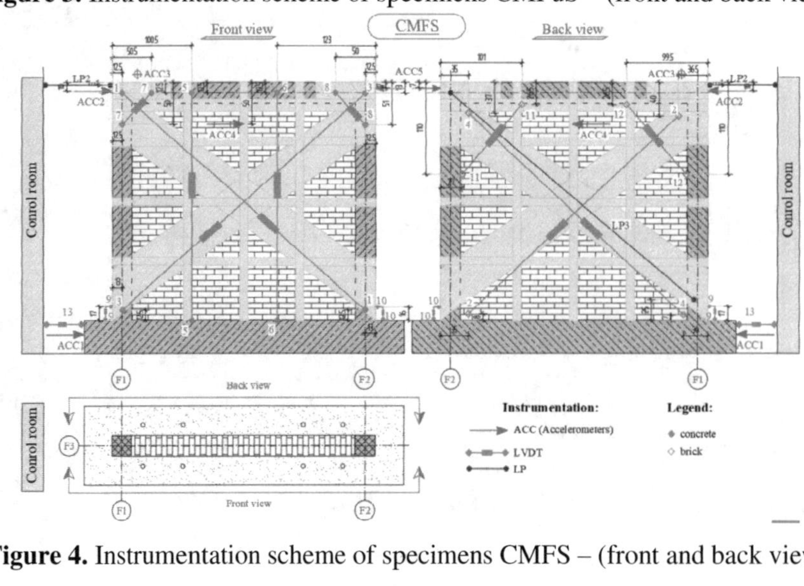

Figure 4. Instrumentation scheme of specimens CMFS – (front and back view).

3. General methodology for testing of the specimens – phases, ground motion and instrumentation

The testing procedure consisted of two main phases. Phase 1: Tests for definition of the dynamic characteristics of the models, in order to check the stiffness degradation of the model produced by the micro or macro cracks developed during the tests – resonant frequency search tests. Phase 2: Seismic testing by a selected earthquake record until heavy damage. The tests were performed in several steps, by increasing the input intensity of the earthquake, (table 1 and table 2), in order to obtain the response in the linear range, as well as to define the initial crack state, the development of the failure mechanism and the possible collapse of the model – seismic response tests. In such a way, the complete seismic performance of the structures starting from the linear range, the appearance of the first cracks in the walls up to the development of the failure mechanisms was captured. The testing has been performed on 5,00x5,00m 5 DOF (Degrees-of-Freedom) MTS seismic shake table at IZIIS Laboratory – figure 5. Detailed information on the shaking table in IZIIS can be found in [9]. Moreover, for simulating the foreseen real load in the exploitation period of a structure, additional load of 5,20 kN/m2 (4x400kg) was placed on the walls (figure 6). Also, in order in-plane excitation to be provided, special system for lateral support of the models was constructed. Detailed information on the specimens can be found in [2].

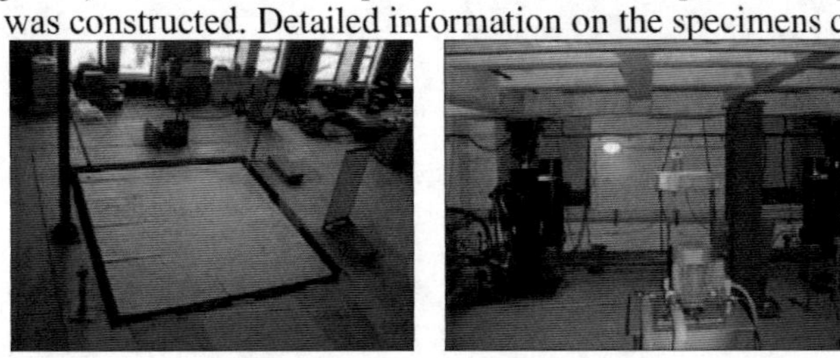

Figure 5. The shaking table (a) and the relevant hydraulic system (b) at the Dynamic Testing Laboratory at IZIIS, Skopje [10].

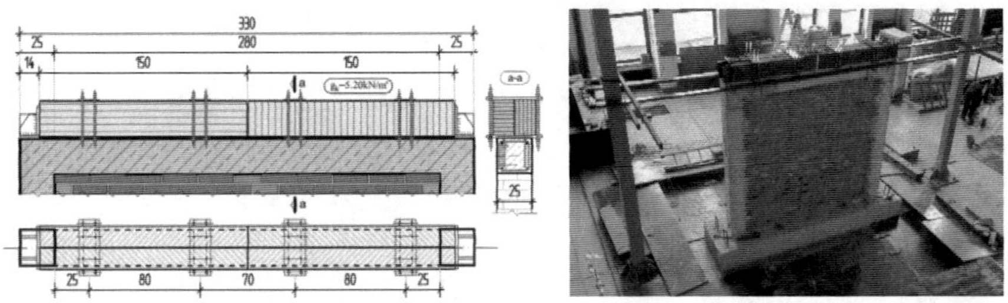

Figure 6. Additional load for the walls (technical drawing) and system for lateral support.

3.1. Definition of the dynamic characteristics of the specimens

Determination of dynamic characteristic of the specimens, in-plane, was done using harmonic sine sweep excitation and random excitation, generated by the seismic shaking table. Peak excitation level is around 0,05g. The change in frequency and stiffness of the specimens CMFuS and CMFS are presented in [11]. The values of the measured natural frequencies are given in Table 1 and Table 3.

3.2. Definition of the seismic behaviour of the specimens under earthquake/dynamic excitation

To determine the seismic response of the specimens under earthquake excitation, in-plane, uniaxial tests were performed. As representative real earthquake excitation, El-Centro (Imperial Valley, California, 18 May, 1940) time history record was used. Since the El-Centro time history record has different dominant frequencies compared with the natural frequency of the specimens, it had been frequently scaled with factor 8.0. The tests were conducted by gradually increasing the intensity of the excitation. Because the specimens could not be significantly damaged with the representative earthquake, both with the original and the scaled one, the models were further excited with random and sine sweep excitation in a previously specified frequency range in a correspondence with the natural frequencies for each model separately. Firstly, random vibrations in defined frequency range were performed, with increase in the intensity of the excitation. Afterwards, sine sweep excitation was carried out with frequency range changed almost in every test, following the decrease in the natural frequency. The intensity of the sine sweep excitations was also progressively raised. Following are the time history records for the representative earthquakes and artificial excitation as well as the Fourier amplitude spectrums of the excitations (figure 7).

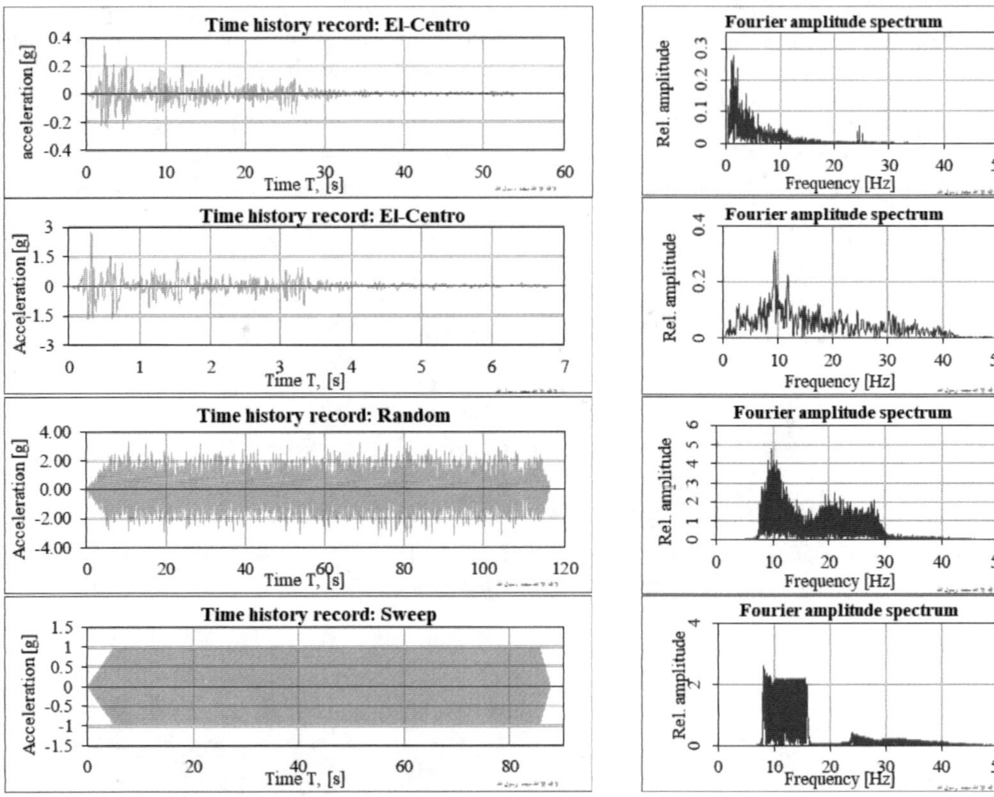

Figure 7. Time history record and Fourier amplitude spectrum.

4. Shaking table tests for seismic behaviour of the specimens

The real nonlinear behaviour of the wall under seismic influences could be established from the experimentally determined hysteresis curves. Under alternating load, the wall gradually reduces its stiffness as cracks form in the masonry (and the tie-columns) depending on the direction of the load. Upon reaching a certain limit deformation in the wall, a significant degradation of the stiffness is observed, which is reflected in a subsequent significant deformation and a decrease in the total bearing capacity of the wall. In order to clarify the nonlinear behaviour of the wall under alternating impact, such as an earthquake, based on the conducted experimental studies and the results obtained for shear force at the base and relative displacement at the top of the wall, the hysteresis curves are drawn.

4.1. Specimen CMFuS – confined masonry wall before strengthening

The specimen was subjected to a sequence of incremental dynamic tests. A series of table motions of increasing intensity were applied with the objective of assessing the ultimate capacity and failure modes of the specimen. On March 6, 2019, a specimen CMFuS was tested at IZIIS-Skopje. A total of 20 tests were performed, of which 7 were to determine the fundamental frequency and the remaining 13 were to investigate seismic behaviour on the wall. Table 1 presents a list of performed tests for specimen CMFuS.

Figure 8 shows the hysteretic base shear-top displacement response of the specimen CMFuS. The base shear has been obtained by summing the product of each accelerometer installed on the specimen time the related mass portion. The proportion between the two axes of all the plots is the same. In this way the eventual progressive specimen stiffness degradation and the consequent fundamental period elongation may be appreciable.

Table 1. List of performed tests for specimen CMFuS.

Test Number	Name of the test	Excitation type	Input: Frequency range (Hz)	Input: Acceleration	Measured fundamental frequency (Hz)
TuS 01	Test01_Random_002	Random	1÷50	0,066.g	/
TuS 02	Test02_Sweep_002	Sine sweep	1÷50	0,030.g	24,63
TuS 03	Test03_Elcentro _50	Time history	/	0,147.g	/
TuS 04	Test04_Elcentr_100	Time history	/	0,340.g	/
TuS 05	Test05_Elcentro_50_X8	Time history	/	0,142.g	/
TuS 06	Test06_Elcentro_200_X8	Time history	/	0,568.g	/
TuS 07	Test07_Elcentro_400_X8	Time history	/	1,088.g	/
TuS 08	Test08_Elcentro_800_X8	Time history	/	1,739.g	/
TuS 09	Test09_Random_005	Random	1÷50	0,169.g	21,18
TuS 10	Test10_Random_015	Random	8÷28	0,537.g	/
TuS 11	Test11_Random_05	Random	8÷28	1,464.g	/
TuS 12	Test12_Random_09	Random	8÷28	2,274.g	/
TuS 13	Test13_Sweep_005	Sine sweep	1÷50	0,062.g	18,89
TuS 14	Test14_Sweep_03	Sine sweep	15÷22	0,331.g	/
TuS 15	Test15_Sweep_100	Sine sweep	14÷20	1,089.g	/
TuS 16	Test16_Sweep_005	Sine sweep	1÷50	0,062.g	12,99
TuS 17	Test17_Sweep_100_8-16	Sine sweep	8÷16	1,247.g	/
TuS 18	Test18_Sweep_005	Sine sweep	1÷50	0,061.g	11,65
TuS 19	Test19_Sweep_03_5-12	Sine sweep	5÷12	0,388.g	/
TuS 20	Test20_Sweep_005	Sine sweep	1÷50	0,062.g	10,60

The hysteretic base shear-top displacement response of the specimen CMFuS immediately before and after the formation of significant damage is presented in figure 9. It very well illustrates the significant degradation of stiffness, which is reflected in the subsequent significant displacement as a result of damage to the specimen. The hysteresis curve before the damage is marked in red and in purple after the damage. Both hysteresis curves are from Test17 with an acceleration of 1,247g (table 2). The maximum shear force before damage is 185.81kN (point A) and -178.35kN (point B) with the corresponding top-displacements -1.92mm and 1.70mm. The maximum shear force after damage is reduced to 102.91kN (point E) and -139.59kN (point F), and top-displacements are increased to -1.86mm and 2.15mm, respectively. The ratio between the maximum shear force at the base before and after damage is from 1.28 (at negative values) to 1.81 (at positive values). The maximum horizontal top-displacement before damage is -2.96mm (point C) and 2.85 mm (point D) with corresponding forces of 80.02kN and -77.06kN. The maximum horizontal top-displacement after damage increases to -4.23mm (point G) and -4.14mm (point H), and the forces decrease to 30.78kN and -13.77kN, respectively. The maximum horizontal top-displacement after the damage compared to before the damage increases between 1.43 times (at negative values) and 1.45 times (at positive values).

IOP Conf. Series: Materials Science and Engineering **951** (2020) 012018 doi:10.1088/1757-899X/951/1/012018

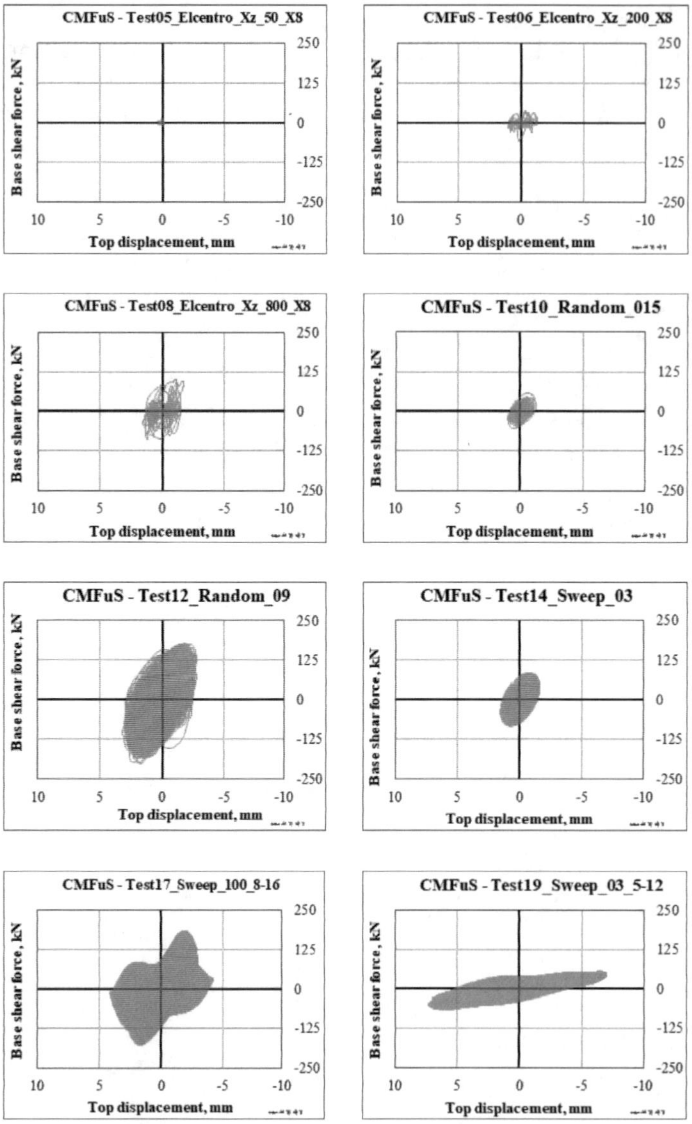

Figure 8. Hysteretic curve – specimen CMFuS (confined masonry wall before strengthening)

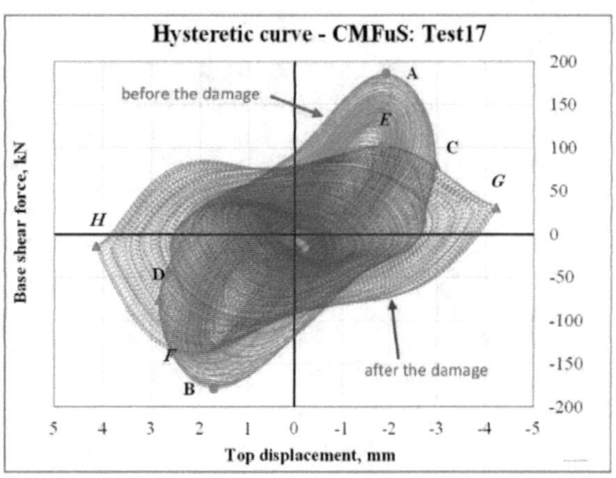

Figure 9. Hysteretic curve – CMFuS: Test17.

Table 2. Characteristic points in the hysteresis curve – CMFuS: Test17.

Pont	Base shear force, [kN]	Top displacement, [mm]
A	185.91	-1.92
B	-178.35	1.70
C	80.02	-2.96
D	-77.06	2.85
E	102.91	-1.86
F	-139.59	2.15
G	30.78	-4.23
H	-13.77	4.14

4.2. Specimen CMFS – confined masonry wall after strengthening

The specimen was subjected to a sequence of incremental dynamic tests. A series of table motions of increasing intensity were applied with the objective of assessing the effect of strengthening, the ultimate capacity and failure modes of the specimen. On March 26, 2019, a specimen CMFS was tested at IZIIS-Skopje. A total of 12 tests were performed, of which 7 were to determine the fundamental frequency and the remaining 5 were to investigate seismic behaviour on the wall. Table 3 presents a list of performed tests for specimen CMFS.

Table 3. List of performed tests for specimen CMFS.

Test Number	Name of the test	Excitation type	Input: Frequency range (Hz)	Input: Acce-leration	Measured fundamental frequency (Hz)
TS 01	Test01_Random_005	Random	1÷50	0,191.g	19,78
TS 02	Test02_Sweep_005	Sine sweep	1÷50	0,063.g	/
TS 03	Test03_Random_005	Random	1÷50	0,175.g	/
TS 04	Test04_Random_005	Random	1÷50	0,117.g	/
TS 05	Test05_Elcentro_100	Time history	/	0,333.g	/
TS 06	Test06_Elcentro_400_X8	Time history	/	1,007.g	/
TS 07	Test07_Elcentro_800_X8	Time history	/	1,733.g	/
TS 08	Test08_Random_005	Random	1÷50	0,142.g	18,44
TS 09	Test09_Random_09	Random	12÷22	2,058.g	/
TS 10	Test10_Sweep_005	Sine sweep	1÷50	0,063.g	13,91
TS 11	Test11_Sweep_1g	Sine sweep	10÷16	1,179.g	/
TS 12	Test12_Sweep_005	Sine sweep	1÷50	0,063.g	12,48

Figure 10 shows the hysteretic base shear-top displacement response of the specimen CMFS. The base shear has been obtained by summing the product of each accelerometer installed on the specimen time the related mass portion. The proportion between the two axes of all the plots is the same. In this way the eventual progressive specimen stiffness degradation and the consequent fundamental period elongation may be appreciable.

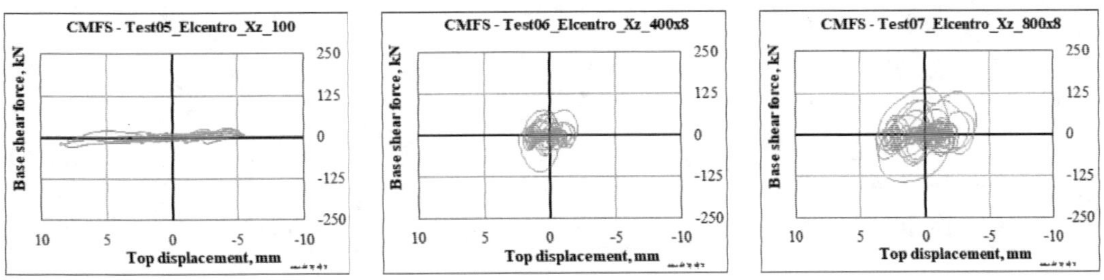

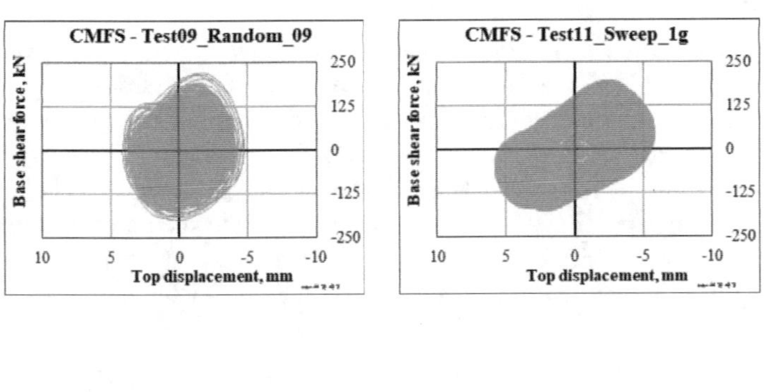

Figure 10. Hysteretic curve – specimen CMFS (confined masonry wall after strengthening).

IOP Conf. Series: Materials Science and Engineering **951** (2020) 012018 doi:10.1088/1757-899X/951/1/012018

Figure 11 shows the hysteretic base shear-top displacement response of the specimens CMFuS and CMFS. The proportion between the two axes of all the plots is the same. The hysteresis curve before strengthening is marked in red and in green after the strengthening.

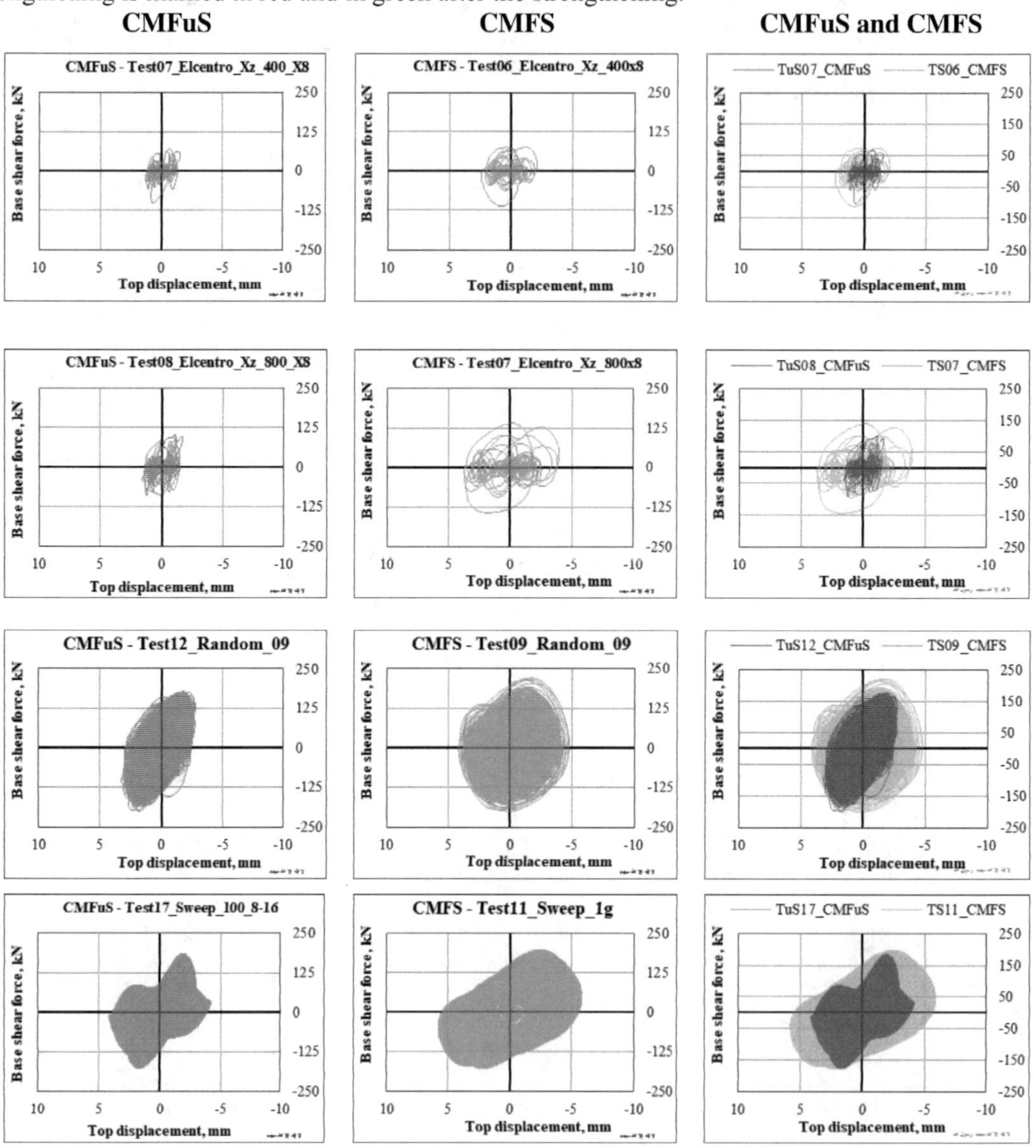

Figure 11. Hysteretic curve – specimen CMFuS and specimen CMFS.

5. Conclusions

The dependence between base shear force and horizontal top-displacement was obtained experimentally. The hysteresis curves are obtained, and their area is a measure of the energy dissipated by the structure. The hysteresis curves were obtained before and after strengthening of the specimen with glass fiber reinforced polymers (GFRPs). According to the understandings in [3] of two constructions, the one that dissipates more energy is better protected from earthquakes. Figure 11 shows a comparison of the hysteresis curves obtained before and after strengthening of the specimen. The tests results indicate that the GFRP system restores and increases the lateral load capacity of strengthened walls. Moreover, GFRP helps to increase the area of hysteresis curves, leading to increased seismic energy dissipation. In the specimen CMFS, which was tested after strengthening, the dissipated energy is about 1.80 times higher than in the specimen CMFuS before strengthening. This is obvious with increasing dimensions of the curves. Dissipated energy is due to the development of inelastic deformations. The use of externally bonded GFRPs materials exhibit considerable potential as an alternative solution to traditional strengthening techniques.

The strengthening also contributes to increased energy dissipation in specimens with an opening (CMWuS, CMWS, CMDuS and CMDS). Further on, the investigations of walls with openings are ongoing and will be presented in other future publications.

Acknowledgments

This research project under contract D-111/2018 was supported financially by the Centre for Research and Design at the University of Architecture, Civil Engineering and Geodesy (UACEG) (www.uacg.bg). The authors would like to express their deepest appreciation to the staff of the Centre for Research and Design.

The authors would like to express their deep gratitude and appreciation to the team of the following companies for their support for the successful implementation of the experiments: A&K Engineering Ltd. (www.akengeneering.org), Isomat International Ltd. (www.isomat.bg), Baumit Bulgaria Ltd. (www.baumit.bg), Shpici stroi Ltd. (www.shpici-stroi.com), Arcon Construction Ltd. (www.arcon.com.mk), and Emil Kroumov & Co. Ltd. (www.kroumov.com).

The authors would also like to thank all participants who contributed to the realization of the experimental tests and to the staff of the Dynamic Testing Laboratory at IZIIS, Skopje (www.iziis.ukim.edu.mk).

The authors express their heartfelt gratitude to all participants who contributed to the implementation of the experimental research.

References

[1] Krstevska L and Poposka A 2019 Shake table test of masonry wall specimens in scale 1:1 - IZIIS – REPORT 2019-34 IZIIS - Dynlab, Skopje.

[2] Abdulahad E and Mahmud E 2019 Implementation of Samples and Preparatory Activities before Dynamic Tests Annual of the University of Architecture, Civil Engineering and Geodesy, vol. 52, no. 3, pp. 891–900, Sofia, Bulgaria.

[3] Eurocode 8: Design of structures for earthquake resistance, European Committee for Standardization, 2005.

[4] Brzev S and Meli R 2012 International Guideline for Seismic Design of Low-Rise Confined Masonry Buildings in Regions of High Seismic Risk The 15th World Conference on Earthquake Engineering (15WCEE), Lisbon, Portugal.

[5] Jain S, Brzev S, Bhargava L, Basu D, Ghosh I, Rai D and Ghaisas K 2015 Confined Masonry for Residential Buildings. Gandhinagar: Indian Institute of Technology.

[6] Singhal V and Rai D 2014 Seismic Behavior of Confined Masonry Walls When Subjected to In-Plane and Out-of-Plane Loading Tenth U.S. National Conference on Earthquake Engineering. Anchorage, Alaska.

[7] Singhal V and Rai D 2014 Role of Toothing on In-Plane and Out-of-Plane Behavior of Confined Masonry Walls Journal of Structural Engineering.

[8] Eurocode 6: Design of masonry structures, European Committee for Standardization, 2005.

[9] Eurocode 2: Design of concrete structures, European Committee for Standardization, 2005.

[10] Corbi I and Rakicevic Z 2013 Shaking Table Testing for Structural Analysis International Journal of Mechanics, vol. 7, no. 4, pp. 459–466.

[11] Mahmud E and Abdulahad E 2020 Shaking table tests for investigation of the change of fundamental frequency and stiffness before and after strengthening of a confined masonry Annual of the University of Architecture, Civil Engineering and Geodesy, vol. 53, no. 3, Sofia, Bulgaria.

IOP Conf. Series: Materials Science and Engineering **951** (2020) 012019 doi:10.1088/1757-899X/951/1/012019

Rapid structural assessment of buildings – experience after the 26th November 2019 Earthquake in Albania

D Blagov, Tz Georgiev, D Kisliakov*, P Pavlov, B Petrov and M Traykova

National Earthquake Engineering Centre (NEEC), Univ. of Architecture, Civil Engineering and Geodesy (UACEG), 1 Hr.Smirnenski Blvd., 1046 Sofia, Bulgaria

*E-mail: kiss_fhe@uacg.bg

Abstract. The authors represent a team who visited the areas affected by the strong earthquake of 26th November 2019 as part of the Bulgarian state support to Albania. The task of the Bulgarian experts comprised a rapid visual assessment of buildings in Tirana, as members of small international groups of structural engineers, local engineers and representatives of municipal authorities. The visit further aimed at gaining organisational and professional experience regarding ensuing operations and at assessing damaged buildings in the epicentral area of the city of Durrës. This summary report presents an overview of the assessed structural systems of the visited buildings in Tirana and the most frequent structural and non-structural damages there. Additional information is presented on certain characteristic damages of buildings in Durrës, and a brief analysis is provided of the reasons for building collapse or structural and non-structural damages. Finally, conclusions are drawn, and tasks for future expert, engineering and management activities towards strong earthquake pre- and post-disaster preparedness for Bulgaria are formulated.

1. Introduction

In the early morning of 26th November 2019, an earthquake of M 6.4 occurred in Albania. The event resulted from a thrust faulting near the convergent boundary of the African and Eurasian plates [1] The source depth was estimated as 20 km [1] to 10 km [2] The epicentre of the earthquake was approximately 30 km west of the capital city of Tirana between the coastal city of Durrës and Thumane inland. 51 casualties and more than 3000 injured were counted, predominantly in Durrës and in the surrounding epicentral area.

An immediate reaction by both the Albanian government and the international community followed. An urgent request for assistance was issued from Albania to all related international organisations [3]. As a result of this request and out of their own initiative, 30 countries, incl. the European Union, offered assistance and supported the immediate relief activities. Some of the latter grew into a long-term support of crucial relevance for the restoration of the affected regions and infrastructure components.

As is common under such circumstances, also the international professional community of civil engineers and all relevant experts in the field of Earthquake Engineering duly responded. The initiated activities addressed a variety of aspects. Further below, two of these will be outlined in more detail. The first one focuses directly on the structural condition of buildings and other civil engineering facilities immediately after the seismic impact. The answer of the question whether a building or facility can be safely used after the strong ground motion has finished is of crucial importance both for the residents and for the operation of the particular civil structure. An adequate answer of this question can be given only by means of a rapid structural assessment of the affected building. Such an assessment is to be undertaken immediately after or parallel to the rescue activities due to the necessity of extreme clarity on the possible further use of the building. Moreover, the structural assessment certainly has to be performed by experienced structural engineers with specialised knowledge of Earthquake Engineering.

The second aspect to be highlighted here refers to the completeness of record and the dissemination of the experience gained during such missions. The professional community of civil engineers worldwide maintains a long tradition in this regard. All major earthquakes at least in the last century were carefully described in terms of geological and seismological parameters, detailed presentation of the damages of buildings, infrastructure components as well as social impact. Currently, large modern observation networks provide considerable amounts of real-time high-resolution information on strong motion events worldwide, and teams of all renowned Earthquake Engineering research centres provide on-site assistance and perform investigations as soon as possible after such events.

Regarding the November 2019 Albanian earthquake, reports by the participating teams from different countries were released for internal use [4]. Unfortunately, possibly the only professional studies publicly available are the comprehensive reports [5] [6]. However, their findings address all relevant aspects of the earthquake impact on the country.

In the days following the earthquake, a team of Bulgarian structural engineers, predominantly from the National Earthquake Engineering Centre (NEEC) at the University of Architecture, Civil Engineering and Geodesy (UACEG) in Sofia embarked on a short-term support mission in Tirana as part of the assistance offered by the government of Bulgaria. The mission was immediately followed by a longer one organised by the Bulgarian Chamber of Engineers in the Investment Design [7]

The present work attempts a strictly professional report by the authors towards sharing their personal experience as members of international engineer teams for rapid assessment of the structural condition of residential buildings in Tirana. These teams were joined by Albanian colleagues and representatives of the respective municipality. The activities were organised by the local operational control centre in Tirana. The international expert teams wherein the Bulgarian engineers participated were assigned to the central part of the city of Tirana. Buildings were visited for which the owners had filed notices to the municipality on damages caused by the earthquake. Moreover, a task force was formed within the Bulgarian group of engineers to apply specific laboratory equipment, such as: laser roulette, binoculars, micrometric magnifier for crack width estimation, devices for non-destructive determination of the probable compressive concrete strength and of the corrosion level of reinforcement bars, devices for localisation and size determination of reinforcement bars and of the concrete cover as well as equipment for determination of the level of concrete carbonisation. Thus, important material parameters could be determined on site.

Finally, the authors visited also Durrës for completion of their professional records. Further below, all findings are shortly discussed, and the lessons learned are presented, which, as we hope, should help the management of similar future events.

2. Structural systems of residential buildings in Tirana

As typical for cities, various structural systems for residential buildings in Tirana exist due to historical development and bearing the imprint of the time when constructed. The main building types in Tirana can be summarised in four groups allocated to mainly five time periods [6]

During the rapid structural assessments of buildings performed by the authors, they examined residential buildings of basically three types: unreinforced masonry buildings up to five storeys, buildings of large prefabricated panels, and newer buildings with reinforced concrete (RC) frame skeleton and infill masonry.

The unreinforced masonry structures prevailed. A typical structural system for the city of Tirana consists of masonry load-bearing walls and partition walls, made of solid bricks. The thickness of the walls on the first and second floors is 50 cm, on the third is 38 cm and on the fourth and fifth ones –

25 cm. Partition walls are of the same type of bricks, but their thickness is 12.5 cm. The floor slabs of the older buildings are of hollow bricks and connected with reinforcement bars passing through the brick holes. This makes the floor slabs relatively soft in their plane, so that they do not act as diaphragms, which results in a poor interaction between different masonry bearing walls. The floor slabs of the more recent buildings were often of strip-like prefabricated panels built on the walls. For better connection between the walls and the floor slabs, reinforced concrete tie-beams are constructed. The measured compressive strength of the concrete floors is 12-15 N/mm^2.

The Albanian masonry buildings and their seismic vulnerability have lately been subject to intensive research. In this regard, the comprehensive study [8] is particularly to be considered. There, the typical structural systems of the masonry buildings from the 1960s till 1990 are discussed in detail together with results from experimental studies towards determination of key material parameters for computational modelling. Special attention is also paid to the resistance of such buildings to near and far-fault ground motion. In general, the observations presented further below correspond to the conclusions in [8]

The load-bearing systems of the contemporary RC buildings in Tirana are predominantly of two types: large-panel stiff walls, mostly built in the 1980s and 1990s, and skeleton moment-resisting frames, as a rule – newer ones, since the 1990s.

3. Observed problems

3.1. Earthquake-induced structural issues

The 26th November 2019 earthquake in the city of Tirana area was classified as strong according to MMI (Modified Mercalii Intensity), with Instrumental Intensity VI, causing slight structural and non-structural damages. The recorded PGA in Tirana (41.3351N 19.606E, Stiff soil, epicentral distance 33.647 km) was 1.08 m/s^2 (0.11 g), and the duration of the main ground motion was ca. 30 to 40 seconds [6]. It was not a highly destructive earthquake, but the social tension and fear in the population were clearly perceptible in the meetings of our teams with building owners and inhabitants. The rapid structural assessment was performed according to a version of [9] developed by the local authorities.

The most frequently observed damages in the masonry structures were cracks in the corners between load-bearing walls and partition walls, or diagonal cracks in the walls. Cracks in the ceiling also were observed in a number of buildings due to the practice of using prefabricated panels for floor slabs. Cracks had formed straight-like lines between the panels, caused by the inevitable shear occurring in the floor diaphragms and caused some shear deformations between the single panels. All these damages could be classified as slight, and in very rare cases slight-to-moderate. None of the buildings visited had to be classified as dangerous, so to be abandoned by its residents.

One of the possible reasons for crack formation in the masonry bearing walls was that relatively flexible floor slabs did not help walls distributing loads between them. On the contrary, they concentrated part of the loading to a particular wall causing its overloading. In some floors, cracks indicating tension stresses were formed on the top side of the floor near the damaged wall. In some buildings, longitudinal cracks between the stone foundation and the masonry walls also were observed, which means that the connection between the foundation and the bearing wall was lost. Another reason for crack formation in the masonry walls was the poor quality of the mortar, i.e. a very low cement content in the latter. Such mortar literally turns to dust when even touched with fingers.

Among the most recent generation of buildings constructed with RC frames, only slight damages were observed as cracking or separation of partition or infill walls in the contact area with the beams and columns. In both types of building structures – masonry ones and skeleton RC structures, the more visible and significant damages were concentrated in the lower storeys. This can be explained by the larger drifts in the lower storeys of structures responding in shear type lateral displacement mode typical for RC frames and masonry structures. Structural damages, jeopardising the overall condition of a building – figure 1 and figure 3 (all photos in this work taken by the authors), were relatively rarely observed by the authors during their mission in Tirana.

Large-panel buildings showed nearly zero damages, neither structural nor non-structural ones. In certain cases, slight openings of joints between the panels were identified. The measured compressive strength of the concrete in contemporary RC buildings was 20-25 N/mm^2.

The most common damages in the residential buildings in Tirana, however, were non-structural: cracks (mostly diagonal or horizontal) in non-bearing brick walls in the masonry and RC-frame buildings – figure 2, around doors and windows – especially due to missing lintels with adequate structure and bearing capacity, between slabs and walls, between stairs (mostly made from low quality concrete in the masonry buildings) and walls, collapsed chimneys.

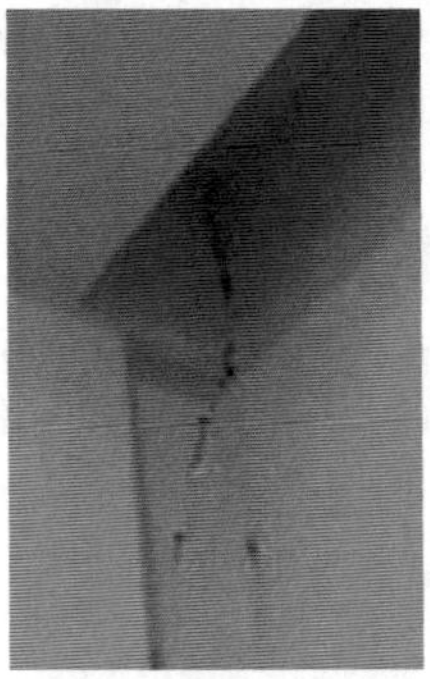

Figure 1. Crack at the beam-to -shear wall joint.

Figure 2. Separation of the corner brick wall from the inward RC frame.

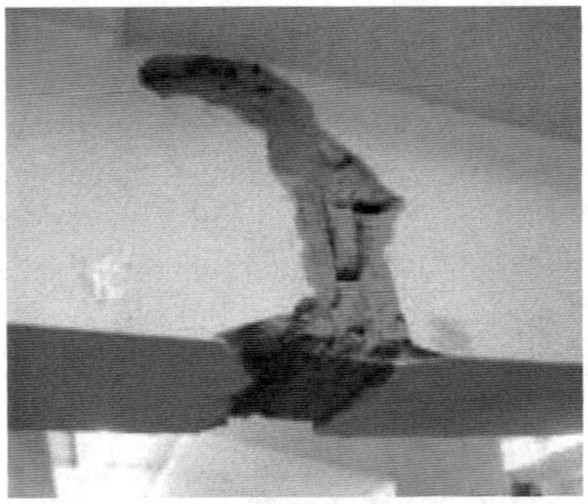

Figure 3. Damaged beam connection in prefabricated RC frame.

3.2. Safety-related structural issues not caused by the earthquake

The city of Tirana is marked by the urban development characteristic for the last ca. 30 years. The identified structural issues, although not caused by the earthquake of 26th November 2019, directly affect the structural quality and integrity of the buildings as well as their seismic vulnerability, respectively. These issues are based on the observations of the authors and are summarised below.

Perhaps the problem most closely related to the seismic performance of the buildings was represented by various reconstructions. It was not uncommon to detect a diversity of building appendixes, encased balconies and removal of internal walls aimed at increasing living space. In some cases, removals have been undertaken on the first floor, and moreover, without adequate design and construction of a lintel. The lack or the actual structure of a lintel was a general problem observed during the mission in Tirana. Furthermore, even additional floors on panel buildings were observed – figure 4. In certain cases, the reconstructions were based on a structural design, however, without seismic performance analysis of the whole building. Unfortunately, this was no standard, when, e.g. a subsequent cutting of a column (including reinforcement) at the foundation was undertaken to enable a pipe connection. A similar case was observed as an additionally created door opening in a wall panel to enable an independent entrance in a flat.

All these reconstructions and upgrades increase the vertical loads on the load-bearing elements. The mass of the building is increased, too, causing bigger horizontal seismic forces. Also, the regularity of the structure in both horizontal and vertical directions is often violated, which additionally affects its seismic performance by causing torsion and other negative effects.

Another common problem was the lack of any seismic joints between adjacent buildings constructed at different times, of different materials and as different structural systems. As a result, local damages of the facade walls were observed around the contact area with the slabs of the neighbouring buildings.

It is also worth highlighting a social aspect of this technical problem. After the earthquake of 26th November 2019, the concerned citizens grew aware of the high vulnerability of seismic non-resistant buildings, and how the illegal interference to the building structure by one owner jeopardises the safety of all residents and of their shared property. This inevitably results in confrontations and social tension both within the neighbourhood in the "Kolektive" residential buildings and in their relation to the authorities.

In the authors' opinion, some typical design solutions were identified during the performed rapid structural assessment in Tirana, which need discussion. Relatively frequently encountered damages of the non-structural elements should be outlined as a weak point in the seismic performance of RC frame buildings. It also is worth reminding that the earthquake was not so strong in the region of Tirana. Buildings with such damages may need to be stiffened, if possible, by adding RC shear walls to them and connecting the existing RC frames to these walls. Undeniably, such an intervention can be undertaken only after a detailed seismic structural analysis. Furthermore, an applied beam-column connection with a large eccentricity also can be classified as unfavourable design – figure 5.

Another problem was the observed design of external columns not in the corners but inward with built non-bearing infill walls at the corners. This resulted in separating the infill walls from the RC frame in the corners caused by the earthquake – figure 2, but such separation could occur without ground motion impact, too. A similar finding was identified at a single flat situated in two adjacent parts of a panel building, separated by a joint. Due to this, there were significant although mostly non-structural, internal damages in the apartments, including large cracks in the infill walls and opening a gap between the parts of the apartment belonging to the different buildings. Here again, severe cracks could occur without seismic excitation, e.g. due to different settlements of the two separate buildings.

An observed problem related to both design and maintenance of the buildings is shown in figure 6. On one hand, the use of non-ribbed reinforcement without sufficient overlapping for the column at the foundation was not an appropriate design solution. On the other, a relatively large spacing between transverse reinforcement in the critical zone of column-to-foundation reinforcement splice is not allowed by modern codes. Furthermore, the obvious lack of adequate maintenance has led here to a deterioration of the concrete, which could further enable more or less intensive corrosion of the reinforcement.

Figure 4. Building of type "Kolektive" with added penthouse on the top.

Figure 5. Unfavourable beam-column connection.

Figure 6. Deteriorated column with non-ribbed reinforcement.

Other maintenance-based problems with unfavourable consequences were the observed permanently inundated cellars in basement floors of large panel buildings as well as the almost universally destroyed rain pipes on their facades. The surface impact of water on the facade panels and their connections, and especially the permanently flooded foundations, seriously affect the overall structural performance of the buildings with particular regard to their seismic vulnerability.

3.3. Impressions from Durrës

In the afternoon of 7th December 2019, the authors visited Durrës to gain professional impressions of the earthquake-induced damages in the city. The observed buildings were located along and near Pavaresia street within a sector of ca. 0.9 km length between the streets David Selenica and 3 Vellezerit Thanasi and the beach in the South-West. There were mainly relatively new buildings with RC beam-column structural framing system and infill masonry walls, characteristic for this part of the city, which underwent a rapid development thanks to the touristic boom after the changes in Albania. The height of the buildings ranged from 3 to 10 storeys. The majority of the buildings were apartment and family hotels, and residential buildings.

During the short mission to Durrës, the engineer team observed in general three types of damages, such as: 1) fully collapsed or few cases of severely damaged buildings, i.e. damaged load bearing building structures, 2) significantly damaged non-structural components, such as facade and partition walls in many buildings, and 3) certain occurrences of soil liquefaction.

The three fully collapsed buildings were the Vila Verde Beach Hotel (four-storey), Hotel Mira Mare (five-storey), and the five-storey Lubjana Hotel. At the time of the visit, all buildings said were completely destroyed, and the sites were already cleaned up. Based on available photographs, the most probable reasons for their collapse were the soft and weak first storeys due to the large open spaces in the restaurant and hotel lobby areas, and possibly an inadequate foundation in an area with a ground containing liquefiable saturated sands. This emphasises the vital relevance of the entire development chain of architectural planning, structural design and construction process and the equally important, inter-related compliance with the basic rules for design and construction in seismic regions.

A case of a seriously damaged building was shown to the team by local colleagues. The site was already inspected, and all residents evacuated. It is a judgment of the authors that the building was severely damaged and did not possess any residual strength. Most probably, it would be due to be demolished in the near future. A visible residual drift and many signs of nonlinear deformations were identified, mainly present in column-beam joints of the first storey, and a collapse of masonry infills due to a frame-wall interaction – figure 7 and figure 8.

Figure 7. Deep crack in a column-beam frame joint.

Figure 8. Completely damaged infill masonry walls.

The most often observed damages were those of the infill walls – figure 9. There were partially as well as fully collapsed walls, cracks, and other damages in the facades and partitions. Generally, due to the climate on the Adriatic coast of Albania, the facade walls of the new buildings are of either double-layer brick masonry with insulation layer in the middle, or of single layer masonry with insulation system outside the wall. In both cases, hollow bricks placed on their narrow side are employed. Generally, all facade walls were non-confined, relatively long and slender and sensitive to storey drifts. It is worth noting that, as a rule, there were no horizontal or vertical confining elements (tie-beams or tie-columns) or any other type of reinforcement. This specific feature of the structural system based on beam-column frame skeleton, performing good ductility but deformable, in combination of non-confined brick masonry infills results in many damages of the infill walls observed at every second or third building within the inspected district. Another severe structural problem observed was a damaged column in the 1st storey with built-in interruption of the vertical reinforcement bar, figure 11.

Figure 9. Completely destroyed facade brick walls.

Figure 10. Soil liquefaction.

Figure 11. Damaged column in the 1st storey.

As said above, the visited district of Durrës is close to the beach, and the construction ground very likely contains saturated sands or even mud [6]. Under such conditions, the potential of soil liquefaction is high. Visible soil deformations were identified around some buildings – figure 10. In the authors' opinion, such effects were relatively rare only thanks to the main earthquake having lasted for ca. 30-40 seconds. In the event of an earthquake of longer duration, the soil liquefaction effects would have been much stronger with much more serious consequences, respectively.

4. Lessons learned and conclusions

As a result of the mission performed by the authors, the following conclusions can be drawn with respect to the safety of residential buildings in earthquake-prone regions:

a) Inappropriate structural design solutions for the building should be avoided. This means that already at the design stage, all load cases comprising decisive seismic excitation should be accounted for. As a result, the load-bearing structural elements and their connections should be designed according to the obtained demands for them, respectively.

b) The "soft first storey" should be avoided. A common problem observed in masonry buildings was the additional weakening of the first floor caused by removing walls for shops and offices. Thus, a "soft first floor" occurred, which is well-known as one of the most frequent reasons for collapse of buildings during strong earthquakes. Although this was not the case in Tirana thanks to the not sufficiently strong seismic impact there, it occurred to numerous buildings in Durrës. Unfortunately, the problem of the "soft first floor" is almost always of economic nature, and therefore difficult to avoid.

c) Any inappropriate structural changes of the building should be avoided, i.e. any change of the structural system of the building should be assessed by a qualified structural engineer with respect to the overall seismic safety of the building.

d) The use of modern codes is of crucial importance for the seismic resistance of the buildings. On one hand, these represent contemporary design codes for structures in seismic regions. Their application enables the appropriate selection of the structural system, design of the bearing capacity of the structural elements, their connections and detailing. On the other, these define the standards for construction materials, ensuring that the latter's parameters correspond to the design ones.

e) Particular attention should be paid to the safety of the non-structural building elements. As said above, facade and internal infill masonry walls in buildings with RC frame structure should be adequately confined. Their structural integrity under ground motion excitation is of crucial importance for the safety and health of both the inhabitants and citizens near the building. In this regard, the use of adequately designed and constructed wall confining horizontal and vertical members and lintels above the openings in such walls substantially contributes to the building safety and reduces the costs of post-earthquake maintenance. Unfortunately, in both Tirana and Durrës, lack of masonry confining members or poorly constructed ones for this purpose were often observed.

f) The November 2019 earthquake in Albania confirmed the necessity and the relevance of the rapid post-earthquake structural assessment of buildings and facilities. For its successful performance, however, the presence of an adequate methodology and corresponding ready-to-use forms for the experts is inevitable. In this regard, the preparation and the operation of the local authorities in Tirana and Durrës can serve as a highly commendable example.

g) Apart from the purely structural issues, the role of the overall organisation of the rescue, supply, assessment and restoration works, the management, control and logistical support, of the coordination and communication of the various task groups after an earthquake event of such gravity definitely requires a very high level of emergency preparedness of all involved authorities. In this regard, the related activities in Tirana and Durrës observed by the authors can also be considered best practice.

The authors trust that the experience shared by this outline may help structural engineers and colleagues involved in the various aspects of Civil Engineering towards preventing as far as possible building deficiencies and damages in future earthquake events.

References

[1] https://earthquake.usgs.gov/earthquakes/eventpage/us70006d0m/executive (last visited June 04, 2020)

[2] https://emergency.copernicus.eu/mapping/list-of-components/EMSR412 (last visited June 04, 2020)

[3] Euro-Atlantic Disaster Response Coordination Centre (EADRCC) Urgent Request for Assistance, Earthquake – Albania 2019 Message No.OPS(EADRCC)(2019)0009

[4] Pavlov P 2020 The 26.11.2019 *Albanian Earthquake– First Order Estimation and Analysis of the Observed Damages*, Public Information Report, Disaster Risk Reduction Council to the Ministry Council of the Republic of Bulgaria, Sofia

[5] Albania, Post-Disaster Needs Assessment, Volume A & B Report 2020 Council of Ministers of Republic of Albania, European Union, United Nations Albania, The World Bank, Tirana

[6] Lekkas E, Marvoulis S, Papa D and Carydis P 2019 *The November 26, 2019 Mw 6.4 Durrës (Albania) earthquake*. Newsletter of Environmental, Disaster and Crises Management Strategies, 15, National and Kapodistrian University of Athens, Greece

[7] https://www.kiip-sofia.com/index.php/bul/nachalo?jsn_setmobile=no&start=12 (in Bulgarian) (last visited June 09, 2020)

[8] Bilgin H and Hysenlliu M 2020 *Comparison of near and far-fault ground motion effects on low and mid-rise masonry buildings*, J. of Building Engineering 30 (2020) 101248

[9] Baggio C et al (Authors), Pinto A V and Taucer F (Eds.) 2007 *Field Manual for post-earthquake damage and safety assessment and short term countermeasures (AeDES)*, EUR 22868 EN, Joint Research Centre Institute for the Protection and Security of the Citizen

Seismic assessment of consequences in the building stock in Bulgaria based on statistical calibration of public data

A Kaneva

National Institute for Geophysics, Geodesy and Geograpgy, Bulgarian Academy of Sciences, Akad. G.Bonchev str., Block 3, Sofia 1113, Bulgaria

E-mail: kaneva_ant@abv.bg

Abstract: A generally accepted step to reduce the consequences of strong earthquakes and be prepared to face them, is to assess the possible damages and undertake countermeasures proactively. The aassessments are made based on analyses of seismic hazard and vulnerability of building structures. To collect the data about the buildings requires a lot of financial resources and time especially for a large-scale study. In this paper an assessment of the possible consequences in the building stock of a town in Bulgaria is performed by using public data. Some problems arising in the use of the available public data for the building stock in Bulgaria collected by the National Statistical Institute are discussed. A statistical approach is applied to calibrate this data for the buildings to classify them according the building typology and period of construction, which are main factors that determine their vulnerability. The consequences to the building stock has been evaluated with public data for several large cities in Bulgaria. Different approaches for the relation between the earthquake intensity and the level of damage are applied. Comparison of the calculated damages in the building stock of the town of Pernik and the observed damages during the 22 May 2012 earthquake is made. The results are discussed in view of applying public data for the building stock to estimate possible consequences from a strong earthquake in large urban areas.

1. Introduction

For successful protection and prevention of disasters, the municipalities in the country prepare plans in which the possible disasters and their consequences have to be assessed. These assessments envisage the relevant preventive actions, assess the ability of municipalities to deal with the consequences (their readiness), organization and management in case of disaster. Risk assessment based on detailed data with important engineering parameters for each building in the municipality is difficult and in most cases impossible. Collecting data is a long and expensive process. This is even more true for risk assessment for large cities across the country. For a quick and rough assessment of the consequences of the earthquake in the building stock, it is necessary to use the available public summary data, which are collected by the NSI (National Statistical Institute) during the census of the population and the building stock. These data impose significant limitations and can lead to misleading estimates. In the article the consequences for the city of Pernik are assessed using public data, EMS98 [1] methodology and mean fragility curves defined in the Syner-G European project [2] for masonry buildings and in RISK-UE WP13 [3]. The results are discussed regarding the observed effects of the earthquake in 2012, May 22, near Pernik.

2. Available public data, constrains and missing data in view of assessment of structural performance during earthquakes

In Bulgaria data about the building stock is collected in the National Statistical Institute (NSI). The NSI Census 2011 data is organized on territorial units, 28 administrative districts in Bulgaria. The data on the buildings is characterized by three factors: type of building (inhabited, non-inhabited, with common house hold.), type of structural system, which includes 4 categories (large-panel buildings, RC buildings, the so called "massive" buildings and others), and the period of construction in decades starting from a period "built till the end of 1949). This public data has several shortcomings in view of the seismic risk assessment.

1. Only residential buildings are included in the NSI data. The non-residential buildings with different functionality may comprise a great deal of the building stock. For example, according the Census2011 data [5] in 2001 there are 2334 residential buildings in one of the capital municipality, "Triaditza". The total number of buildings in "Triaditza" in 2001 is 3677 (data from the cadastre, RISK-UE project [3]), or the ratio of non-residential buildings to the total number of buildings is 0,36. In the present paper only the residential buildings are considered in the estimates due to the constraints in the available data.

2. The building typologies for the districts included in the Census 2011 are four, less than the typologies in Census2001. These are: Large-panel buildings, RC buildings, „Massive", and "other buildings". The RC buildings are described in the methodological notes [6] as „buildings which bearing system and the floor constructions are built of reinforced-concrete and the walls are made of panels, brick masonry or other materials". Here both frame and wall systems are in one group. The so called „Massive" buildings are the masonry ones "with bearing walls of brick and stone masonry and the belts, the beams and the floor construction are made of RC but have no reinforced concrete columns". In the Census survey list the masonry buildings are described by two types: with RC floors and with wood/steel beams floors. In the public data masonry buildings are one group only.

In the group of "other buildings" are included the structures that are built of stones, sun-dried brick, wood and other materials.

3. Data for the year of construction is considered in decades that do not coincide with the changes of design codes for constructions in seismic areas. Construction period "till the end of 1949" is the first one in 2011 census.

4. The data in Census 2011 for the structure type is not correlated to the period of construction. The only such correlated data is available in the Census 2001 and it is generalized for the buildings in the whole country. Two more tables of such data are presented for the buildings in the towns/cities and the buildings in the villages.

5. Data for the buildings concerns the administrative units, districts (total 28 in the country) and municipalities (total 265 in Bulgaria). For the towns, only total number of buildings is available.

All these constraints and missing data are factors that can lead to great uncertainties in the estimates.

3. Assumptions to calibrate the data for the buildings to overcome missing data

One of the most applied methodology to assess the possible damages in the building stock is the EMS98 [1], and it is used in this study too. The limitations in the public data are also a factor to apply this approach. The building characteristics that are considered in a quick but rough estimate are the structure type and period of construction in accordance to the level of the earthquake resistance design code. The relation between the structure type and period of construction in EMS-98 defines the vulnerability class (from A to F). This implies the data for the type of structure to be related to the period of construction (level of seismic code) of the buildings. Such data is missing in the last census. As already mentioned this data is related only in the Census 2001 (figure 1) where the type of structure is correlated to the period of construction in generalized form for the whole country (for cities and for villages). The data in figure 1 is for all residential buildings in the country. In figure 2 is presented the distribution of buildings by type and period of construction for the cities.

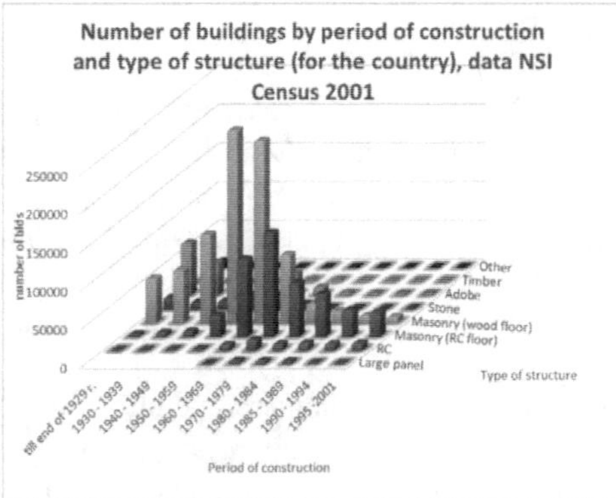

Figure 1 Buildings by period of construction and type in Bulgaria

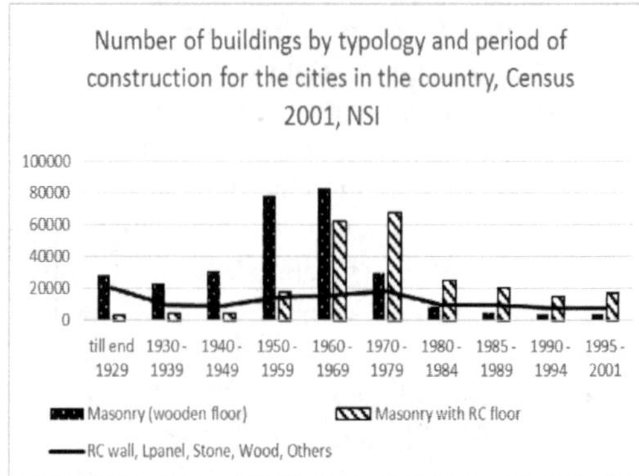

Figure 2 Buildings by period of construction and type in Bulgaria

To illustrate the available public data for the Municipality of Pernik [4], the data for 2011 is presented in Table 1 and 2.

Table 1. Number of buildings by construction period in Municipality of Pernik data [4]

Construction period	Number buildings
till end of 1949	2250
1950 - 1959	4377
....	
2000-2011	1212

Table 2. Number of buildings by building type in the Municipality of Pernik, data [4][5]

Building structure type	Number of buildings
RC-Large panels	409
RC- skeleton	1271
Massive - Masonry	22775
Others	701

For each municipality the total number of buildings is available, grouped in the 4 typologies (Table 2). The typology identification is the main factor that determines the building vulnerability. The missing data is the number of buildings in each construction period of the same typology. In this study it is assumed to calibrate the data based on the statistical distribution from the 2001 Census separately for the towns and for the villages (by proportional distribution respective to the data from Census 2001).

Next assumption concerns the sub-categories of the masonry buildings which are the prevailing typology (figure 1) in the country and in the towns (figure 2). The masonry buildings considered in [3] are of two sub-categories depending on the horizontal load-bearing system: with deformable steel or wooden floors and with non-deformable reinforced concrete floors. So is the data for masonry buildings in Census 2001. In this study the number of masonry buildings in each sub-category is assumed proportional to the distribution of these sub-types of structures in the Census 2001.

To make the correspondence between the period of construction according the census decades and period of seismic code in Bulgaria, review of the code periods is taken into account. In Bulgaria the first regulations for seismic design of the buildings are from 1947 [7] (no map in the regulations attached). Before 1947 Italian regulations were accepted to be applied [8]. Next change of the regulations is in 1957 [9] when a seismostatistic map of seismic areas is included, reflecting the observed intensities of past earthquakes. In the revision in 1964 the map was re-evaluated with zones of VII, VIII and IX degree MSK. In 1977 and 1987 the changes in the code included new data and the map was re-evaluated. In 2007–2009 a new hazard map was elaborated in accordance to the requirements of EN1989. In order to take advantage of the Census 2001 interconnected data for the period of construction and the structure typology the following correspondence between the code period and Census period of construction is assumed:

Code period	Census period of construction
1. Very old - before 1929	< 1949
2. 1929-1957	1950-1959
3. 1957- 1964	1960-1969
4. 1964-1977	1970-1979
5. 1977-1987	1980- 1989
6. 1987-2009	from 1990 -
7. > 2009	

The period of construction is an important factor in the vulnerability estimate of a building as it defines the level of seismic design of the building and level of seismic loading. For example, the town of Pernik was in seismic zone VIII according the 1957 map, the static seismic force according this code is determined as 1/20 of the weight and vertical loads of the building. In the period 1964- 1987, the town of Pernik is not in seismic zone ≥VII MSK. A comparison of the seismic zones areas in Bulgaria shows that in the period 1961-1977 only 22% of the territory of Bulgaria is in zones with intensity ≥VIIMSK, in 1977 – 1987 this percentage is 40% and from 1987 it increases to 98% [10].

4. The towns selected for evaluation

Several towns are chosen for assessment, Pernik, Blagoevgrad, Plovdiv and Triaditza administrative unit of Sofia.

4.1. The town of Pernik

Pernik is a town about 20 km SW from the capital of Bulgaria, Sofia. The population of the town is 75246 inhabitants and of the Pernik municipality is 97181 inhabitants in 2011, NSI data, Census 2011. The total number of buildings in the town of Pernik is 10980 and in the municipality is 25156 [5].

The town is chosen as an earthquake of $M_s = 5.8$ ($M_w = 5.6$) hit Western Bulgaria, near Pernik on May 22, 2012. In the town of Pernik the maximum intensity was VII-VIII MSC scale [11]. The same intensity was observed in Divotino, a village in the Pernik municipality. Many damages of VII degree of MSC in other villages in the municipality were observed [11]. In [12] Intensity VIII is observed in the central part of the town of Pernik and in the village of Divotino. The number of buildings in Divotino is 1400 and in the town of Pernik, 10980.

In order to make rough estimate of expected consequences based on EMS-98 methodology, the total number of buildings for the town of Pernik is proportionally distributed among the typologies according to the data for the municipality (Table2). Then the buildings of each typology are distributed in function of the construction period based on the data of Census 2001. The results for the buildings in the town of Pernik in function of type and period of constructions after the calibration are presented in Table 3. The same procedure is applied to the data Census 2011 for Blagoevgrad, Plovdiv and Triaditza administrative unit of Sofia.

Table 3. Distribution of buildings in function of type and period of construction in the town of Pernik, calibrated to data from Census2001

Construction period	Large panel	RC	Masonry (RC floor)	Masonry (wood floor)	Others
till end of 1949	0	23	232	1541	170
1950- 1959	0	20	327	1465	57
1960- 1969	12	85	1168	1553	37
1970- 1979	53	139	1264	557	18
1980- 1989	90	133	844	250	13
1990 -	23	155	603	138	10
total	179	555	4436	5504	306

4.2. The town of Blagoevgrad, Plovdiv and Triaditza administrative unit of Sofia
Blagoevgrad is a town in the SW part of Bulgaria with population of 68679 at the end of 2019 and the number of residential buildings 4630 in 2011 [5]. It is situated in seismic zone with one of the strongest shallow earthquake in Europe in the last two centuries occurred in the Krupnik-Kresna area on April 4, 1904. Its magnitude is estimated as 7.8 [13].
Plovdiv is the second largest town after Sofia in Bulgaria. The population of the town is 347851, the number of the residential buildings is 18828 [4], [5]. The earthquakes in 1928 in this zone were assessed as of X and XI Forel - Mercalli scale (figure 3 and 4) [14].

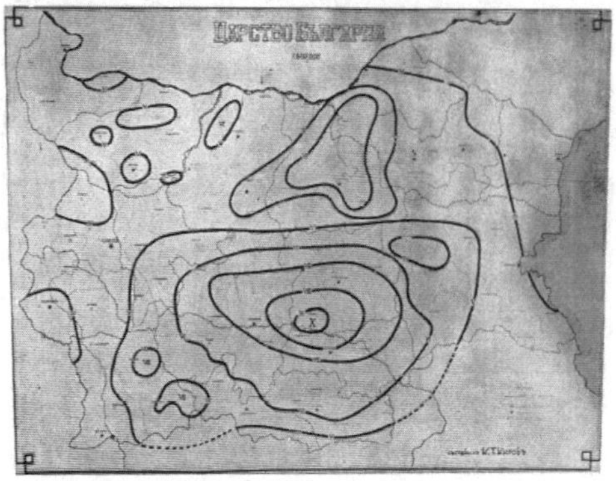

Figure 3 Isoseists of the April 14, 1928 earthquake [14]

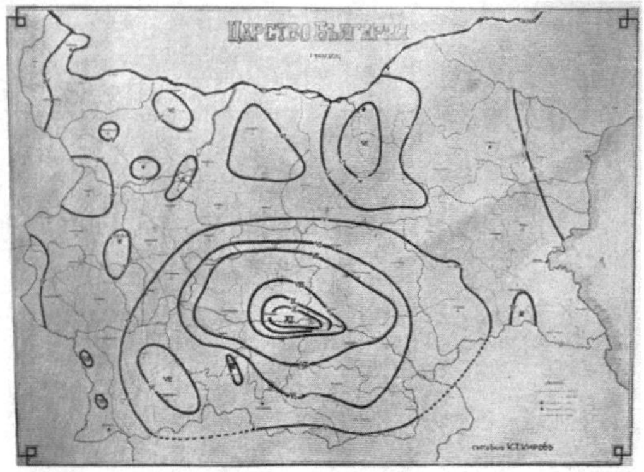

Figure 4 Isoseists of the April 18, 1928 earthquake [14]

5. Results of the assessments and analysis
All estimates are performed under the following assumptions:

Each town is considered as a point in the seismic map with equal intensity.

When assessment is performed applying EMS-98 methodology the intensity measure for each town is from the seismic zone map for 1000 years return period [15]. These are for Pernik and V. Tarnovo - VIII MSK –intensity (PGA= 0,15g); for Blagoevgrad, Plovdiv and Sofia- IX MSK –intensity (PGA=0,27g). When assessment is performed applying fragility curves with intensity measure type PGA, the values of PGA are according the hazard map 475 years return period of [16] (figure.5).

The definitions of the range of damages in EMS98 are assumed fixed as: "few" is minimum equal to 0 and maximum value is 15 %, "many" is 15 and 55% for minimum and maximum and most is from 55 to 100%. The vulnerability class based on EMS98 in function of design code period is according [19].

In this study the selected fragility curves are for the masonry buildings from the set collected in [2]. The curves are for low-rise buildings with flexible and rigid floors, typical for the conditions in Bulgaria and the type of intensity measure is PGA. The fragility curves for "masonry" buildings in [3] are applied.

5.1. Results for the town of Pernik and comparison to observed damages and other estimates

The resulting picture of the expected damages to buildings in the town of Pernik based on the EMS98 definition of VIII degree is as follows: From 0 up to 3% (295 buildings) of all the buildings in Pernik are assessed to be in grade 5 (Table 4). From 3 to 14% are in grade 4. From 4 to 22% are in grade 3 and from 6 to 25% are in grade 2. In summary from 13% to 63% (6929 buildings) of all residential buildings in Pernik are in grade 2 to 5 including.

Information for the impact of this earthquake in Pernik can be find in the plan for disaster protection of the Municipality [18]: „The social effects of the earthquake were that 40-50 inhabitants were slightly injured and sought medical help; 270 families left homeless; 8203 dwellings suffered some degree of damages; 273 buildings were marked as dangerous and the mayor ordered to be fully demolished". These demolished buildings can be interprited as buldings in damage grade 5 and 4 and fall in the lower limit of the estimate made in this study for VIII MSK intensity measure. Hadjiiski et al [12] underline that the intensity VIII MSK is observed in the central part of Pernik and the heavy damages are in the old buildings without seismic design. This fact outlines that assuming one intensity for the whole studied area is not the best practice to make estimates but gives idea of the limits of number of buildings in each expected damage grade.

The results for the damaged buildings in Pernik obtained in this study are quite rough estimate. However, the observed damages fall within the calculated range of damages calculated using public data for the buildings, calibrated in a manner to be classified according building typology and period of construction/level of seismic design.

5.2. The towns of Blagoevgrad, Plovdiv and Triaditza administrative unit of Sofia.

The procedure to assess the consequences in the residential buildings in Pernik was performed for several towns in Bulgaria. The intensity measure of the earthquake used to calculate the damages to the buildings is given in brackets after the name of the town (**Table 4**). The resulting number of buildings suffering different damage grades are listed for each town in the table.

Table 4 Number of buildings in each damage grade for several towns in Bulgaria, estimates with public data and EMS98 methodology

EMS98 damage grades	Town of Pernik (VIII MSK)	Town of Blagoevgrad (IX MSK)	Town of Plovdiv (IX MSK)	Town of Sofia-Triaditza (IX MSK)
Grade 5	0 - 295	134 - 676	500 - 2568	61 – 317
Grade 4	295 – 1544	185 - 955	736 - 3835	95 – 493
Grade 3	436 – 2380	278 -1118	1135-4616	145 –581
Grade 2	683 - 2710	97 - 357	453 - 1661	50 – 182
In damage grade 5 to 2	1441 to 6929	694 to 3106	2824 - 12680	351 - 1573

In RISK-UE project [3] the buildings in Triaditza were studied. The data was collected from the cadastre plan and it contained the following information for each building: building typology, year of construction, functional usage, number of floors, exact built area. The estimate was for 3647 buildings, residential and non-residential. In this paper the number of studied residental buildings is 2334 as in the Census data for 2001 The ratio between the residental to all the buildings in Triaditza is 0.64. The estimates of damaged buildings in Triaditza in this study are in very good agreement with the results achieved in the RISK-UE project Level 1 [3] having in mind the ratio of buildings studied (Table 5). The use of public data calibrated to fill missing information give results for Triaditza comparable to the results with detailed cadastre data for the buildings.

Table 5. Number of buildings in different damage grades Level 1 in Triaditza [3]

Damage grade	None	Slight	Moderate	Heavy	Very heavy	Destruction
Number	236	991	933	948	465	74

5.3. Assessment based on fragility curves

In the figures below (figure 5 and 6) the fragility curves for masonry buildings in [3] and mean curves in [2] for the masonry low-rise buildings are shown. In Syner-G project fragility curves from many studies have been collected and compared. The dispersion among them is in quite wide range (mean and standard deviations) for a class of buildings. From all fragility curves for low rise bearing wall masonry type, mean curves for two limit states (yielding and collapse) are defined (figure 5 and 6) in the project.

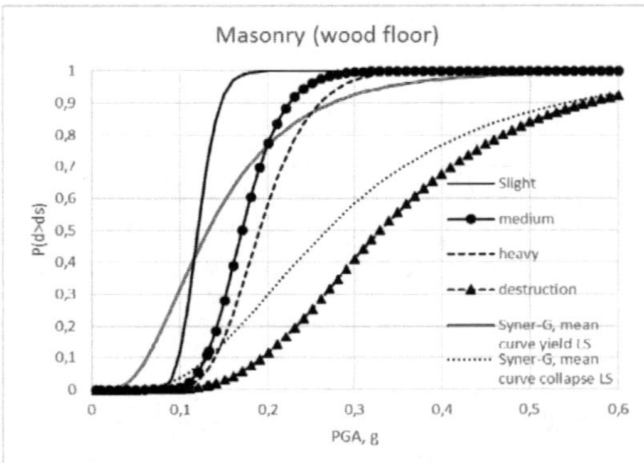

Figure 5 Fragility curves for masonry (wood floor) low-rise buildings [2][3]

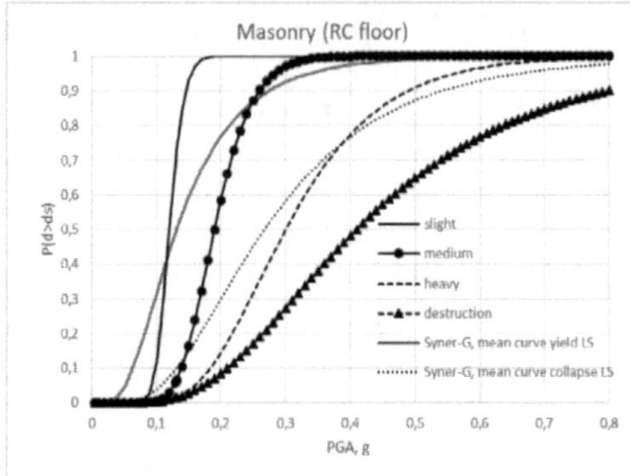

Figure 6 Fragility curves for masonry (RC floor) low-rise buildings [2][3]

These two sets of fragility curves are used to assess damages in the masonry buildings in Pernik. This type of buildings was chosen because in district Pernik 80.8% of the buildings are type masonry [4]. The reference peak ground motion for the town of Pernik is a_g =0,15g, seismic hazard map for design for 475 years return period [17]. The Syner-G project mean curve for the masonry low rise buildings defines that 15% of that type of masonry buildings would reach or be greater than limit state "collapse" for PGA =0.15 g. The RISK-UE WP13 [3] curves define about 20% masonry building with flexible floors in damage states "heavy damages and destruction" and about 3% masonry building with RC floors in damage states "heavy damages and destruction". These assessments give results that vary from 15% to 23%. Acceleration time histories of the earthquake are recorded in several stations in Sofia with maximum acceleration with amplitude of NS component 98.55 cm/s^2 [12] but no record of the PGA in the town of Pernik is available of the 22 May 2012 earthquake to make comparison of the results.

It is worth to mention here one of the conclusions about the masonry buildings of the Syner-G project [2], i.e. that „though in Europe the number of masonry buildings is significant, the number of studies with focus on this typology are quite limited and for this reason the team of the project could identify only two main classes of the typology, low rise and mid rise".

6. Conclusions

For quick but rough assessment of possible consequences from a strong earthquake in the residential buildings in our country public data for the building stock available from the census can be applied. In the assessment the period of construction of the buildings and the building typology should be taken into account. The leading characteristic is the typology. The period of construction can be calibrated proportional to the census 2001 data. The period of construction is the second important factor to define the vulnerability of the building. In just 60 years, the maps of seismic zones for design have undergone several changes as well as the level of seismic design. The areas of the seismic zones of different intensity have changed significantly over the years. For rough assessment, the range of expected damages from a probable earthquake can be obtained based on the EMS98 methodology.

Most of the collected fragility curves in Syner-G project are analytically obtained, some are from observed damages. These fragility curves reflect the specific characteristics of the building type in the investigated regions and should be used based on careful informed selection. When the fragility curves of a certain building type for a certain limit state/ damage grade are plotted, large dispersion of probabilities for the damages can be observed. The large dispersion for the masonry buildings estimates is due to a number of factors [20] including the level of seismic design. On the other hand the 12 degree Intensity EMS98 scale is not the measure in the hazard maps in the contemporary codes. The seismic intensity measure is PGA and that supposes increasing the number of the seismic zones. Fragility curves then would be more applicable approach for estimates of possible consequences.

Analyzes for the response of masonry buildings in seismic conditions, which reflect the specific conditions of construction practice in our country are still too limited.

References

[1] Grünthal G 1998 European Macroseismic Scale 1998 EMS-98 (Luxembourg: Centre Européen de Geodynamique et de Séismologie)

[2] Kaynia A 2013 Guidelines for deriving seismic fragility functions of elements at risk:Buildings, lifelines, transportation networks and critical facilities, Syner-G reference report 4 (JRC scientific and policy reports, Italy)

[3] Kostov M, Vaseva E, Kaneva A, Koleva N, Varbanov G, Stefanov D, Darvarova E, Solakov D, Simeonova S and Christoskov L 2004 WP13: Application to Sofia. RISK-UE project Final report (Nice)

[4] NSI Census 2011 2012 Reference book district Pernik, https://www.nsi.bg/ (in Bulgarian)
Reference book district Blagoevgrad (in Bulgarian)
Reference book district Plovdiv (in Bulgarian)
Reference book district Sofia (NSI, Sofia) (in Bulgarian)

[5] NSI Census 2011 2012 *Districts Book* ***14*** *Pernik* Vol. **3** (NSI, Sofia)
Vol 3 *Districts Book* **1** Blagoevgrad (NSI, Sofia)
Vol 3 *Districts Book* **16** Plovdiv (NSI, Sofia) (in Bulgarian)
Vol 3 *Districts Book* **23** Sofia (NSI, Sofia) (in Bulgarian)

[6] NSI Census 2011https://www.nsi.bg/en/content/5788/housing-fund

[7] Regulations for design and construction of buildings and structures in seismic regions in Bulgaria 1947 *State newspaper* **98** (Sofia) (in Bulgarian)

[8] Ignatiev N, Sotirov P History of the Bulgarian codes for seismic design of structures and transition to the Eurocode system of analyses http://www.petkovstudio.com/documents/7.pdf (in Bulgarian)

[9] Code for design and construction of buildings and structures in seismic regions in Bulgaria 1957 *Notifications of the Presidium of the National Assembly* **65** (Sofia) (in Bulgarian)

[10] Solakov D and Simeonova S 2007 Earthquakes and seismic hazard in Bulgaria *Insurer Press newspaper* **11** (Sofia) (in Bulgarian)

[11] Botev E, Dimitrov D, Georgiev I and Protopopova V 2013 Investigations of the 22 May 2012 Earthquake in Pernik Region, Bulgaria Conf. Proc. 7th Congress of the Balkan Geophysical Society, European Association of Geoscientists & Engineers DOI: https://doi.org/10.3997/2214-4609.20131708

[12] Hadjiiski K, Simeonov S, Botev E 2012 Seismic impact of the earthquake of 05/22/2012 for Sofia, http://www.niggg.bas.bg/en/ (in Bulgarian)

[13] Georgiev I, Dimitrov D, Pashova L, Shanov S , Nikolov G and Botev E 2006 Kinematics and geodynamics of the Krupnik-Kresna seismogenic zone *J. Geodesy* **17** (Sofia: BAS) http://clg.cc.bas.bg/17thvolumeGeodesy/Georgiev6.htm

[14] Stoyanov K 1931 Report on the work done from April 25, 1928 to November 1, 1931 Directorate for assistance and reconstruction of the earthquake area of 1928 (Sofia: State printing house)

[15] Order РД-02-20-2 for design of buildings and structuresin seismic regions 2012 (Sofia MRRB)

[16] BDS EN1998-1, National annex

[17] http://gis.mrrb.government.bg/mrrgisv1/mrr_tematicmaps/geoportal/inmapnorms_local.html

[18] Plan for earthquake protection of the population Draft project Municipalityof Pernik (in Bulgarian) retrieved May 2020, official site Pernik Municipality

[19] Simeonov S, Solakov D, Georgiev I, Vatzeva R, Dimitrov D, Stefanov D, Simeonova S, Trifonova P, Vaseva E, Cherkezova E et al. 2018 Development of Method for Analysis, Assessment and Mapping of Seismic Risk in Republic of Bulgaria *Bulletine Construction and Architecture* **5** (Sofia: ITUS) (in Bulgarian)

[20] Lagomarsino S and Cattari S 2014 Fragility functions of masonry buildings Chapter 5 of *Syner-G: Typology definition and fragility functions for physical elements at seismic risk* Pitilakis K, Crowley H, Kaynia A (Springer)

IOP Conf. Series: Materials Science and Engineering **951** (2020) 012021 doi:10.1088/1757-899X/951/1/012021

Technical solution for frame rail viaduct that can guarantee durability and safety exploitation

A K Jiponov

Road and Transport Structural Facilities Department, FTE, UACEG,1 Hr. Smirnenski Blvd, Sofia 1046, Republic of Bulgaria

E-mail: ajiponov@gmail.com

Abstract. Bridge constructions with frame static scheme are proven solution for railway viaducts, which guarantees durability, trouble – free maintenance, appropriate behaviour ensuring the assumption and transmission of vertical and horizontal efforts, redistribution of deformations occurring during the exploitation. The lack of bearings and expansion joints, makes them extremely reliable and adaptable to both the loads and the deformations that occur during the exploitation. An example of such a construction, with a seamless period of operation for more than 30 years, is the bridge in railway area of Station – Poduyane. It is a 3 spans frame structure, with a total length of 105m, with 18 track lines at the Railway Yard Area. In the current article, 2 railway viaducts will be presented – part of the Modernization of the Railway line Sofia – Plovdiv, railway section Elin Pelin – Ihtiman. Specific issues of the design process have been considered, with particular attention to the element stiffness proportion ratio, details of the joints, the effects of temperature, the shrinkage and creeping of the concrete, isolating the abutments from the soil effects / pressure.

1. Structure and foundation types for frame rail viaduct - overview

1.1. Regarding the superstructure

Usually a prefabricated-monolithic superstructure is used, where the beams are from plain reinforced concrete. After their mounting, in-situ concrete is casted over the supports, which embeds the beams in the respective sectors. This way it is possible to achieve spans' lengths which are significantly longer than the beams' length. The beams themselves are unified in transversal direction in a grillage structure with the help of a monolithic slab [1], [2], [3], [4].

1.2. Regarding the foundation

The foundation can be of direct / spread type and pile type.

Direct foundations are modelled with the help of springs, which stiffnesses' values depend on the soil layer properties, on which the foundation is positioned. It is not correct to place a spread foundation directly onto a solid rock layer. In this case at least 0,50 m of the solid rock layer are removed and subsequently replaced with coarse sand layer. The obtained sand bed is functioning as an elastic environment with its respective properties [5] [6].

Pile foundations, when needed, could be implemented with friction piles or standing piles. The calculation model which defines their structural behaviour is again achieved with the help of springs, which stiffnesses' values depend on the soil layers properties, through which the piles are penetrating.

2. General information about the present two structures

As a static scheme the viaducts are frame structures. The 8 no. of spans consist of prefabricated reinforced beams with length of 22m and weight of 35t each, which are unified with cast in-situ concrete slab thus creating a common structure. In longitudinal direction the beams are embedded with cast in-situ concrete over the supports and abutments, thus creating the frame structure. All the loads and actions on the structure are according to the Eurocodes. The abutments themselves are part of the frame structure. They are not subjected to passive and active soil pressure, because are protected by a retaining wall, which is founded on the same pile cap – **Figure 10**.

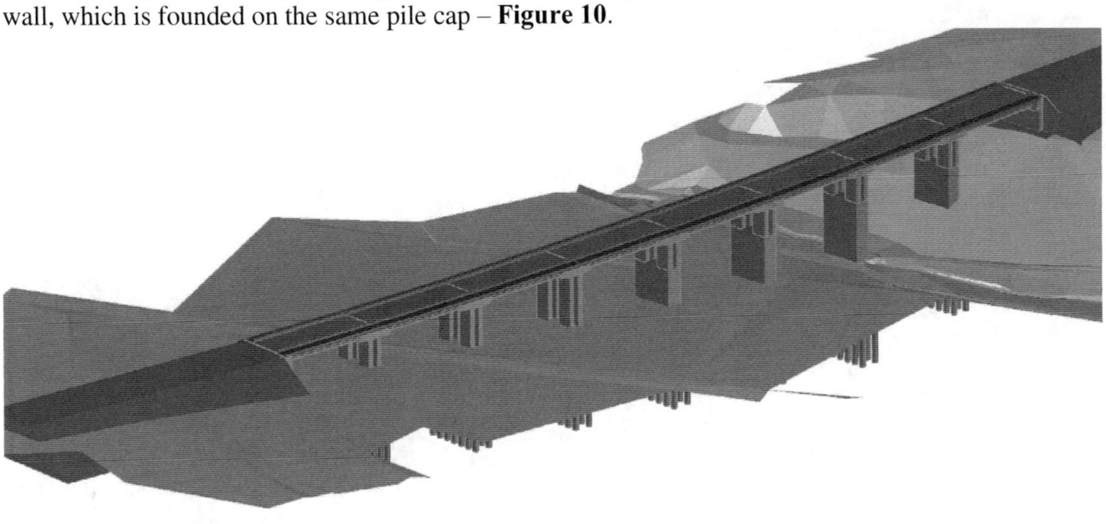

Figure 1. Visualization.

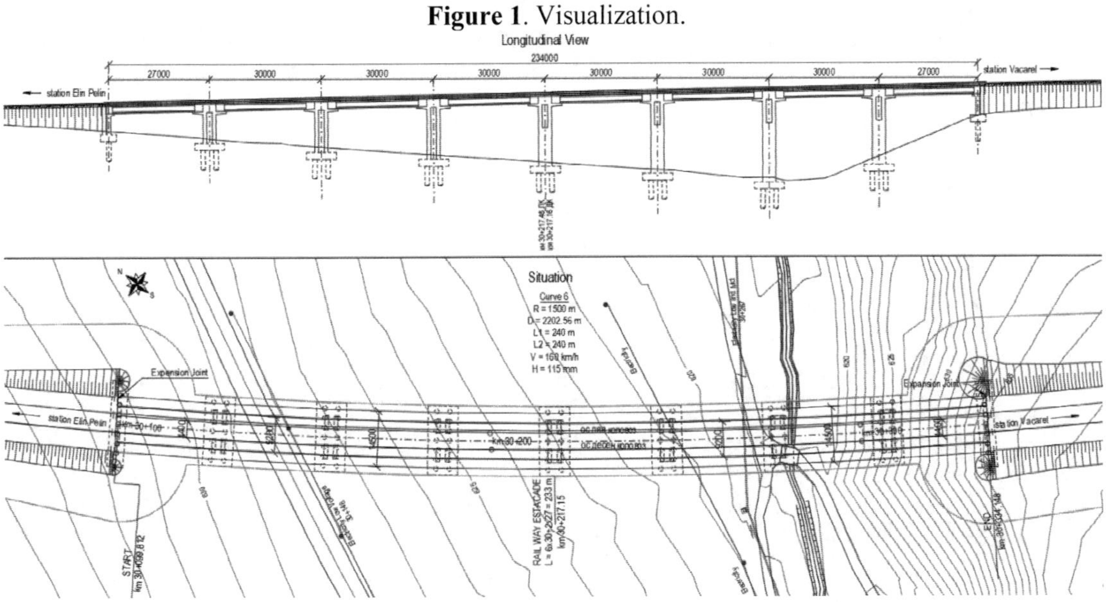

Figure 2. General plan.

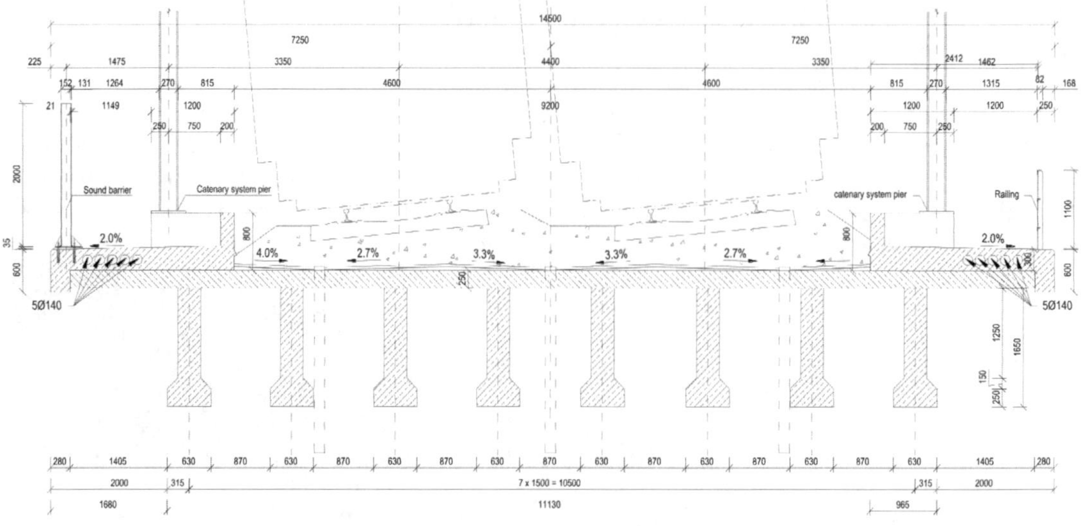

Figure 3. Cross section.

2.1. Viaduct at km 30+217

The railway viaduct consists of two tracks with 6 no. of middle spans with length of 30m each and 2 no. outermost spans with length of 27m each. The overall length is 234m. The viaduct is positioned in a curve with R=1500m for the left track in direction of Ihtiman and R=1504m for the right track. It crosses both a coulee, for which a correction is envisioned, and 2 no. of 20kW power lines, one of which requires a dislocation.

The dilatations between the frame structure and the retaining wall are with overall value of about 8 cm (from -5 to +3 cm).

2.2. Viaduct at km 30+765

The railway viaduct consists of two tracks with 12 no. of middle spans with length of 30m each and 2 no. outermost spans with length of 27m each. The overall length is 414m. The viaduct is positioned in a curve with R=1500m for the left track in direction of Ihtiman and R=1504m for the right track. The dilatations between the frame structure and the retaining wall are with overall value of about 15 cm (from -6 to +9 cm).

2.3. Construction technology

The following technological order will be maintained during the construction of the facility:

2.3.1. Casting the in-situ piles with diameter of 120cm and length of 20m. During the casting, the following should be done:

- conducting tests for bearing capacity of a single pile of the pile groups positioned at the beginning, middle and end of the viaduct;
- running a check and recording an ultrasound scan of the achieved concrete density of all piles, which should remain as a project documentation.

2.3.2. - Formwork mounting, reinforcement placing and concrete casting for all the middle supports' and abutments' pile caps;

2.3.3. Formwork mounting, reinforcement placing and concrete casting for the abutments to level top edge of predalle and all piers to level 25cm beneath top edge monolithic in-situ slab;

2.3.4. Mounting of the prefabricated and pre-transported to the construction site reinforced concrete beams with length of 22m and weight of 35t each;

2.3.5. Formwork mounting, reinforcement placing and concrete casting for monolithic sectors and in-situ slab. During the reinforcement placing the overlapping of both horizontal and vertical bars to be achieved with front contact welding or with rebar couplers;

2.3.6. Mounting of the dilatation joint between the viaduct and the retaining wall;

2.3.7. Concrete casting for the sidewalks' sections, with leaving of openings for the handrail sceptres;

2.3.8. Mounting of the gullies and casting the concrete for dewatering inclination, which is with maximal thickness of 5cm. After a while, depending on the type of insulation used, the hydro insulation of the slab along with the protective layer is implemented;

2.3.9. At both ends of the viaduct, at the connection with the backfill, two types of connections could be implemented:

- first variant – implementation of two curtain walls, which has sleepers installed subsequently before and after them;
- second variant – implementation of transitional devices to prevent falling out of the ballast bed onto the dilatation joint.

The advantage of the second variant is that it prevents the lifting of the heavy machine for sieving of the ballast bed during operation and maintenance of the railway [7].

3. Specific points of the design for the presented rail viaducts

3.1. Elements' stiffness proportion ratio

With long viaducts the separate elements' stiffness proportion ratio is crucial for optimizing the structural behaviour. For the correct proportions to be obtained the respective analyses should be made. The final goal is to get a structure with optimal behaviour in all of the standardized limit states. The parameters for the limits of the desired structural behaviour are the minimal and maximal stiffness for specific elements. In the superstructure the minimal stiffness is required, based on the desired deformation in horizontal and vertical direction to be in standardized limits. These factors can be influenced with the cross-section to span length ratio. The goal is a structure, which is not susceptible to severe deformations. Suitable and easy method for determining the dimensions is established by years of practice and the indicative proportions of structural height to span length are varying from 1/8 to 1/24, depending on the type and purpose of the structure [3], [4], [6].

During the design of the connection for the joints between superstructure and substructure elements in the particular case a proportion is sought where the superstructure has greater bending stiffness than the substructure elements, which are connected to it.

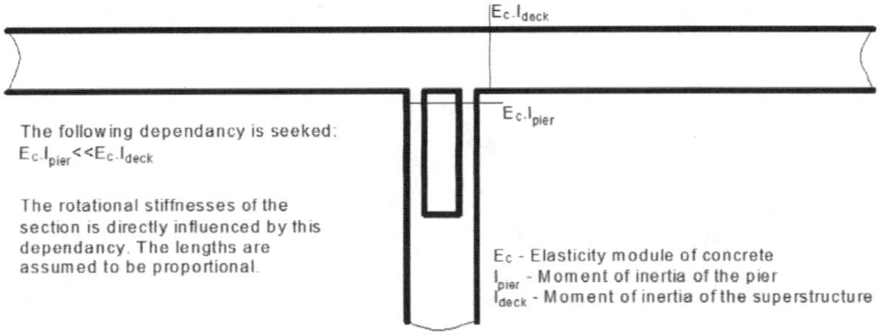

Figure 4. Stiffness proportions

This way a behaviour is provided for the superstructure beam, which is very close to the classic continuous beam behaviour. Thus two connected structures are obtained which don't affect drastically each other. The final result is that the bending moments caused by the superstructure shortening / lengthening deformations are drastically decreased both in the substructure elements themselves and in the joints' equilibrium. Other important aspect of this design is that loadings on the bridge superstructure don't create critical bending moments for the substructure and vice-versa. For determining the correct ratio several options for the joints have been analysed, all with different geometric proportions of the substructure elements. The last critical aspect is the equalization of the stiffness properties of the substructure elements.

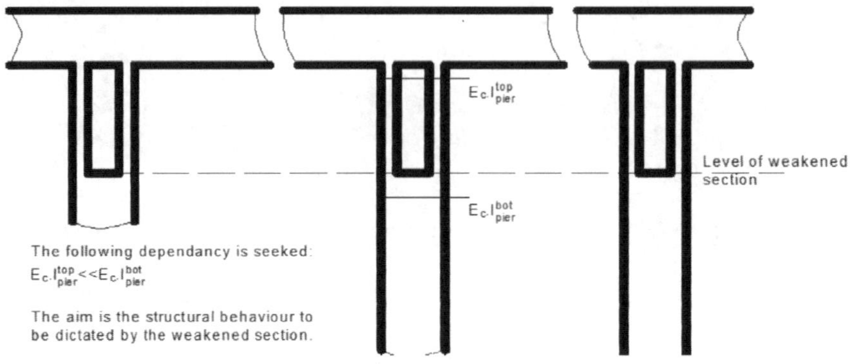

Figure 5. Weakened section

The main factor participating in this is the height of every single pier. Leaving different supports' heights without taking into consideration their different stiffness properties could lead to critical situations. Increased second order effects, uneven dynamic loadings distribution and a number of other problems could prove to be critical. Such parameters are often avoided and rarely are being left to be leading for the design, because of their time consuming and often unreliable analysis. This problem could be solved with a number of methods, like using a typical cross section with reliable stiffness in horizontal / transversal direction – a box, for example.

In the particular case a decision has been made for all the piers to be designed in a way to guarantee high stiffness below a particular level. For determining the level a quick proportions determining is enough as a start and it is updated after initial analyses. The determining criterium is uniform behaviour of the structure under dynamic effects / horizontal loadings, which eliminates the overloading of certain elements.

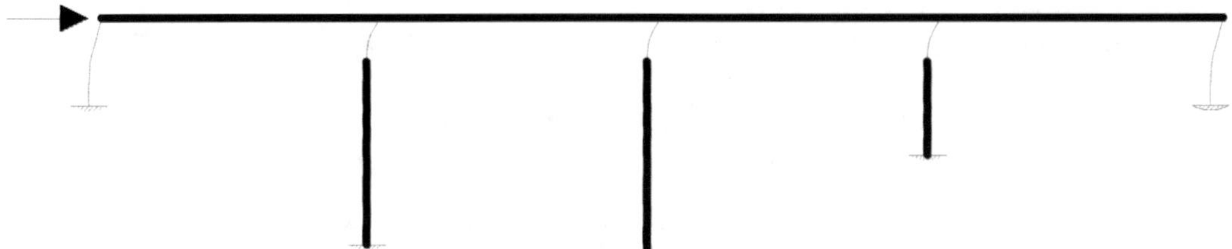

Figure 6. Static scheme of the structural behavior under horizontal loading

The stiffness should be in the limits – from minimal stiffness, which is dictated by allowable deformations outside the structural plane, to maximal stiffness, which leads to critical states from superstructure shortening / lengthening effects in horizontal longitudinal direction.

IOP Conf. Series: Materials Science and Engineering **951** (2020) 012021 doi:10.1088/1757-899X/951/1/012021

3.2. Joint detailing for substructure – pier connection

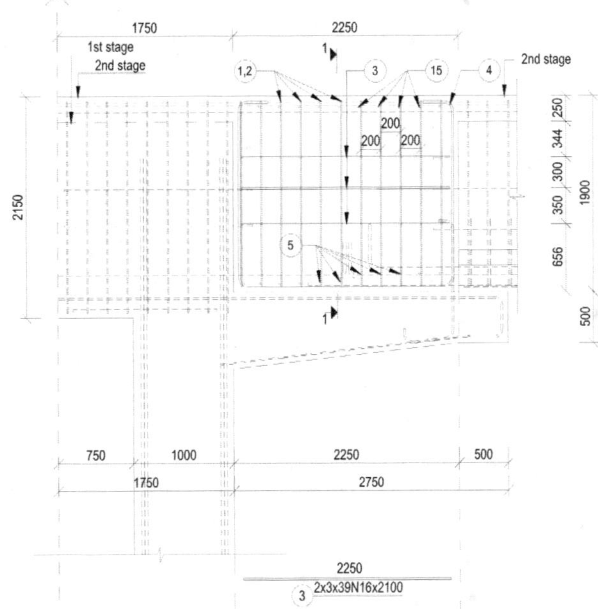

Figure 7. "Reinforcement basket" of the frame joint superstructure - pier

The joints' detailing and their implementation are important for the desired behaviour of the structure. In the particular case a prefabricated-monolithic solution is sought. For this purpose, a detail has been designed, which provides the possibility for mounting the beams in their design positions without the help of temporary supporting elements. For all elements' reinforcement quantities have been predetermined, which need to be correctly anchored in the joint. For the reinforcement baskets to be correctly assembled, all of the waiting reinforcement is selected to be anchored in free zones with open access for work.

3.3. Isolating of the abutments from the backfill

A vertical reinforced concrete retaining wall is positioned behind each of the abutments for protecting the endmost support from direct contact with the backfill.

The aim is the effects from soil pressure on the structure to be eliminated. Due to the large uninterrupted structure length the displacements at the end supports are expected to be quite big. Such displacements of the abutment could lead to varying of the active and passive soil pressure, which respectively are influencing the structure. Such varying on the other hand could lead to interrupted contact between structure and backfill [8].

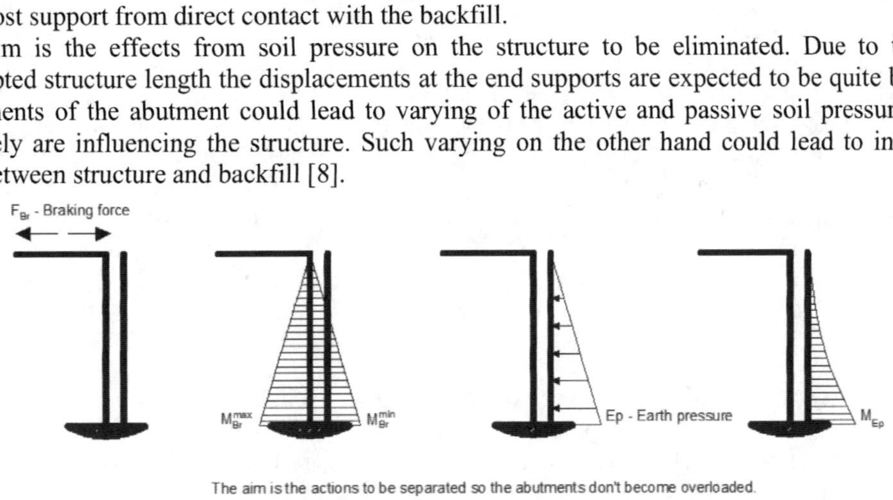

Figure 8. Behavior of the abutments and the retaining walls behind them

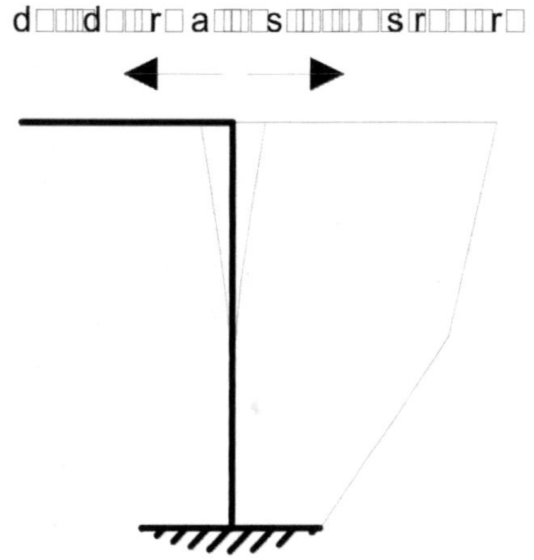

Figure 9. Abutment deformations due to forces inducing horizontal displacement

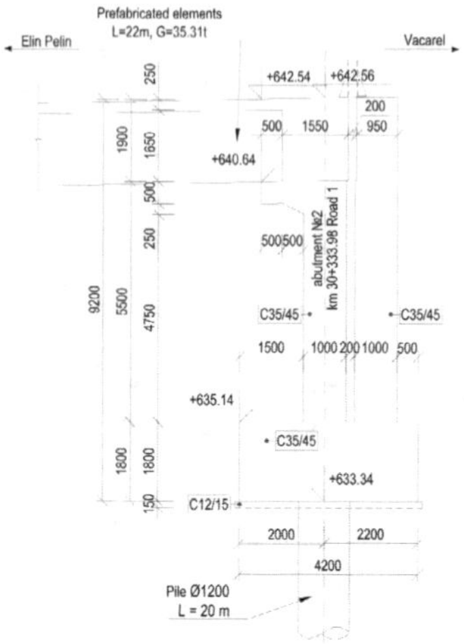

Figure 10. Form of the abutment

This interruption is critical for the structure's durability, due to it causing soil settlements, active corrosion zones and overall problems with the structure behaviour, as well as considerable stresses in the superstructure itself. For the particular case, a retaining wall is designed within direct proximity to the endmost bridge supports. The two structures are stepping on the same foundation. When designing such a structure it is necessary the possible displacements in both elements to be taken into account and the needed distance between them to be provided. Also, the common foundation should be designed for the cooperative behaviour of the two elements.

3.4. Thermal effects

The thermal effects should be taken into account with their full capacity since they are important critical effect for continuous structures. They usually lead to significant deformations and can dictate the division of the structure to separate blocks. With the correct stiffness proportion ratio, the thermal effects could be lowered to their minimal values, but they are still critical for certain elements. Load combination with dominant thermal effects is taken into account for every structural element.

3.5. Creep and shrinkage of concrete

The creep and shrinkage of concrete are effects which are difficult to be defined precisely. Because of this every design normative has them included with their maximal values. These values in most of the cases are on the conservative side and can be reduced by more precise analyses and building conditions, time for construction, care for concrete, quality of the materials. Due to the susceptibility of the continuous structures for the particular case is decided in favour of the Eurocode values, which are on the conservative side. The design of the structures is made with the maximal values of creep and shrinkage and their effects are added in every load combination for which they prove to be unfavourable. The deformations due to creep and shrinkage are critical parameter for determining the maximal and minimal dilatations and are also taken into account when defining the joints' dimensions and their location.

4. Advantages of the presented technical solution to other common methods

The presented structure has considerable amount of positive and negative compared to the traditional simply beam style static scheme, common for railway bridges. On the side of the design, a frame structure provides significant complexity to the structure behaviour. The static flow of the forces often cannot be determined as directly as in the simply supported structures. Effects like a creep, shrinkage, thermal deformations have a significant impact on the frame structure, compared to the simply supported static conditions, where those effects are significantly simplified or straight out ignored. However, the frame structure provides option of forces transfer in directions that the standard solutions do not. Having stiff connections between the substructure and the superstructure provides better mechanism for horizontal force distribution. In the traditional solutions, horizontal forces, bridge stability, and durability are provided via the usage of bearings, construction joints, dampers, and other devices. Those devices are expensive, especially the ones that can guarantee a good quality and low maintenance, negating the price difference between the frame and the traditional beam structures. Even more so if the devises used, have to be overengineered to withstand and maintain stability in case of accidental horizontal loads, such in earthquakes or severe wind action. The frame structures do put out more strain on the substructure itself. However, this trade of is often more than ideal as substructures of bridges tend to have design that can be utilized significantly better and be able to withstand the additional forces easily. In the frame static solution significant strain on the structure can occur due to the soil-abutment interaction, however a good solutions for isolation of the structure from the soil are optional, like the presented double wall abutment.

5. Conclusion

The use of frame structure reduces a lot of the critical key point, that are common to failure, high maintenance, and excessive costs. The modern solutions for bearings, joint connections are becoming more and more advanced, increasing the lifespan of the element, but still lack the durability index of the frame structures. The increased lifespan such elements is achieved by increased complexity, reflecting on their cost for the product itself and its application is inevitable. In the modern bridge development, most of the setback of the frame structure are avoided by the state of the construction quality. The modern constructor has arsenals of utilities and tools to easy the issues like complexity, quality of elements and connection, logistics, scaffolding, framework, and reinforcement. For most construction, the above-mentioned setbacks of the frame structures are irrelevant as the methods used are not significantly affected by them. Even more so, application of good design solution for isolation of the abutments from soil pressure, like the double walled abutments, further improve the general behaviour of the structure. This leads to the costs for a frame structure to be quite comparative compared to traditional solution where quality joints, bearings, and dampers must be used.

References

[1] The European Standard EN 2007 Eurocode 1992-1 - General rules and rules for buildings
[2] The European Standard EN 2007 Eurocode 1992-2 - General rules and rules for bridges
[3] Narayanan R S and Beeby A 2005 Designers` guide to EN 1992-1-1 and EN 1992-1-2. Eurocode 2: Design of concrete structures. General rules and rules for buildings and structural fire design ISBN: 07277 3105 X Thomas Telford
[4] Hendy C R and Smith D A 2007 Designers` guide to EN 1992-2. Eurocode 2: Design of concrete structures. Part 2: Concrete bridges ISBN: 978-0-7277-3159-3 Thomas Telford
[5] Martin P, Bruke Jr 2009 Integral and semi-integral bridges, ISBN: 978-1-4051-9418-1, Wiley-Blackwell
[6] Krizek J 2011, Integral bridge soil-structure interaction
[7] Chen W., Duan L., 2014 Handbook of International Bridge Engineering, ISBN: 978-1-4398-1030-9, CRC Press
[8] Topurov K., 2012 Reinforced concrete bridges, ISBN: 978-954-8873-93-2, ABC Technic
[9] Nikolov P 2013 Concrete road bridges, Designers` guide ISBN: 978-619-90057-2-9 Konsult 2000
[10] Nikolov P 2012 Modelling of structural and non-structural elements of bridges International Jubilee Conference UACEG2012: Science & Practice
[11] Topurova I, Nikolov P and Topurov K 2016 Highway Viaduct Renovation Procedia Engineering Bridges in Danube Basin 2016 – New trends in bridge engineering and efficient solution for large and medium span bridges Žilina, Slovakia, published by Elsevier
[12] Nikolov P 2012 Limitation of the translation for overpasses with integral abutments International Jubilee Conference UACEG2012: Science & Practice

IOP Conf. Series: Materials Science and Engineering **951** (2020) 012022 doi:10.1088/1757-899X/951/1/012022

Main features and issues in the design of tall buildings in Bulgaria

A Traykov

Department of Structural Mechanics
University of Architecture, Civil Engineering and Geodesy
1046 Sofia, Bulgaria

E-mail: alex_fce@uacg.bg

Abstract. In recent years, construction of tall buildings has been flourishing in many places around the world. In a considerable number of cases it is very much a matter of demonstrating financial and technological capabilities. Bulgaria is not isolated from these processes, mainly designing and building office, residential and hotel buildings with reinforced concrete structures. This requires systematization of information and experience in the design of tall buildings, as well as discussing the basic features and problems in the structural design of tall buildings and making recommendations based on existing experience and trends observed in the development of analysis methods and design. Basic issues related to height, architectural solutions, structural systems and other aspects that affect design solutions and problems related to these issues are discussed. The regulations related to the design of structures of tall buildings were commented on, with the main issues discussed mainly based on structural Eurocodes. The most important aspects of computer modeling of tall building structures and numerical analysis are discussed. Common problems related to the dimensioning and detailing of basic structural elements are commented. Finally, general recommendations are made for the research and design of tall buildings with reinforced concrete construction.

1. Introduction

Historically, the construction of tall buildings and facilities is not an end in itself. In the case of equipment - chimneys, towers, masts, the height is related to meeting certain technological requirements.

In the case of buildings, the construction of high-rise buildings is mainly related to the price of land in the center of large cities, and the implementation is also related to specific favorable conditions for construction - mostly appropriate geotechnical conditions. In all cases, it is necessary to have sufficient material and technical resources to carry out the construction. It is no coincidence that the very concept of skyscrapers for many years is associated mainly with cities in the United States. Even in the first few decades after World War II, very tall buildings were an exception, even in Europe's major cities.

In recent years, however, there has been a significant change in this regard. In many parts of the world, the construction of tall buildings is booming, becoming a kind of competition. In a significant number of cases, this is clearly a demonstration of financial and technological capabilities rather than a real necessity. Participation in such projects is a matter of prestige and a "business card" for investors, construction companies and design bureaus.

Bulgaria is not isolated from these processes. As a country with a relatively small population, which has historically been relatively evenly distributed in a large number of settlements, there was virtually no need to build tall buildings. Even in the capital Sofia until 20 years ago, there were very few tall buildings. The very concept of a tall building has evolved over time. In the 1960s, 12-storey buildings were considered tall, then separate higher buildings began to appear. These were landmark buildings for the city, some of them with an experimental construction - the trade union house on Macedonia Square. In the 80s of the twentieth century, high-rise buildings in Sofia and other cities in the country can be specified – in Sofia these are the building of the trade unions, the Ministry of Transport, Energoproekt, Hotel Rodina, Park Hotel Moscow, Hotel Hemus. An iconic building is also the so-called then a "Japanese" hotel – New Otani designed by the famous architect Kurokawa. Three high-rise residential buildings have been built in the Iztok district for the needs of the diplomatic corps, three more at the beginning of Mladost district, as well as several other. In Varna these are the buildings of the Black Sea Hotel and the Municipality building, in Plovdiv - the Maritsa Hotel, in Burgas - the Bulgaria Hotel. Among the other tall buildings, the main examples are also hotels - "Kuban" in Sunny Beach, "International" in Golden Sands, "Dobrudja" in "Albena".

With the expansion of construction opportunities after the changes in 1989, there is a tendency to gradually increase the number of storeys of some newly built buildings, but until 2010 the criterion for a high-rise building remains in the range of 15-18 floors, i.e. up to about 50-60 m height.

2. Purpose of the paper

After 2012 and the revival of the real estate market after the economic crisis began the design and construction of buildings over 85-100 m, in some cases higher. It should be noted that the design developments in different phases of development of a project are probably currently about 10 times more than the buildings that are built or under construction. It should be noted the fact that some of these developments are adaptation projects developed in other countries and other conditions. The reasons for the lack of continuation and realization as construction of most of the project developments are complex, but the main ones are the following:

- difficult market realization with unclear prospects of properties in buildings of this size, whether for sale or rent;
- developer's awareness of the scale of investment in the project development process;
- the relatively higher unit price of a built-up area in a tall building compared to that of "normal" construction, in some cases significantly higher;
- urban planning problems related to obtaining a construction permit for such buildings, especially in residential areas with high building density;
- problems related to transport access and maintenance of multi-occupant buildings;
- complicating and increasing the cost of installations.

The aim of the work is to systematize the information and experience gained in the design of tall buildings, as well as to discuss the main features and problems in the design of high-rise buildings and formulate recommendations based on experience and trends in the development of research methods and design.

3. Main aspects and factors that influence the design of tall buildings

The following main aspects are important in the design of tall buildings

3.1. Height

There is no clear and common definition of a tall building. Basic document related to urban planning [1] gives only general guidelines for building height. It defines as high construction all buildings taller than 15 m. As it is shown in [2] there are different criteria for the defining and measuring of tall buildings. One such criterion is the height relative to context that means that a building of given number of floors could be considered as tall in provincial city taken into account the local urban norm while the same building may not be considered a tall building in many big cities. Another criterion could be the aspect ratio. A slender building could give the appearance of a tall building even if it is not very high. The same source maintains that number of floors is not good indicator for defining if a building is tall

IOP Conf. Series: Materials Science and Engineering **951** (2020) 012022 doi:10.1088/1757-899X/951/1/012022

because of different floor height in different buildings and due to different functions of floors. In spite of it a building of 50m or 14 or more floors could be considered as tall building [2]. The number of floors includes the main floors above ground, including the level of the ground floor and the floors of mechanical equipment and mechanical mezzanines with a significant area and does not include the floor levels with a significantly smaller area than the area of the main floors. Mechanical rooms above the general roof area should not be counted, too.

According to [3] tall buildings are those which height is 73 m (240 feet) or more. The Bulgarian Chamber of Engineers in Investment Design (KIIP) has created Technical counsel that works together with Sofia Municipality with the task for preliminary review and preparation of a technical statement about the structural design of tall buildings. As tall buildings are considered all buildings over 74m and those over 50m with irregular configuration in plan.

Heights and construction - in accordance with the traditional method of construction and height - heights from 90 to 130 m, and there are projects for 150, 180 and 200 meters of buildings.

3.2. Architectural designs

Experience shows that the main types of buildings that are designed in Bulgaria are office buildings and hotels. Those buildings are usually tower type.

Layouts of that type of buildings are presented on figure 1.

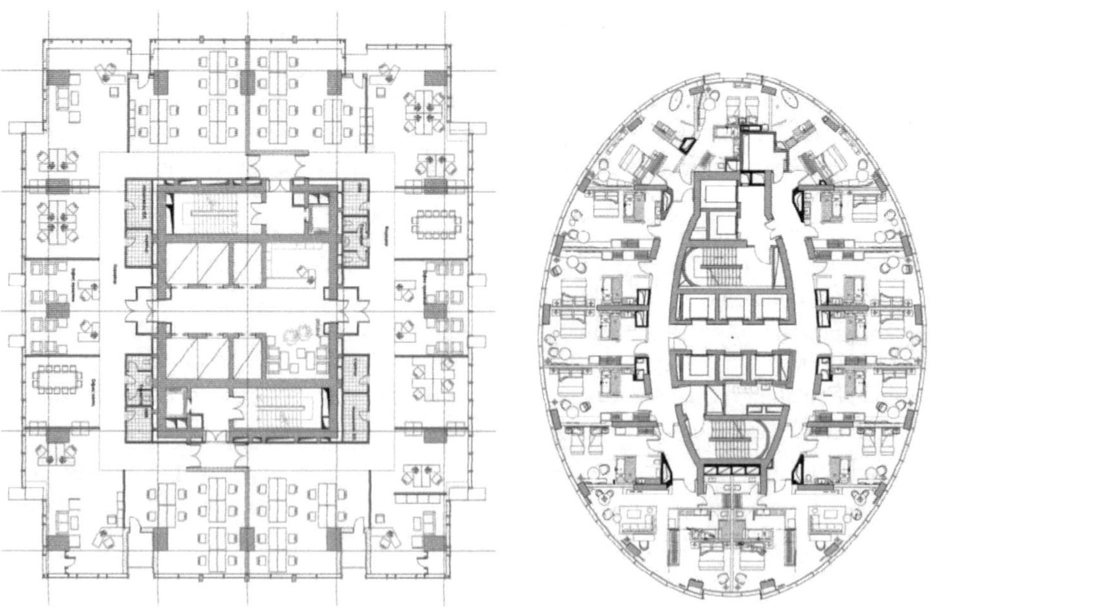

Figure 1. Floor layouts of tower buildings.

Residential buildings are designed with more variety. Different layouts are presented on figure 2.

IOP Conf. Series: Materials Science and Engineering **951** (2020) 012022 doi:10.1088/1757-899X/951/1/012022

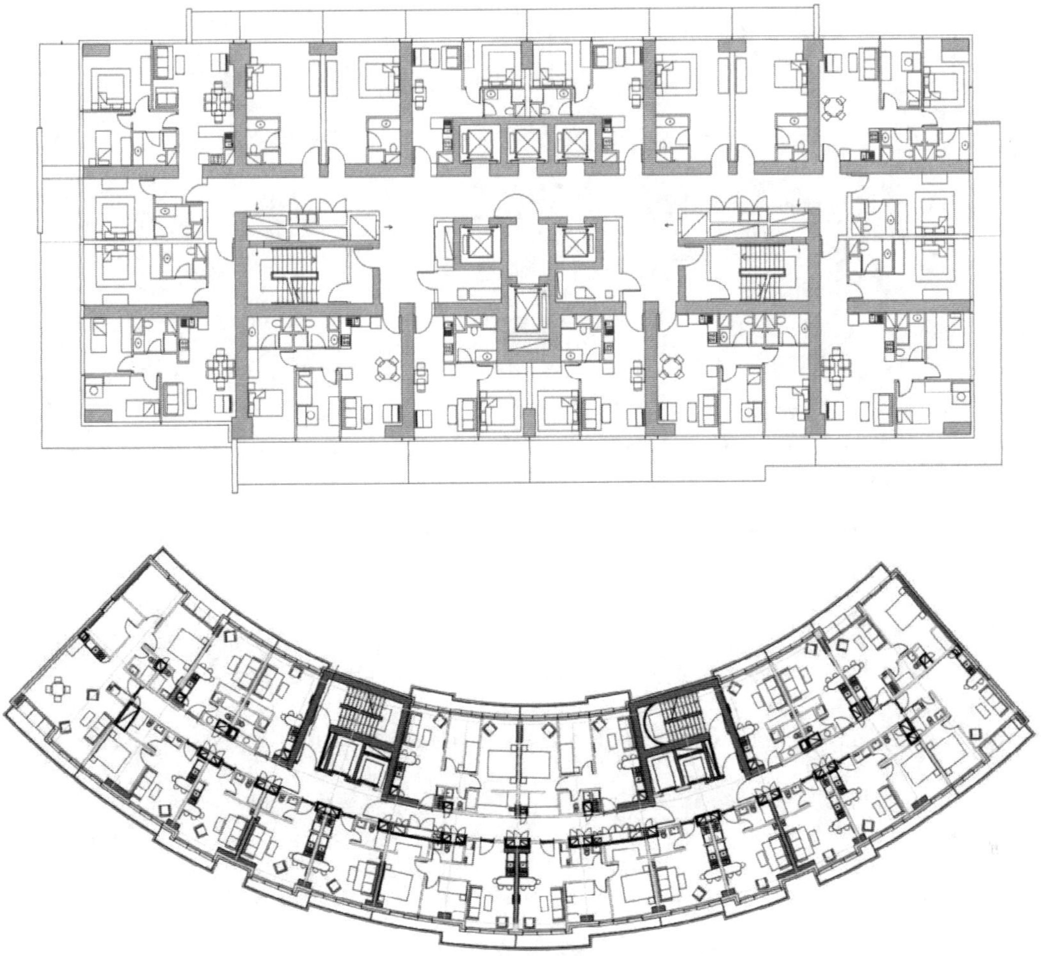

Figure 2. Floor layouts of residential buildings.

3.3. Structural systems

In high-rise buildings, it is necessary to make a clearer distinction than in conventional buildings of structural systems for resisting vertical and horizontal forces.

Usually in office buildings there is an aim for large bays as determined by the purpose, which implies a preference for flat slabs. Unfortunately such structures cannot be considered appropriate for tall buildings with regard to the horizontal loads.

The system resisting horizontal loads is of main interest when discussing the design of tall buildings. The height of buildings that are designed and constructed in Bulgaria so far defines the earthquake as the most important horizontal load. The main elements that provide the earthquake resistance of tall buildings are usually a combination of cores, shear walls (earthquake resistant walls). Some buildings do not include all those element types.

Floor structural layout of tower type of buildings is/are shown on figure 3

IOP Conf. Series: Materials Science and Engineering **951** (2020) 012022 doi:10.1088/1757-899X/951/1/012022

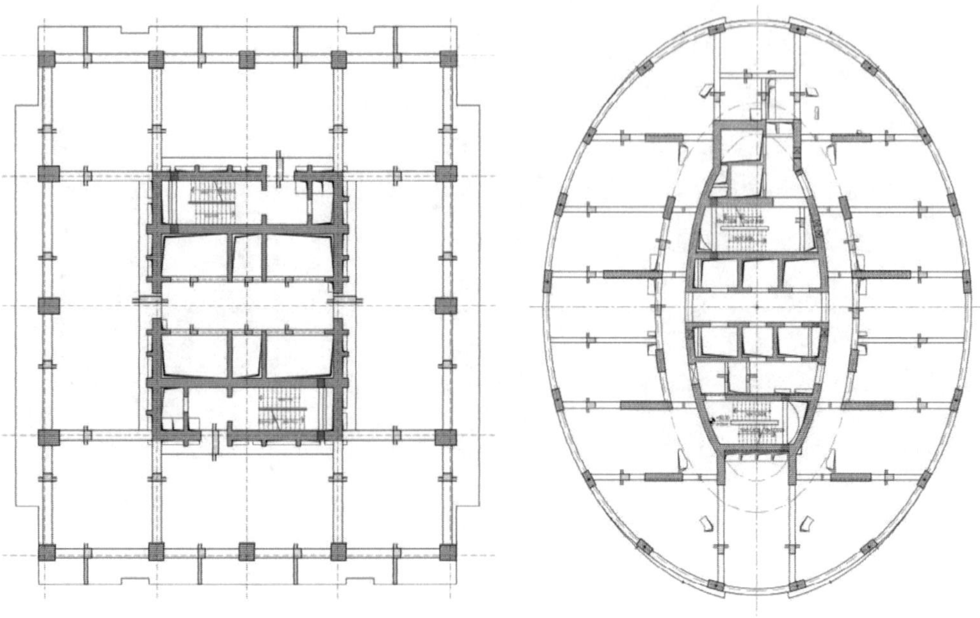

Figure 3. Floor structural layouts of tower buildings.

Floor structural layout of residential buildings is/are shown on figure 4

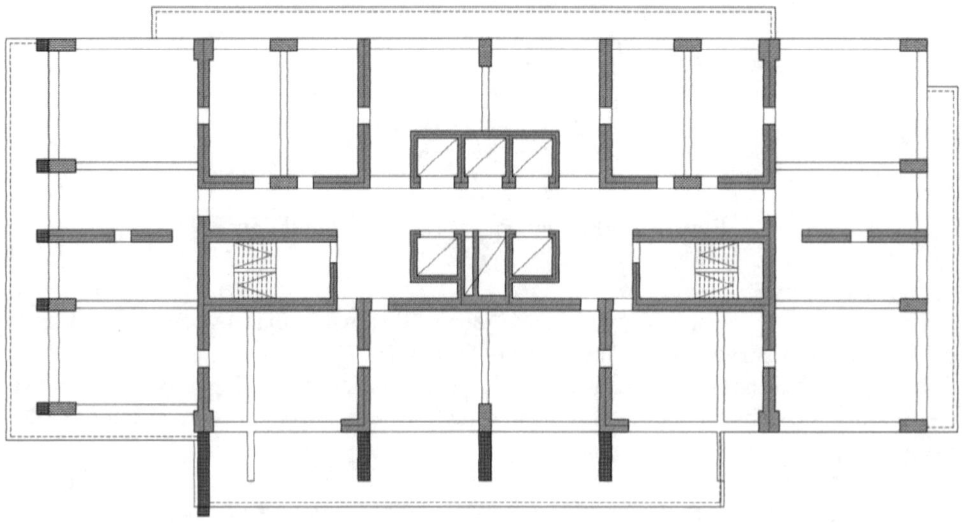

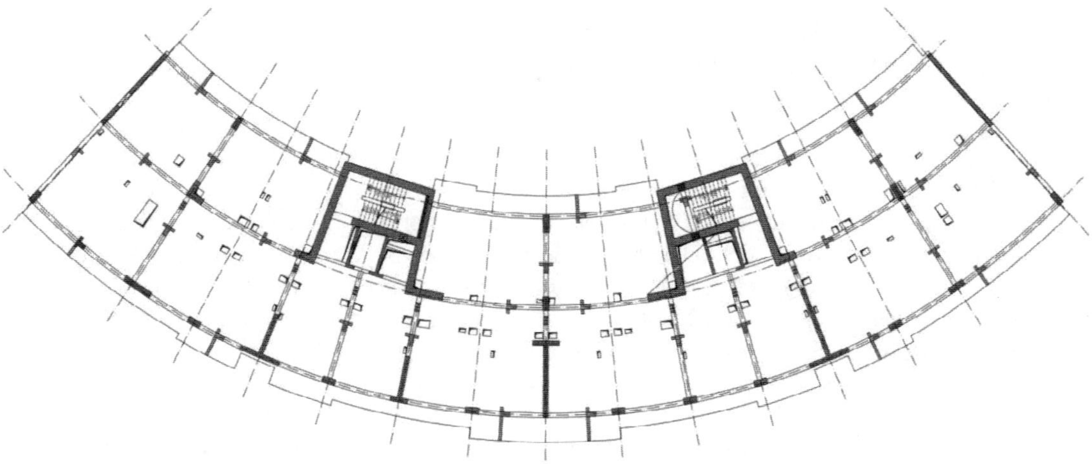

Figure 4. Floor structural layouts of residential buildings.

Flat slabs are the floor structure, usually preferred by architects and developers for different reasons. All those reasons are not related to structural design considerations. On the contrary – from structural point of view flat slabs are not suitable for tall structures – they are thicker than slabs combined with beams and for that reason induce higher seismic forces at the floor level. Flat slabs are not ductile and they do not provide adequate stiffness, especially at the buildings' contour. It should be noted that it is important to design beams at the facades, especially in tower type buildings.

The following key issues can be noted in this regard:

- application of traditional architectural solutions inconsistent with the height of the building
- insufficient peripheral stiffness associated with seismic behavior and building torsional stiffness
- application of prestressed slabs in order to reduce slab thickness.

Such design solutions are not recommendable because of structural and/or fire protection and durability considerations.

3.4. Legislation

Legislation is an important issue that affects the design of tall buildings.

In Bulgaria, as in most countries, there are no special regulations related to the design of tall buildings. Tall buildings usually fall into the category for which the design is required to be carried out in accordance with Eurocode system. The Eurocode itself was not developed with specifically for tall buildings. Rather, some specific guidelines in this regard can be found in Bulgarian norms [4]. For example buildings high "over 20 floors" or "over 30 floors" are considered in certain context. This is one of the reasons in practical applications to seek recommendations based on conclusions from comparisons of different standards, especially seismic ones, which are made in various literature sources, for example [5] и [6].

One of the tasks of the established Technical council to the Chamber of Engineers (KIIP) Sofia-city with the support of the Ministry of Regional Development and Public Works and Sofia Municipality is to deal with existing gaps in the legislation. A good example in this direction can be taken from [7].

It should be taken into account that the technical norms considering the design of tall buildings are not limited only to structural design. There are a lot of additional requirements that are imposed by architecture, greenery, roads, access to the building, fire protection, supply systems and other that can affect significantly the structure itself and shall be taken into consideration.

IOP Conf. Series: Materials Science and Engineering **951** (2020) 012022 doi:10.1088/1757-899X/951/1/012022

3.5. Computer modelling

There are a lot of important aspects of the computer modelling of buildings in general that are of particular importance for the modelling and analysis of tall buildings. Some of them are briefly mentioned below.

First of them is to choose the appropriate software. In spite of that not all the software for structural engineering is suitable for the analysis of tall buildings there are at least several popular commercial packages that are reliable and are used widely for the design of such structures. There are a number of important aspects of the creation of the computer model. It is important to start the modelling by carrying out thoroughly the regularity checks. The results from those checks as well the correct definition of the structural system of the building affect the value of the seismic forces that are used in the analysis.

Analysis and design of tall buildings must include developing a number of various computer models. In general, they can be described as follows:

- computer models for structural analysis of structures for vertical loads - these models are used for the design of slabs and for beams and columns that do not participate in the earthquake resistant system. It is recommended to develop and use both 2D and 3D computer models for the study of floor structures. This makes it possible to ensure the safety and the absence of defects in the elements of the floor structures, as well as to reduce the risk of damage due to deflections of slabs and / or beams.
- computer models for structural analysis of structures for horizontal loads - these models are used for the design of walls, cores and frames that create the earthquake resistant system.

The way the building is modeled in the computer model has a major impact on the obtained results. This applies to any building, but is very important in modeling the structure of tall buildings. Since the structure should be analyzed in consistent way a lot of important results and design checks need to be monitored simultaneously. Among the most important quantities to be controlled are the story drifts, the normalized axial forces in walls and columns and the quantities participating in the torsional flexibility checks. Design checks of at least the main structural elements should be carried out at every iteration. The question of choosing a critical area remains open, especially in cases of lack of part of the basement walls.

The usual method for seismic response assessment, which is considered as reference analysis method, is by spectral response and modal superposition. In fact, a linear analysis is performed, which has many drawbacks.

The regulations recommend the use of various nonlinear methods, but in reality, the possibilities for using nonlinear analysis are limited. That application of nonlinear methods for the analysis of buildings is not sufficiently developed in the regulations from practical point of view, in particular with regard to the tall buildings. Anyhow it is highly recommended to carry out nonlinear analyses in order to compare the results that are obtained by different analysis methods and to compare those results in order to gain information and experience. Studies taking into account specific effects related to the calculation of structures of tall buildings and the application of appropriate nonlinear methods for them were performed in [8], [9] and [10].

3.6. General problems related to dimensioning and detailing

The problems related to dimensioning and detailing of the structural elements of high-rise buildings are the result of the high values of the internal forces in them, as well as the deficiencies or incompleteness of the technical norms.

Iterative analysis related to dimensioning and detailing of:

- reinforced concrete cores
- seismic resistant walls of rectangular, L, U and other complex cross section shape
- columns and beams as frame elements

is necessary in order to achieve compliance to the requirements of the regulations and to achieve safe and consistent design.

Foundation of tall buildings is another and relatively independent topic that is of particular importance for tall buildings, which goes beyond the scope of the article. Specific for tall buildings is the raft foundation, often combined with piles. There are a lot of specific issues regarding the computer modelling, detailing and construction of such foundations.

4. Conclusions and recommendations

Based on the observations and analysis, the following conclusions and recommendations could be formulated:

- tall building design and construction is area of intensive development;
- almost all tall buildings in Bulgaria are office or residential, and the structures are reinforced concrete. Specific studies and dissemination of knowledge about the analysis, design and construction of that type of buildings is important to improve the quality of engineering and buildings performance. Exchange of international and local experience and publication of recommendations based on scientific research and good practices is essential.;
- there is insufficient experience in design and construction of tall buildings. The combination of local knowledge and traditions with modern methods of analysis, design and construction is the right approach to achieve better design and construction decisions. It is necessary to pay specific attention on right decision of the structural system, computer modelling and detailing;
- monitoring is important both in terms of the behavior of the building, and with regard to the accumulation of knowledge to enrich professional practice in the future;
- application of nonlinear methods for analysis should be expanded and comparison of the obtained results with those obtained by means of the reference methods should provide information both about the structural behaviour and also about the calibration of the parameters of the computer models used.

References

[1] Spatial Planning Act 2001 *Republic of Bulgaria SG,* amendments issue 21 2020
[2] Criteria for the defining and measuring of tall buildings 2016 *CTBUH (Council on Tall Buildings and Urban Habitat)* Illinois Institute of Technology Chicago, USA
[3] Moehle J, Bozorgnia Y and Yang T Y 2007 The Tall Building Initiative, *Structural Engineers Association of California SEAOC 2007 Convention proceedings* Squaw Creek, California, USA
[4] Ordinance № RD-02-20-2 of 2012 on the design of buildings and facilities in earthquake areas 2012 *Republic of Bulgaria SG*, issue 13
[5] Critical comparison of major seismic codes for buildings, Technical report 2013 *fib Bulletin* **69**, CEB-FIP
[6] Santos S H C, Giarlelis C, Traykova M, Bucur C, Zanaica L and Lima S S 2017 Comparative study of a set of codes for the seismic design of buildings, *39th IABSE Symposium – Engineering the Future* Vancouver, Canada
[7] Guidelines for performance-based seismic design of tall buildings 2017 TBI (Tall Buildings Initiative), ver. 2.03 *PEER Report 2017/06*, California, USA
[8] Traykova M, Boiadjieva-Marinova R and Traykov A 2015 Strengthening of existing RC frame in case of construction of additional floor, with account for the construction stages *IABSE Conference, Geneva 2015: Structural Engineering: Providing Solutions to Global Challenges – Report* pp. 1243-1247
[9] Traykov A and Boiadjieva R 2015 Computer Modelling of Structures with account of Construction Stages and Time Dependent Material Properties, *J. of Civil Eng. Architect. Res.* **2** (11) 1076-1085
[10] Bui D, Traykov A and Traykova M 2017 Shear lag effect and its effect on the design of high-rise buildings, *High Rise Constructions 2017*, Samara, Russia, E3S Web of Conferences 33, 02001

Analysis of the load bearing capacity of scaffoldings used in Poland

E Błazik-Borowa *, J Bęc, P Jamińska-Gadomska, B Kawecki, T Lipecki, M Pieńko, A Robak, J Szulej and P Wielgos

[1] Lublin University of Technology, Faculty of Civil Engineering and Architecture, Nadbystrzycka 40, 20-618 Lublin, Poland

* E-mail: e.blazik@pollub.pl

Abstract. The paper presents static-strength analyses of 120 scaffoldings used on construction sites in Poland in the period 2016-2018. The geometry of each analysed scaffolding was taken from measurements made on the construction sites, and the scaffolding loads were determined on the basis of the standards requirements. The results of scaffolding inventory, surveying measurements, measurements of axial forces in standards, and measurements of scaffolding accelerations during vibrations were used to create numerical scaffolding models. For each scaffolding, buckling load analysis and linear static analysis were performed for four load cases which took into account self-weight, parallel and perpendicular wind actions and service load. On the basis of numerical analyses, it was found that 14 scaffoldings could fail due to loss of stability. For 56 scaffoldings, the ultimate limit state condition was not met in at least one group of components, therefore 47% of tested scaffoldings should not be approved.

1. Introduction

Scaffoldings are temporary structures that are used on most construction sites. They ensure safety of people working at height or in places with difficult access. The specificity of scaffoldings use means that their users are more prone to accidents than other construction workers. From 2011 to 2014, 479 people fell from scaffoldings located in Poland [1]. The direct causes of falls from scaffolding are mainly: leaning on handrails, shear of deck hooks or lack of decks, but also failure of scaffolding and its falling over together with people.

The necessity of performing static and strength calculations for scaffoldings was demonstrated, among others, by Żółtowski [2]. The paper [2] describes failure of scaffolding mounted inside the distillation column with a diameter of 5.3 m and a height of about 57 m, located at Petrochemia in Płock. The scaffolding was made using an older type of tubular system. The tubular elements had an outer diameter of 48.3 mm, but the walls of the elements had various thickness: 2.9; 3.7 and 4.0 mm. Furthermore, significant steel corrosion was also found in the elements. During the use of the scaffolding, the standards buckled, joints were damaged and eventually horizontal load-bearing elements were damaged and the scaffolding collapsed. As a result, two people were killed and three were injured. Static-strength calculations performed after the disaster showed the bearing capacity of the elements was significantly exceeded.

Another example of scaffolding failure occurred in 2010 at the historic building in Gostyń [3]. The scaffolding was used to enable maintenance works for building towers at 38.5 m. During the storm, the

IOP Conf. Series: Materials Science and Engineering **951** (2020) 012023 doi:10.1088/1757-899X/951/1/012023

access tower of scaffolding was detached from the building and the working platforms were moved by the wind to the roof and outside the building. In the result, building's historic tiles were destroyed. The direct cause of the scaffolding failure in Gostyń was a strong wind, but a detailed static and strength analysis of the structure showed that the structure did not meet the ultimate limit state requirement. The failure did not take place earlier, because the scaffolding was just assembled and no construction works were carried out on it before the storm.

According to the National Regulation [4] scaffoldings that are different from those described in the producer's documentation should be assembled on the basis of individual projects. The projects of atypical scaffoldings should be prepared in accordance to the standards EN 12810-2 [5] and EN 12811-1 [6] recommendations. Such designs have not been prepared for the two scaffoldings which failures were described above. In order to check whether other scaffoldings in Poland would meet the requirements of EN 12810-2 [5] and EN 12811-1 [6], static-strength analyses of 120 scaffoldings used at construction sites in 2016-2018 were performed. For each scaffolding, the geometry used in calculations was corresponding to the one measured at the construction site, whereas scaffolding loads were determined on the basis of standard requirements.

2. Description of research methods

2.1. In situ measurements

The research on scaffoldings was performed by five research teams, as a part of the ORKWIZ project (full title: „Modelling of risk assessment of construction disasters, accidents, and dangerous incidents at workplaces using scaffoldings"). The subject of the study were frame scaffoldings, mostly set up at buildings facades. The total number of 120 scaffoldings was tested during the research. Scaffoldings were located in various parts of Poland. Each scaffolding was assigned a symbol, consisting of a letter denoting the given research team (D, E, L, W and P), and a subsequent number in the tests of the given team (e.g. W15).

On each of the construction sites where scaffoldings were tested, the measurements lasted one working week. The methodology of the studies, the results and their analyses are presented, among others, in works [7]-[20]. The results of a detailed inventory of scaffoldings, geodetic measurements, measurements of normal forces in standards and accelerations measured in various points of scaffoldings during vibrations were used in the research described in this paper.

As part of the inventory of each scaffolding, the following information was collected: type of scaffolding system; scaffolding schemes with marked: base jacks, standards, decks, handrails, toe-boards, bracings, anchors, ladders and other scaffolding elements; type of platforms; presence of protective covering and its type; dimensions of mudsills; presence of earthing and lightning protection mounted on a scaffolding. Photographic documentation included photos with all the elements listed above. Then, on this basis, 3D drawings were developed in the Autocad. The documentation prepared this way enabled accurate mapping of each scaffolding. In case of scaffoldings, seemingly small changes have a great impact on the effort of individual elements of the structure.

Scaffoldings geometry measurements were supplemented with geodetic surveying (Figure 1a). As part of these measurements, the location of pairs of points which lie on the circumferences of the circular cross-sections of standards was determined. On the basis of these points, the cross-sectional centres of standards were determined, and the scaffolding geometry was developed based on their locations. The method of measuring and creating scaffolding geometry based on the results of these measurements has been described in detail in [11].

Measurements of normal forces in standards were made using an instrument designed especially for this purpose, shown in Figure 1b. The device consists of two parts and each of these parts is comprise of a rigid frame and a force sensor placed on it. Before the measurement, each part was positioned under one stand of a scaffolding frame. The frame of the device was positioned with the screws so that the force sensor indicated zero. Then the base jacks were unscrewed and the scaffolding rested on the force sensors. After that, readings of normal forces values in the standards of one frame were taken. For each scaffolding three frames at the lowest level of scaffolding were chosen to be tested: one with attached bracing, one without it and one at the end of the structure The method of measurement and analysis of the results has been described in detail in [10].

The study of dynamic parameters consisted of measuring accelerations of selected points below the highest level of decks. The signal analyser and accelerators from Brüel & Kjær were used. Based on the acceleration changes measured over time, the frequencies of free vibrations and scaffolding damping parameters were determined. The Fast Fourier Transform (FFT) analysis was performed for each of the measured signals. Based on these results, the frequencies of free vibrations and damping parameters were determined. The research methodology and analysis of the results were thoroughly presented in [7].

(a) (b) (c)

Figure 1. Measuring stations [11]: (a) geodetic measurement, (b) measurement of forces in standards, (c) measurement of accelerations during scaffolding vibrations

2.2. Numerical analysis

Numerical models and calculations were made for each scaffolding using the Autodesk Simulation Multiphysics 2013. The models took into account geometric imperfections defined as the distances between the location of standards connections in real and ideal geometry of the structure. Figure 2 shows an example of numerical model for scaffolding W12. Static scheme of the remaining scaffoldings are included in [8]. Geometric characteristics of elements cross-sections and characteristic yield limits – f_{yk} (Table 1) were adopted on the basis of information from manufacturers or from the own research. The calculated yield stress was determined from the formula $f_y = f_{yk}/\gamma_M$, where the partial safety factor γ_M was equal to 1.1.

To determine the frequencies of natural vibrations for each scaffolding, static calculations including only the self-weight of the scaffolding were carried out. Based on the resultant frequencies and forms of natural vibrations, the supports modelling anchors were calibrated and the rigidity of elastic supports modelling cooperation between the analysed scaffolding and adjacent scaffoldings was selected. For some scaffoldings, mainly low ones with low self-weight, cooperation with the ground was also modelled with the use of elastic supports. Then, normal forces in standards from static calculations for verified static scaffolding schemes, were compared with measured values. These results showed for some standards, a possible lack of full support on the ground, which was verified on the basis of soil tests and information from scaffolding inventory. During scaffolding tests on construction sites, it was checked if the standards were loaded. The methodology of boundary conditions selection in scaffolding models has been described, among others, in [21], while the impact of the adopted static schemes was checked in [22]and [23].

IOP Conf. Series: Materials Science and Engineering **951** (2020) 012023 doi:10.1088/1757-899X/951/1/012023

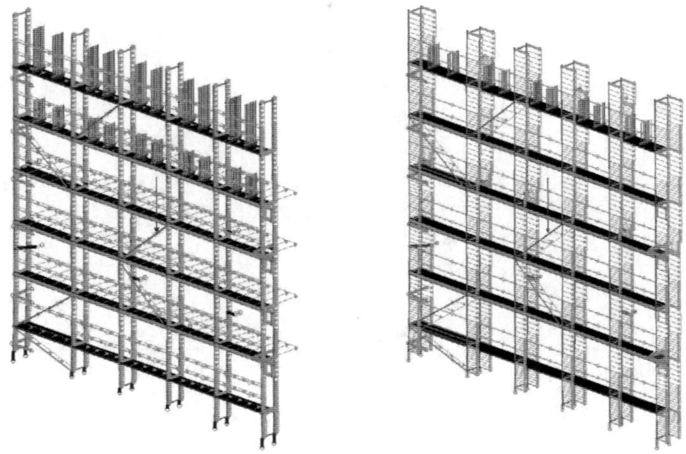

(a) (b)

Figure 2. Static scheme of W12 scaffolding for cases: (a) LC1, (b) LC4. Load cases description summarized in Table 2

Table 1. Characteristic yield strength f_{yk} [MPa]

Scaffolding system	The number of scaffoldings	Standards	Base jacks	Transoms	Other components
Altrad Mostostal	36	315.0	235.0	315.0	315.0
Plettac	33	320.0	235.0	235.0	235.0
Layher	23	320.0	320.0	320.0	320.00
Delta	8	315.0	235.0	235.0	235.0
RUX	5	235.0	235.0	235.0	235.0
Hunnebeck	5	235.0	235.0	235.0	235.0
RR-0,8	3	450.0	300.0	300.0	300.0
BAL	2	460.0	270.0	400.0	270.0
Termosprzęt	2	235.0	235.0	235.0	235.0
PERI	2	235.0	235.0	235.0	235.0
Omega	1	235.0	235.0	235.0	235.0

For verified scaffolding models stability and linear static analyses were performed. The calculations were made for 4 load cases, in accordance with EN 12811-1 [6] and summarized in Table 2. The characteristic wind action was determined from the formula:

$$F_k = c_s \sum A_i c_{f,i} w_{k,i} \,. \qquad (1)$$

where: A_i – the reference area, c_f – the aerodynamic force coefficient, c_s – the site coefficient, w_k – the characteristic wind velocity pressure, i – index of the scaffolding element. A detailed description of the scaffolding wind load determination methodology can be found in [17].

The static-stress analyses were performed as linear calculations, because the obtained results not only confirm the possible exceedance of the load capacity of structural elements, but also show the scale of failure risk, measured as the quotient of normal stress caused by the design load, and the calculated yield stress. Stability analyses were made for scaffoldings with characteristic loads.

IOP Conf. Series: Materials Science and Engineering **951** (2020) 012023 doi:10.1088/1757-899X/951/1/012023

Table 2. Load cases

Load case	Self-weight	Service load			Wind load		
	γ_f	q_k [kN/m^2]	Load on platforms	γ_f	w_k [kN/m^2]	Direction to the facade	γ_f
LC1	1.5	2.0	highest level	1.5	0.2	perpendicular	1.5
		1.0	below the highest level				
LC2	1.5	2.0	highest level	1.5	0.2	parallel	1.5
		1.0	below the highest level				
LC3	1.5	0.5	highest level	1.5	EN 1991-1-4 [24]	perpendicular	1.5
LC4	1.5	0.5	highest level	1.5	EN 1991-1-4 [24]	parallel	1.5

q_k – characteristic value of a service load, w_k – the characteristic wind velocity pressure, γ_f – partial safety coefficient.

3. Calculation results

3.1. Results of buckling analysis

The results of the stability analysis are buckling forms and critical load multipliers α_{cr}. Figure 3 shows the number of scaffoldings for individual ranges of α_{cr}. The multiplier α_{cr} took values from 0.2 to 24.9. The scaffoldings for which α_{cr} values were less than 1 were obtained only for cases with the wind perpendicular to the facade. There were 3 such scaffoldings for LC1 and 14 for LC3 load case. All these scaffoldings had protective coverings. In the performed calculations the wind was assumed as working towards the facade. It was found out that this situation is more unfavourable for the structure than when the wind acts from the façade because when it pushes the scaffolding against the façade anchors stop working. They either buckle or change the geometry so that they do not participate in creating the rigidity of the scaffolding system. This results in the loss of global stability at the level of scaffolding which is not supported horizontally. The issue of low stiffness of anchors in the case of wind loads towards the facade was discussed in the paper [25] and the impact of coverings on scaffolding effort in case of tall buildings was presented in [26].

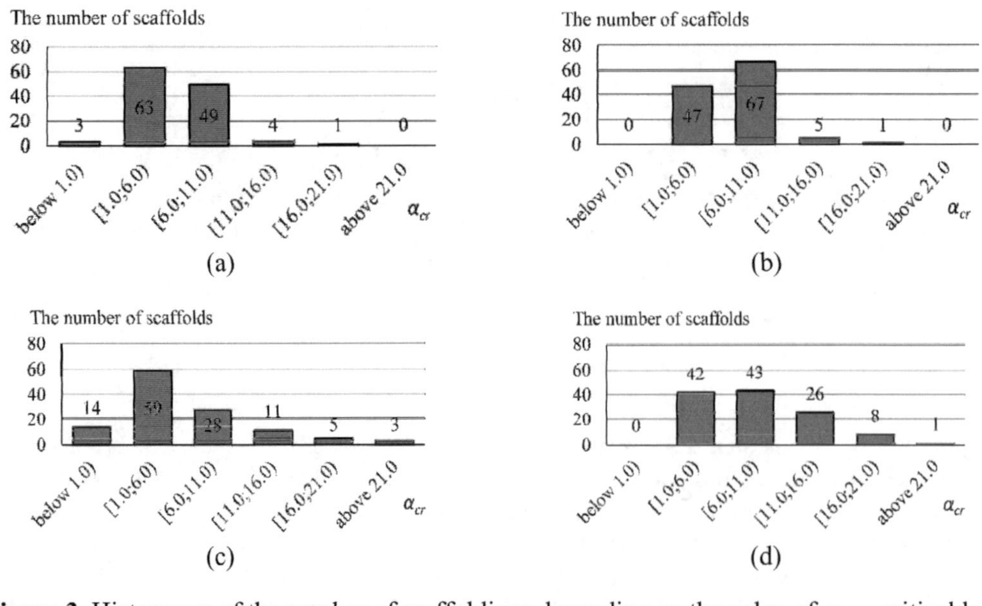

Figure 3. Histograms of the number of scaffoldings depending on the value of α_{cr} – critical load multiplier, for cases: (a) LC1, (b) LC2, (c) LC3, (d) LC4

IOP Conf. Series: Materials Science and Engineering **951** (2020) 012023 doi:10.1088/1757-899X/951/1/012023

3.2. *Results of static analysis*

The results of static calculations were: displacements, internal forces and normal stresses determined from the formula:

$$\sigma = \frac{S}{A} + \frac{M_2}{W_2} + \frac{M_3}{W_3}. \quad (2)$$

where: S – normal force, M2 and M3 – bending moments about the main axes of the cross-section, A – cross sectional area, W2 and W3 – section modules at bending about principal axes.

Normal stresses in the W12 scaffolding elements are shown in Figure 4, whereas Figure 5 presents histograms of the section load utilization coefficient determined from the formula:

$$\kappa = \frac{\sigma}{f_y}. \quad (3)$$

The presented histograms are graphs showing the number of scaffoldings for which κ value was obtained for four groups of elements: base jacks, standards, transoms and the group of other elements (bracings, consoles, girders, etc.). As can be seen in Figure 5, scaffolding may exceed its bearing capacity if it is exposed to loads applied in accordance with EN 12811-1 [6]. The load capacity for base jacks was exceeded in 56 scaffoldings, for standards – in 53 scaffoldings, for transoms – in 35 scaffoldings and for other elements – in 42 scaffoldings.

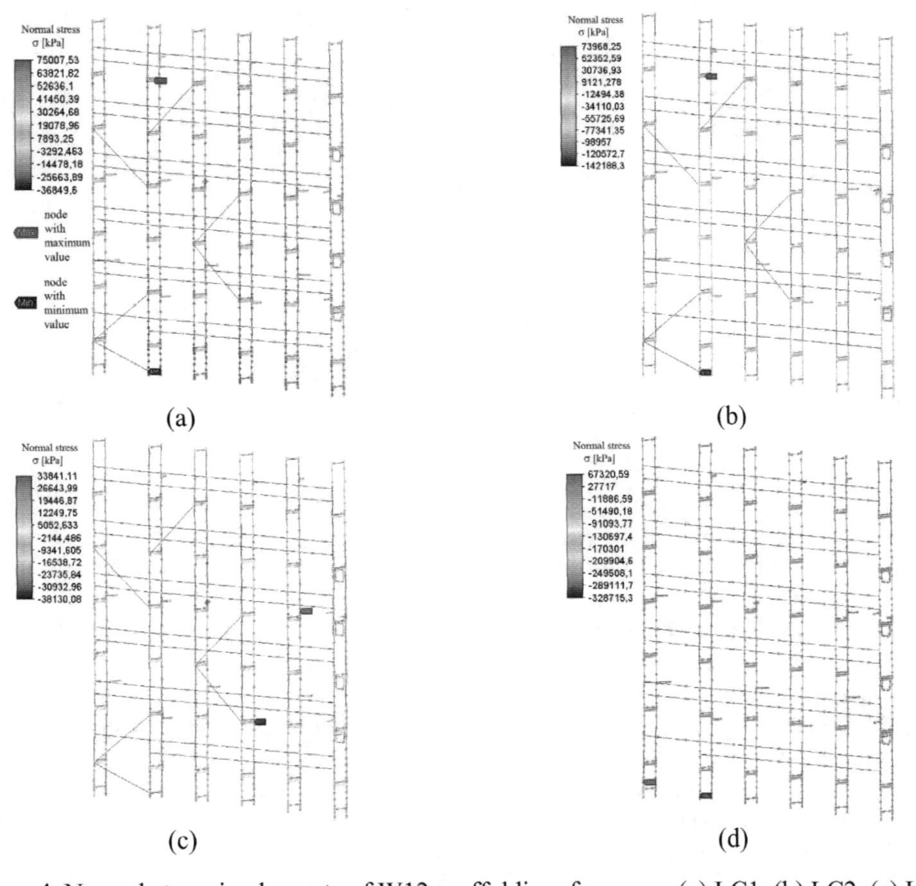

Figure 4. Normal stress in elements of W12 scaffolding, for cases: (a) LC1, (b) LC2, (c) LC3, (d) LC4

IOP Conf. Series: Materials Science and Engineering **951** (2020) 012023 doi:10.1088/1757-899X/951/1/012023

In 56 scaffoldings the load capacity was exceed for one or more component groups. It means that 47% of scaffoldings should not be approved for use. If the possibility of buckling was also taken into account, the number of defective scaffoldings increases to 59. Of the 49 scaffoldings with protective covering, 35 did not meet the $\kappa<1$ condition and the load capacity of their elements was exceeded in all load cases. The highest effort in standards and base jacks was observed for the LC3 and LC4 cases (self-weight and wind, respectively, perpendicular and parallel). Transoms were the elements with the least number of fail rate. The main reason for exceeding load capacity in base jacks is probably the wind load acting perpendicularly to the scaffolding. The other components of scaffolding had various applications, and on average 25% of them could fail regardless of the load case.

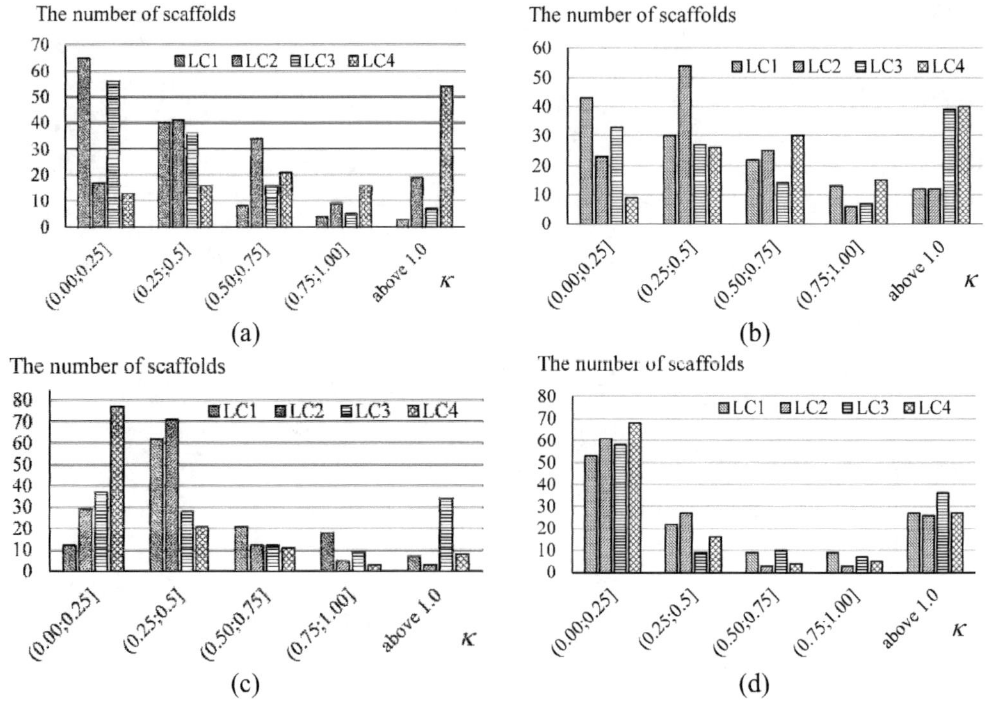

Figure 5 Histograms of the κ coefficient for: (a) base jacks, (b) standards, (c) transoms, (d) other elements

4. Conclusions

Static analyses of 120 real scaffoldings tested in years 2016-2018 were carried out. It was found that if the applied load was equal to the one recommended in standards, 14 scaffoldings would fail due to the loss of stability and in 56 scaffoldings exceeding capacity of one or more component group would be exceed. This means that 47% of tested scaffoldings should not be approved for use.

Scaffoldings, which had been found to be in a state of emergency under the standard loads, did not suffer from disasters, as the probability of such loads is very low. Scaffoldings are structures with a relatively short service life and the occurrence of maximum wind load is low but not impossible. Usually, in the event of atypical use of scaffoldings or significant loads resulting from the type of carried out work, the construction managers commission additional analyses before scaffolding is being used. However, not every scaffolding is placed on a construction site or there is no one who could assess the situation. Furthermore, the standard recommendations were meant to minimize the risk of failure at the design stage. The lack of scaffolding calculations in technical projects or skipping the scaffolding design completely poses a risk to the health and life of scaffolding users.

IOP Conf. Series: Materials Science and Engineering **951** (2020) 012023 doi:10.1088/1757-899X/951/1/012023

Acknowledgement
This paper has been prepared as a part of the project supported by the National Centre for Research and Development within the Applied Research Programme (agreement No. PBS3/A2/19/2015 "Modelling of Risk Assessment of Construction Disasters, Accidents, and Dangerous Incidents at Workplaces Using Scaffoldings").

References
[1] Błazik-Borowa E, Czarnocki K, Dąbrowski A, Hoła B, Misztela A, Obolewicz J, Walusiak-Skorupa J, Smolarz A, Szer J and Szóstak M 2015 *Safety in the Construction Industry*, (Lublin: Publisher of Lublin University of Technology) p 132 (in Polish)
[2] Żółtowski W 2000 *Inż. i Bud.* **7** 359–61 (in Polish)
[3] Błazik-Borowa E 2017 *Proc. 28th Int. Conf. on Structural Failures* Międzyzdroje pp 19–36. (in Polish)
[4] Regulation of the Minister of Infrastructure of 6.02.2003 on health and safety during construction work, Journal of Law 2003, No. 47, item 401, 2003. (in Polish)
[5] EN 12810-2. Façade scaffolds made of prefabricated components. Part 2: Particular methods of structural design, European Committee for Standardization, Brussels, Belgium, 2004
[6] EN 12811-1. Temporary works equipment. Part 1: Scaffolds –Performance requirements and general design, European Committee for Standardization, Brussels, Belgium, 2004
[7] Bęc J and Błazik-Borowa E 2019 *MATEC Web Conf.* **285** 1–8
[8] Błazik-Borowa E 2018 *Loads and Actions on Scaffolding as Engineering Structures* (Lublin: Publisher of Lublin University of Technology) p 346 (in Polish)
[9] Błazik-Borowa E 2018 *Proc. 8th Int. Sym. on Environmental Effects on Buildings and People: actions, influences, interactions, discomfort* Krakow-Tyniec pp 87-8
[10] Błazik-Borowa E, Bęc J, Robak A, Szulej J, Wielgos P and Szer I 2017 *Int. Saf. Heal. Peop. Constr. Con. Towards better Safety, Health, Wellbeing, and Life in Construction*, Cape Town pp 154–63
[11] Błazik-Borowa E, Borowa A, Kawecki B, Kotowicz M and Robak A 2019 *Eng. Struct.* **196** 109360
[12] Błazik-Borowa E, Czarnocki K, Hoła B and Szer J 2016 *Rusztowania* **1** 18–20 (in Polish)
[13] Bojar H, Silveira F, Rebelo M, Czarnocka E and Czarnocki K 2019 *Ann. Agr. Env. Med.* **26** 138–42
[14] Bucon R, Czarnigowska A Kmiecik P and Robak A 2019 *IOP Conf. Ser.: Mater. Sci. Eng.* **471** 112059
[15] Hoła B, Nowobilski T, Rudy J and Błazik-Borowa E 2017 *Tech.Tran.* **7** 31-9
[16] Jabłoński M, Szer I and Szer J 2018 *J. Civ. Eng. Manag.* **24** 437–43
[17] Lipecki T, Jamińska-Gadomska P, Bęc J and Błazik-Borowa E 2020 *Arch. Civ. Mech. Eng.* **20** 1–15
[18] Lipecki T, Jamińska-Gadomska P, Sumorek A, Błazik-Borowa E and Bęc J 2018 *J. Phys.: Conf. Ser.* **1101** 012019
[19] Pieńko M, Robak A, Błazik-Borowa E and Szer J 2018 *World Acad. Sci. Eng. Technol. Int. J. Civ. Environ. Eng.* **12** 72–7
[20] Robak A, Pieńko M, Błazik-Borowa E, Bęc J and Szer I 2019 *MATEC Web Conf.* **284** 1–14
[21] Cyniak P, Błazik-Borowa E, Szer J and Lipecki T 2018 *AIP Conf. Proceedings, Comput. Methods Mech.* **1922** 1–12
[22] Dolejs J and Ilcik J 2017 *Ernst Sohn Conf. Proc. EUROSTEEL* 2017 **1** 4698–703
[23] Misztela A 2012 *Int. J. Appl. Mech. Eng.* **17** 931–40
[24] EN1991-1-4. Eurocode 1: Actions on structures – Part 1-4: General actions – Wind actions, European Committee for Standardization, Brussels, Belgium, 2005
[25] Ilcik J, Arora V and Dolejs J 2016 *Eng. Struct.* **118** 334–43
[26] Hrischev L and Slavchev V 2018 *Annu. Univ. Archit. Civ. Eng. Geod.* **1** 29–44

Influence of wind loads on façade scaffolds covered with different types of nets

L Hrischev[1,*], V Slavchev[2], V Tepeliev[1], D Dinev[1] and R Angelova[3]

[1] University of Civil engineering, architecture and Geodesy, Faculty of Structural Engineering, 1 Hr. Smirnensky blvd., 1000 Sofia, Bulgaria
[2] University of Structural engineering and Architecture "L.Karavelov", Faculty of Construction, 175 Suhodolska str., 1000 Sofia, Bulgaria
[3] Technical University of Sofia, Department of Textiles, 8 Kliment Ochridski blvd., 1000 Sofia, Bulgaria

*E-mail: l.hrischev@abv.bg

Abstract. In the process of the structural design of the working and façade scaffolds, the self-weight of the elements, the service load and the wind load should be taken into account. The wind load on a covered façade scaffold is determined by many factors, including the wind velocity, the solidity ratio of the building's façade, the geometric characteristics of the scaffold' components and the type of cladding – netting or sheeting. The paper investigates the effects of wind loads on both types of façade scaffolding – with and without claddings. A computational finite element model was developed to study the specifics of the scaffoldings in the function of the nets permeability and the solidity ratio of the façade. Load combinations, according to the requirements in EN 12811-1, were considered to obtain the internal forces and stresses of the scaffold elements subject to wind actions. Based on the numerical analyses, it was found that the porosity and air permeability of the nets have a significant effect on the internal forces and stresses of the façade scaffolds elements, which also depend on the anchorage patterns and solidity ratio of the façade. In conclusion, recommendations are given regarding the application of specific structural schemes and anchoring patterns, depending on the permeability of the different façade scaffold nets.

1. Introduction

Working scaffolds are used primarily in the execution of various types of finishing works on buildings and facilities. Their purpose is to provide convenient platforms enabling work at higher levels while addressing fall arrest and protection against falling tools and objects. [1]. Specific requirements for scaffoldings are described in several public documents, guidelines and standards.

According to Regulation № 2 from 22.03.2004 (Bulgaria) [2], to execute construction/installation work at height using a scaffold, it should comply with the instructions from the manufacturer for installation, operation, allowable loads, dismantling and the requirements for health and safety during work. On the other hand, Regulation № 7 from 23.09.1999 (Bulgaria) [3] also stipulates that structural design for the scaffolding is to be carried out to evaluate its stability and bearing capacity should the manufacturer's documentation is missing or the intended structural configurations are not covered in the manufacturer's guidelines. The specifics of individual design solutions for scaffoldings, designed according to European standards, are discussed in many publications and monographs [4-7].

When the design of scaffolding is deemed necessary, the design should be under the requirements of EN 12811-1 [8]. According to the standard, in the structural design of working and façade scaffolds, the self-weight of the elements, the service load and the wind load should be considered. The wind load on a covered/clad façade scaffold is determined by many factors, including the wind velocity, the solidity ratio of the façade of the building, the geometric characteristics of the components of the scaffold and the type of cladding - netting or sheeting.

According to EN 12810-1 [8], the manufacturer must declare a standard set of system configurations between 24 m and 25,5 m in height for all the scaffold components as well as anchoring patterns/schemes [9]. The analysis of the standard sets of façade scaffolding configurations declared by different manufacturers shows that they provide configurations for unclad scaffoldings, scaffoldings clad with permeable nets and scaffoldings clad with impermeable sheeting, for both solid and partially open facades [10]. However, the different manufacturer configurations for scaffoldings with nets do not take into account the permeability of the possible netting options, nor the possible solidity ratios of the facades.

There are numerous papers in the literature related to wind loads on scaffoldings. An assessment of the wind action on a façade scaffolding, based on full-scale data measurements, was presented in [11]. The wind effect on clad (covered) façade scaffoldings was principally investigated, and one of the conclusions was that in general, the wind affects mainly the anchors, as this load-causes the most significant stresses in the stands (standards), located in the bottom frames and their base jacks. The results from wind tunnel experiments, based on a prototype of scaffolding covered with nonporous cladding showed that the wind pressure on the inner surfaces of scaffolding plays a crucial role in wind loads on a clad scaffolding [12]. Irtaza et al. [13] used wind tunnel test to investigate the permeability of two types of nets, called Type A and Type B, where the Type B net was made by double folding the Type A net. It was concluded that the wind forces, acting on scaffolds, transferred from the covering materials to the scaffold structure, depended mainly on the air permeability of the netting and the building. Beale [14] presented a detailed review and summary on the research of scaffold structures. Special attention was paid to the wind impact on clad façade scaffolds. The latest research on wind effect on temporary structures was analysed and discussed in detail in [15].

Some other studies also show that different types of nets, e.g. nets for covering façade scaffoldings, nets for agricultural facilities or other, exhibit different air permeability and therefore cause different wind pressures on the clad structures [16-18].

The study presented in this article aims to determine the influence of nets with different permeability accompanied by different façade solidity ratios on typical façade scaffolds configurations.

2. Materials and methods

2.1. Typical façade scaffolding configurations and load combinations

The research was done on three typical types of façade scaffolds configurations: Scheme 1, Scheme 2 and Scheme 3 (Figure 1). They resemble the configurations according to different manufacturers' schemes. Scheme 1 is recommended for unclad scaffoldings, Scheme 2 is recommended for scaffoldings clad with permeable safety nets, and Scheme 3 is recommended for scaffoldings clad with impermeable sheetings [10]. Computational FE models representing each of the three schemes were developed. Scaffold frames were modelled as frame elements with tubular cross-section ϕ48,3 with a thickness of 2,7 mm. The diagonal elements were modelled as frame elements with tubular cross-section ϕ42,2 and a thickness of 2,7 mm. The characteristic yield strength f_{yk} of the main scaffolds elements was taken as 235 N/mm^2. The scaffold platforms in the FE model were modelled as shell elements - one shell element per platform. The three scaffold schemes showing the monitored points and standards are presented in Figure 1. The denoted points and standards were chosen to represent the most loaded elements of the scaffold structure. It has to be mentioned that due to the differences in the anchoring patterns, the location of the most loaded anchor varied: the most loaded anchor in (a) is at different location compared to (b) and (c).

The calculations were performed for scaffold load class 3 according to the load combinations stated in [8], which comprise: a) Service condition, b) Out of service condition. The wind load was determined, considering the influence of the five different types of scaffolding nets, which characteristics are presented in 2.2.

Linear FE analysis was performed to determine the maximum internal forces in the most loaded elements of the scaffolding in the three schemes at different façade solidity ratios. It was considered only the wind load, perpendicular to the façade plane, acting on the clad surface of the scaffold as well as on its structural elements. The wind design was done for the territory of Sofia, Bulgaria, where the base value of the wind load is 0.43 kN/m^2 at a basic wind speed of 26.1 m/s [19].

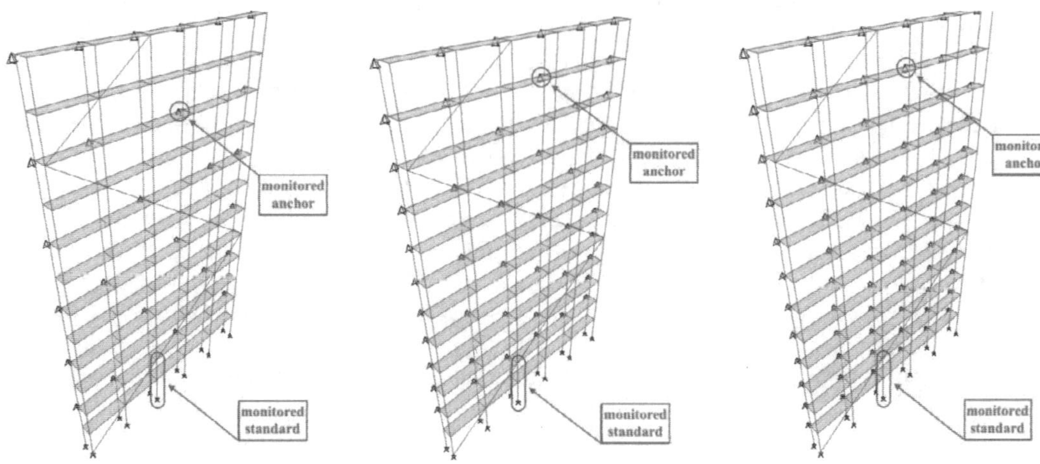

(a) (b) (c)

Figure 1. Static scheme of the scaffolds: (a) Scheme 1, (b) Scheme 2, (c) Scheme 3.

2.2. Characteristics of the façade scaffold nets.

Five types of scaffold nets were studied - type "A" (Figure 2a), type "E" (Figure 2b), type "F" (Figure 2c), type "G" (Figure 2d) and type "L" (Figure 2e). The net, type "F", is designed to cover a façade scaffolding to protect against falling objects. The nets type "A" and "G" are mainly used during an external façade insulation installation to prevent falling object hazards, but also to protect the thermal insulation layers from the weather. The net, type "E", is used for advertisements, which is one of the modern trends in the recent years, i.e. to utilise the available façade surfaces for commercial purposes [20]. The pressure on the nets is determined at different wind speeds using an open wind tunnel. The research methodology is presented in [21].

The porosity of the nets was determined by image analysis, using the method in [22]. The method, based on the processing of binary images, was developed for textile layers with the irregular shape of the pores. The method was successfully applied for the scaffold nets under consideration. The measured porosity (an average value of 10 single measurements) was determined as: 34.5% (net type "A"), 21.8% (net type "E"), 73.6% porosity (net type "F"), 40.2% (net type "G"), and 57.6% (net type "L").

(a) (b) (c) (d) (e)

Figure 2. Samples of nets: (a) net type "A", (b) net type "E", (c) net type F", (d) net type "G", (e) net type "L". Images are indicative and not to scale.

3. Results and discussion

The results from the linear static FEA for determining the maximum forces and stresses in the most loaded anchors and standards of the scaffolding in Scheme 1, Scheme 2 and Scheme 3 are presented below. The study was done without cladding and with a clad with the five net types. The solidity ratio of the façades was taken into account implicitly by the site coefficient c_s [8]. The site coefficients c_s used corresponds to solidity ratios of the façade with values from 0.1 to 1. Only the results of wind acting perpendicular to the façade - from the building towards the scaffolding (suction) are given.

3.1. Influence of the porosity of the nets on the pressure drop

Figure 3 presents a comparison between the porosity and pressure drop on the nets. The results show that there is a high correlation between nets' porosity and the pressure from wind action: the value of the coefficient of determination from the linear regression equation is very high (R^2 = 0,91). Every single thread is an obstacle for the passage of the fluid, thus influencing the pressure drop between the two sides of the net. The shape and the size of the net pores, as well as the thickness of the net, could also influence the pressure drop. Further research would be dedicated to the characteristics of the nets that could influence their behavior under different wind loads.

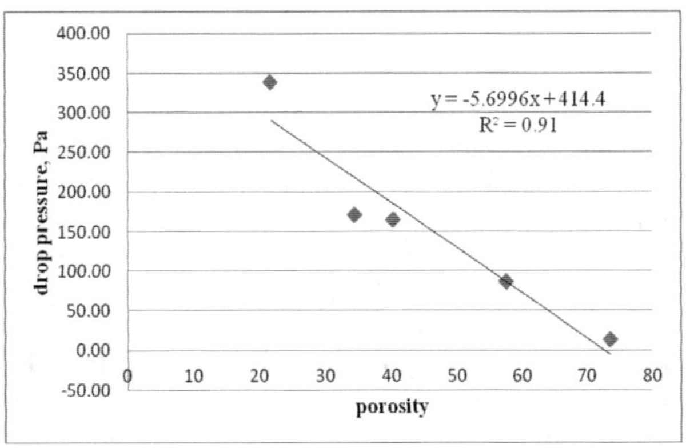

Figure 3. Relationship between porosity and pressure drop on nets at wind speed 26,1 m/s.

3.2. Internal forces in anchors at different anchoring schemes at nets with different porosity

The tensile forces in the anchors are presented in Figure 4 for the most loaded anchors for each scheme for load combination a) Service condition, and in Figure 5 for load combination b) Out of service condition. It has to be noted that the most loaded anchors are different in different schemes. For the current analysis, the maximum pull-out capacity of the anchors is assumed to be 4.5 kN.

The performed analysis shows that larger forces in the anchors occur in load combination b) Out of service condition, which is caused by the higher values of the wind load. As the porosity of the nets used decreases, the tension forces in the anchors' increase. It is observed that the increment is proportional to the porosity of the nets. Also, it is apparent that when the solidity ratio of the façade increases (resulting in decreasing of the site coefficient c_s) the tension forces in the anchors decrease.

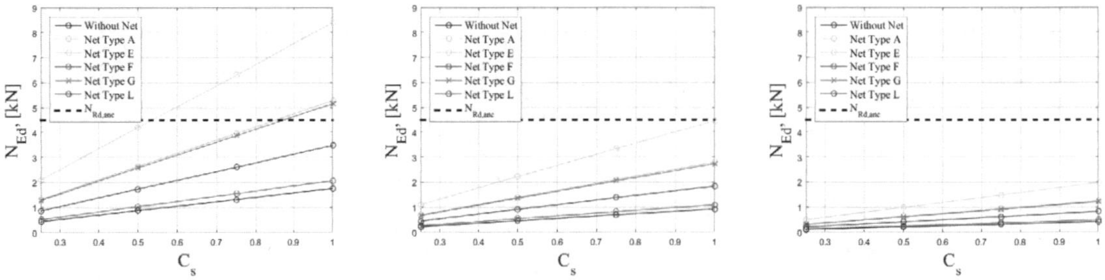

(a) (b) (c)

Figure 4. Axial tension forces in most loaded anchors in load combination a) Service condition: (a) Scheme 1, (b) Scheme 2, (c) Scheme 3.

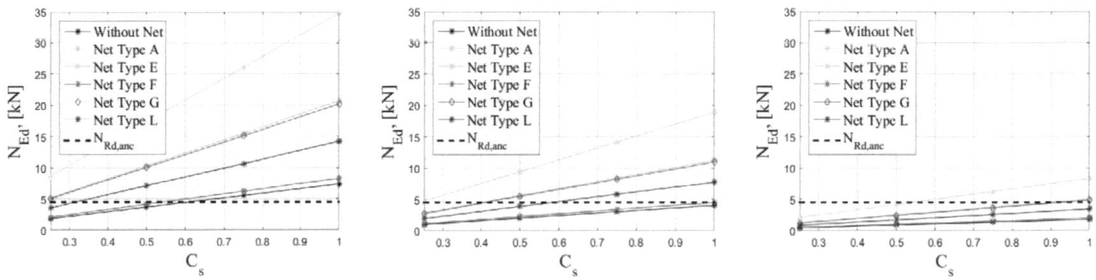

(a) (b) (c)

Figure 5. Axial tension forces in most loaded anchors in load combination b) Out of service condition: (a) Scheme 1, (b) Scheme 2, (c) Scheme 3.

In Scheme 1 at load combination b), Out of service condition, at values of c_s bigger than 0.6, at nets type "A", type "E" and type "G", in the most loaded anchors tension forces are observed with values greater than 4.5 kN. With the pull-out capacity of the anchors thus assumed, a relevant reduction of the spacing between the anchors and/or application of another anchoring scheme is necessary. In Scheme 1 clad with the net with the highest porosity (net type "F"), at c_s with values up to 0,55, it was observed that lesser internal forces than the assumed allowable bearing capacity of the anchors occurred. This implies that at high values of the solidity ratio of the facades (meaning low values of the coefficient c_s), Scheme 1 can be applied only in case of using very low porosity nets. It should also be noted that even in the case of unclad scaffolding and site coefficient c_s values above 0,55, the bearing capacity of the anchors is exceeded and usage of additional anchors or anchors with higher pull-out resistance is required.

In Scheme 2, the use of a high porosity mesh does not lead to exceedance of the bearing capacity of the anchors. However, when the porosity of the nets decreases, and the coefficient c_s increases, the internal forces in the most loaded anchors have higher values than the assumed values for their bearing capacity.

In Scheme 3, the use of nets with smaller porosity and different site coefficient c_s values does not lead to exceeded pull-out resistance of the anchors. Only in the case of the net with the lowest porosity (net type "E") and at values of the coefficient c_s above 0,55, the anchors fail in tension as the forces are about twice the resistance (4,5 kN). In these cases, it is necessary either to use anchors with greater pull-out capacity or to change the structural arrangement of the scaffolding or to change its anchoring pattern. Combinations of the listed measures are also possible, which is considered as a good engineering practice.

The use of different nets with different porosity for claddings leads to an increase in the forces in the anchors compared to the forces in an unclad scaffolding, as follows: in Scheme 1 – an increase from 12,5 % (for net type "F") up to 373,2 % (for net type "E"); in Scheme 2 – an increase from 12,4 % (for net type "F") up to 376,7 % (for net type "E") and in Scheme 1 – an increase from 12,4 % (for net type "F") up to 373,7 % (for net type "E). In general, it can be concluded that several factors have a significant adverse impact on the anchor forces induced by wind loads: namely the anchoring scheme, the solidity ratio of the façade, the net porosity and the air permeability related to it.

3.3 Stresses in the vertical standards at the different schemes clad with nets with different porosity
Figures 6 and 7 show the compressive stress range in the most loaded standards denoted in Figure 1. Load combination a), Service condition, proves to lead to the highest values of compression stresses in the vertical standards. When determining those stresses in the standards, it turned out that in Scheme 1 and Scheme 2, not only the normal forces but the bending moments in the plane perpendicular to the façade increased as well.

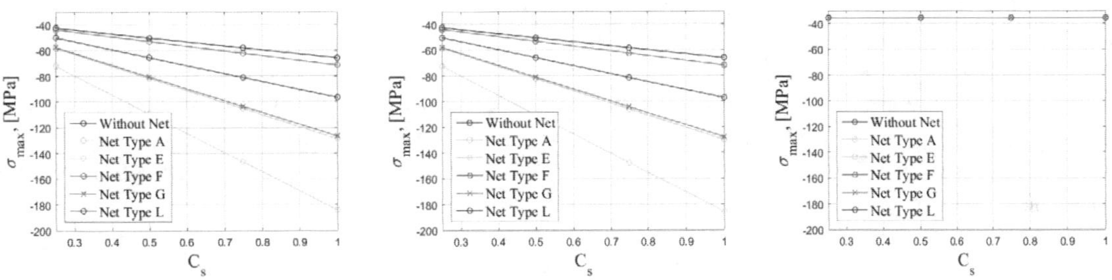

(a) (b) (c)

Figure 6. Maximum normal stress (compression) in most loaded standards in load combination a) Service condition: (a) Scheme 1, (b) Scheme 2, (c) Scheme 3.

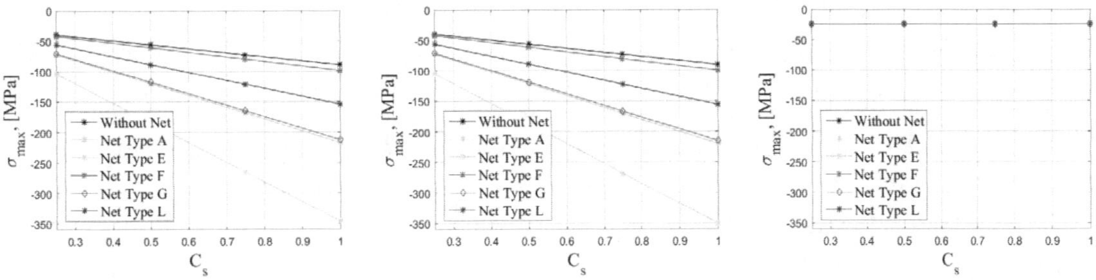

(a) (b) (c)

Figure 7. Maximum normal stress (compression) in most loaded standards in load combination b) Out of service condition: (a) Scheme 1, (b) Scheme 2, (c) Scheme 3.

In Scheme 1 and Scheme 2, it is also evident that when the porosity of the net decreases, the stresses in the vertical scaffold standards increase, but the values remain the same in both schemes. Normal stresses rise significantly at nets with low porosity and at high values of the coefficient c_s.

In Scheme 1 and Scheme 2, the use of nets with lower porosity (e.g. nets type “F” and type “L”) results in stresses, which are lower than the admissible ones for the chosen steel grade. However, when using nets with low porosity and lower values of the solidity ratio of the facades, the observed stresses in the scaffolding standards could happen to be higher than the admissible ones. This can be easily noticed at the nets with low porosity, e.g. nets similar to type “E”. It is observed that the smaller the porosity of the nets is, the greater the increase in stresses becomes. The use of different nets with lower porosity for scaffold cladding leads to a stress increase in the standards compared to those in the unclad scaffoldings, as follows: in Scheme 1 – the increase varies from 6 % (for net type “F”, and C_s = 0,25) up to 288 % (for net type “E”, and C_s = 1,0); in Scheme 2 – the increase varies from 6,1 % (for net type “F”, and C_s = 0,25) up to 288,8 % (for net type “E”, and C_s = 1,0).

In Scheme 3, the permeability of the nets does not lead to a change in the forces in the vertical load-bearing members of the scaffolding (standards), regardless of the values of the site coefficient c_s. The main reason for this is that all anchoring pipes are connected to every scaffolding node and thus in the vertical standards bending moments do not occur, which normally would increase the normal stresses in them.

It can be concluded that the porosity of the façade scaffolding nets affects both the forces in the anchors and the stresses in the scaffolding standards.

4. Conclusions

This paper presents a study of the influence of nets with different porosity on the elements of three types of schemes for façade scaffoldings with a height of 24 m. Scheme 1 is recommended for unclad scaffoldings, Scheme 2 is recommended for scaffoldings clad with permeable safety nets, and Scheme 3 is recommended for scaffoldings clad with impermeable sheetings. The results of the static linear elastic finite element analyses were used for determining the internal forces and stresses of the most loaded elements of the scaffolding – the anchors and the vertical standards.

For the façade scaffold nets used in the present study, the relationship between their porosity and the imposed wind load was established. The analysis of the different types of façade scaffold nets shows that there is a high correlation between the porosity of the nets and the imposed wind loads associated with them. However, further research is needed to take into account all the nets' characteristics that affect their air permeability and thereof the wind loads on the scaffold.

It was found that the use of nets with lower porosity leads to an increase in the anchors' forces regardless of the adopted anchoring scheme. Nets with lower porosity also lead to an increase in stresses in the scaffold standards in Scheme 1 and Scheme 2. The internal forces in the scaffolding elements are particularly large when using nets with low porosity and low values of the solidity ratios of the façade.

In general, it can be concluded that several factors have a significant impact on the internal forces and stresses in the scaffold elements under wind load conditions. Those comprise the anchoring pattern scheme, the solidity ratio of the façade, the net porosity and the air permeability related to it.

As a result of the conclusions and findings of the study, the authors recommend not to apply the manufacturers‘ schemes for anchoring directly, but to prepare structural design project for the implementation of the façade scaffolding.

References

[1] Kirov N 2006 *Formwork and Scaffolds* (Sofia: Daniel SG) p 30.
[2] Regulation No. 2 of 22.03.2004 for minimum requirements on health and safety during construction and mounting work, Official Journal 2004, No. 37, item 87, 2004 (in Bulgarian).
[3] Regulation No. 7 of 23.09.1999 for minimum requirements on health and safety on working places and with use of working equipment, Official Journal 1999, No. 89, item 215, 1989 (in

Bulgarian).
[4] Błazik-Borowa E and Pieńko M 2017 *Scaffoldings* (Lublin: Lublin University of Technology) pp 73-108.
[5] Robak A, Pieńko M and Błazik-Borowa E 2015 Civ. Env. Eng. Rep. **18** 3 115-120.
[6] Kuzhin M and Kuzhin B 2016 MATEC Web of Conferences **86** 07007.
[7] Hrischev L and Slavchev V 2018 *Annu. Univ. Archit. Civ. Eng. Geod.* **51** 10 29–44.
[8] EN 12811-1 Temporary works equipment. Part 1: Scaffolds - Performance requirements and general design.
[9] EN 12810-1 Façade scaffolds made of prefabricated components. Part 1: Product specifications.
[10] Hrischev L, Dinev D and Tepeliev V 2018 *Proc. 10th Jub. Int. Conf. Civ. Eng. Design and Constr.* Varna pp 136-145.
[11] Lipecki T, Jamińska-Gadomska P, Bec J and Błazik-Borowa E 2020 *Arch. Civ. Mech. Eng.* **20:27** 1–15.
[12] Wang F, Tamura Y and Yoshida A. 2013 *Wind Eng. Ind. Aerodyn.* **120** 37–50.
[13] Irtaza H, Beale R, Godley M and Jameel A 2014 *IOSR J. Mech. Civ. Eng.* **4** 54-62.
[14] Beale R. 2014 J Constr Steel Res **98** 188–200.
[15] Beale R and Andre J 2017 Design Solutions and Innovations in Temporary Structures (Hershey: IGI Global) 52-124.
[16] Agarwal A and Irtaza H 2018 *J. Inst. Eng. India Ser. A* **99** 2 311–320.
[17] Castellano S, Starace G, De Pascalis L, Lippolis M and Mugnozza G 2016 *Biosyst. Eng.* **150** 131-141.
[18] Briassoulis D, Mistriotis A and Giannoulis A 2010 J. *Wind Eng. Ind. Aerodyn.* **98** 919–928.
[19] EN1991-1-4/NA Eurocode 1: Actions on structures - Part 1-4: General actions-Wind actions/ National annex.
[20] Ivanova E and Boyadzhieva D 2020 Proc. of Int. Scient. Conf. Tech. Techn. Educ. Saf. Borovets **1** 38-41.
[21] Hrischev L, Markov D, Tepeliev V and Dinev D 2019 *Annu. Univ. Archit. Civ. Eng. Geod.* **52** 3 929–943.
[22] Angelova R 2020 *C. R. Acad. Bulg. Sci.* **73** 1 96-102.

IOP Conf. Series: Materials Science and Engineering **951** (2020) 012025 doi:10.1088/1757-899X/951/1/012025

Analysis of accidents with tower cranes on construction sites and recommendations for their prevention

K Radlov* and G Ivanov

Department „Construction Technology and Mechanization“,
Faculty of Structural Engineering,
University of Architecture, Civil Engineering and Geodesy; 1046 Sofia, Bulgaria

*E-mail: kradlov@abv.bg

Abstract. Analyzing and investigating accidents with tower cranes on construction sites is a particularly important preventative measure to ensure their accident-free work and reduce the number of occupational accidents at the construction site. The urgency of this problem is increasing very much in view of the relatively long operational life of the tower cranes on construction sites. This article presents various types of results from researches related to the analysis of accidents with tower cranes on construction sites worldwide. In addition, an approach for analysis of the absolute probability of a major tower crane accident caused by failures of mechanical components is considered, as well as analysis of possible sequences of failures of mechanical components leading to a tower crane accident, which are analyzed based on Fault Tree Analysis (FTA). An adequate statistical approach is proposed to establish the most significant organizational and technical causes, mainly related to the organization of the construction site and the activities for maintenance of the machine park, which cause accidents with tower cranes. An approach is proposed to calculate the conditional probability (weighting factor) of a specific organizational and technical cause to lead to a serious accident with the tower crane using Bayesian formulas, as well as an approach to determine the total assessment for each organizational and technical cause regarding its potential for causing accidents. Finally, proposals and recommendations are made related to the identification of key technical and organizational measures, the adequate and timely implementation of which could lead to a reduction in the number of occupational accidents with tower cranes on construction sites.

1. Introduction

In construction practice, incidents related to accidents during the operation of tower cranes are not uncommon. Such events not only have the potential to cause great damage to private property and forcible interruption of construction activities, but can also lead to accidents involving injuries to people. Much effort is being made by scientists, researchers and organizations worldwide, which are aimed at conducting a detailed analysis of accidents with tower cranes on construction sites ([1], [2], [3], [4] and others). The report [2] examines accidents with different types of construction cranes, which analyzes their potential to cause major accidents involving serious injuries to people. The results of this research testify to the serious consequences of accidents with construction cranes. The study [3] analyzes the calendar age of tower cranes in which accidents occurred in the period from 1983 to 2014, emphasizing that the average age of tower cranes involved in accidents is 16.9 years, most accidents occurred with tower cranes aged about 14 years, and a large proportion (21.6%) of accidents occurred with tower cranes over 25 years old. The main goal of all researches is to identify adequate recommendations that will lead to timely prevention and/or reduction of the number of accidents and occupational accidents with tower cranes.

IOP Conf. Series: Materials Science and Engineering **951** (2020) 012025 doi:10.1088/1757-899X/951/1/012025

2. Review of the results of previously performed statistical analyzes regarding the actual causes of accidents with tower cranes worldwide

The Health and Safety Laboratory (HSL) has carried out an analysis of worldwide tower crane accidents from 1989 to 2009 [4], in an attempt to establish the causes of each accident. The research report [4] includes only accidents that occurred in the most widely used in construction top-slewing tower cranes, which are from different manufacturers (Liebherr, Potain, Wolff, Favelle Favco, Comedil, Jaso Peiner, Kroll, Raimondi, Kodiak and others). A total of 85 major accidents with these cranes were analyzed (leading to: fall of the entire crane, collapse of the boom, etc.), which were conditionally divided into 7 different categories, depending on the causes for their occurrence:

- accidents occurred during installation, dismantling or extension of the crane tower/boom - 29 accidents (34% of the total number). These are accidents that occur during lifting, turning or assembly/disassembly of large and heavy components of the tower crane structure. The technologies and procedures for assembly/disassembly and extension of the boom/tower of the tower crane in most cases require the implementation of a complex sequence of steps that are associated with the performance of work at height. These complex activities are usually associated with a long duration of operations and require a large number of staff, which creates additional preconditions for human error, leading to severe consequences;
- accidents occurred as a result of external natural influences - 15 accidents (18% of the total). Of these accidents, 2 were caused by earthquakes, and the remaining 13 were due to strong winds. Some of the accidents caused by strong winds are related to the omission to provide the free rotation of the crane boom during the wind load, as well as to the incorrect position of the boom (angle of inclination) in the tower cranes with luffing boom;
- accidents caused by improper use of the crane (mistake made by the operator and/or riggers) - 6 accidents (7% of the total). They are mainly related to: lifting stuck loads, overloading the crane, impact/interaction between two cranes on the same construction site, etc. Possible prerequisites that create conditions for the occurrence of such accidents are: insufficient qualification of the crane operator and/or the riggers, poor quality of the construction supervision on the site, etc.;
- accidents occurred as a result of compromising mechanical/structural components of the tower crane - 4 accidents (5% of the total number). They are mainly related to: broken bolts of the support-rotating device; broken weld in bearing structural component; damage/destruction of an element of the kinematic chain of a lifting mechanism and/or a jib crane mechanism; failure of the brake system in one of the mechanisms of the crane, etc. Possible prerequisites that create conditions for the occurrence of such accidents are mainly the following: omission to perform the mandatory periodic checks/inspections of the mechanical components of the crane; poor quality checks/inspections of mechanical components; poor quality repair works on mechanical components of the crane, etc.;
- accidents occurred as a result of compromising the foundation/supporting structure under the crane - 2 accidents (2% of the total number). They are caused mainly by non-compliance with the manufacturer's instructions during the construction of the supporting structure and the foundation under the tower crane (the so-called "original design");
- accidents caused by compromising electrical components or the crane control and management system - 1 number of accidents (1% of the total number). They are mainly related to a fault in the inverter/controller (PLC) settings of the crane drive system, which in some cases could lead to uncontrolled discharge of the load. The risk of such damage and accidents is greater with modern cranes, which are equipped with more complex control systems. This largely depends on the competencies of the electrician who performs the settings of the control system during the installation of the crane or replacement of spare parts for the electrical equipment of the crane during operation;

- accidents for which the cause is still unidentified (unknown) - 28 accidents (33% of the total number). These are accidents whose causes cannot be defined with sufficient reliability because the available information about them is not sufficiently accurate and/or sufficiently detailed.

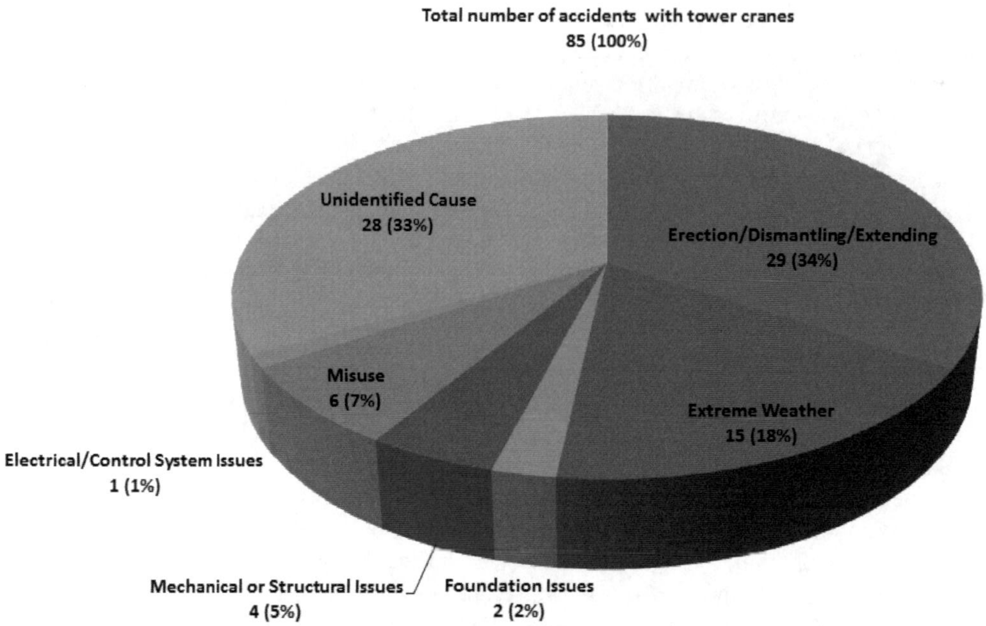

Figure 1. Results obtained for the causes of major accidents with tower cranes for the period 1989-2009 according to the research report [4].

3. Analysis of the absolute probability of a major accident with the tower crane due to failure of mechanical components

Based on the results of research [5] it can be concluded that among the various possible mechanical damages to the tower crane, those that lead to compromise/fall of the tower crane boom are those that cause the most severe accidents on the construction site. Analysis of the absolute probability of severe accidents with the tower crane caused by failure of mechanical components is considered in the study [6]. In this study, sequences of failures of mechanical components of the tower crane are analyzed using the Fault Tree Analysis (FTA) method. Accidental failures of the individual mechanical components in the machine are distributed mainly by exponential law [7], [8]. The calculation of the risk of an accident with the tower crane $Q(T)$ is performed according to the dependence:

$$Q(T) = 1 - P(T)\,, \tag{1}$$

where:
T is the period of operating time of the tower crane, for which the probability of accident-free operation is assessed;
$P(T)$ is the probability of accident-free operation of the tower crane for a period of operating time T.

$$P(T) = \exp[-\int_0^T E(t).dt]\,, \tag{2}$$

where $E(t)$ is the intensity of failures, which is equal to the probability that the accident will occur after a period of accident-free operation t.

In assessing the absolute probability of a tower crane accident caused by failure of mechanical components, the correct definition of the causal links between individual random events (failures of individual components) plays an important role, whose individual probability of occurrence is of different frequency, but they all contribute to and lead to one final event - the occurrence of a major accident. A "Fault Tree" is built for each individual operating state of the crane. Figure 2 shows the "Fault Tree" for operating state z_2 – tower crane in operating state with a load suspended on the hook, in which the tower crane moves and the boom rotates simultaneously [6].

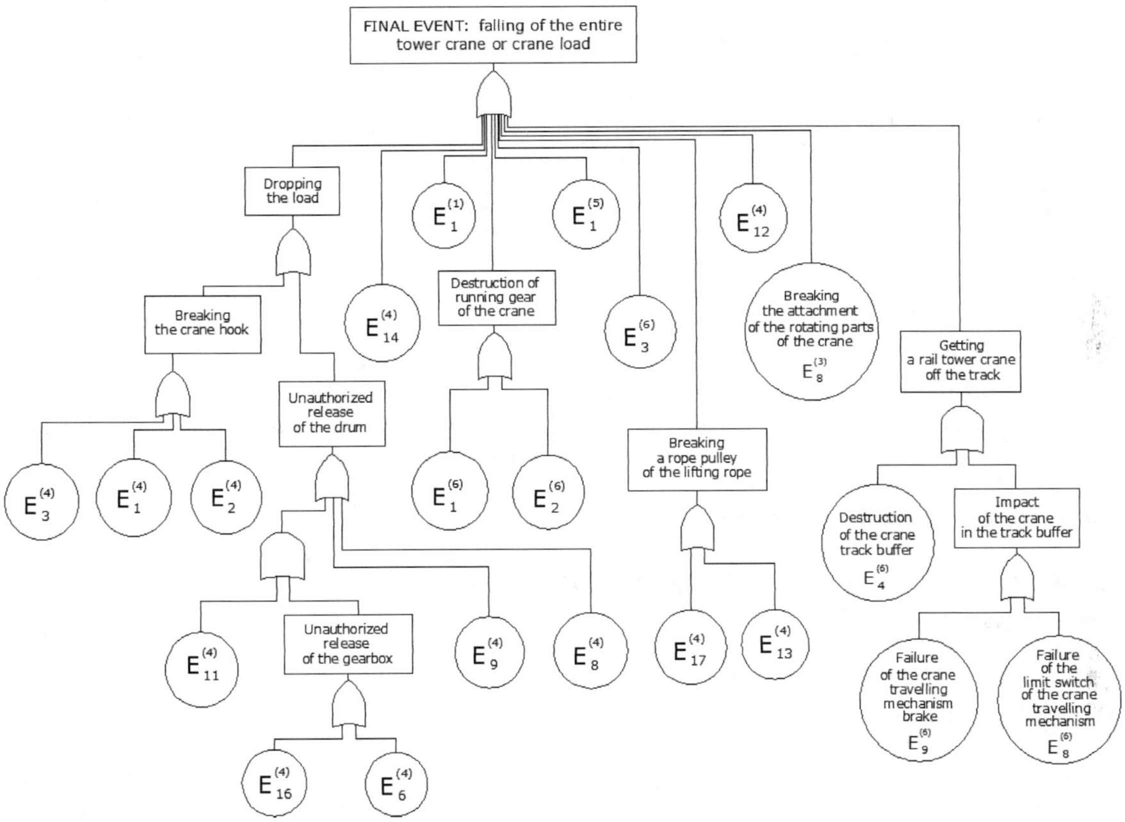

Figure 2. Tower crane "Fault Tree" in operating state z_2 [6].

Logical operators (AND and OR) and event symbols are used to construct the "Fault Tree". Assuming that the selected initial random events (failures of individual components) are independent of each other, then the probability of occurrence of the final event due to several initial random events that are associated with a logical operator "AND" is determined by the following formula:

$$Q(T) = Q(E_1 \cap E_2 \cap E_3 \cap ....E_n) = Q(E_1,T).Q(E_2,T).Q(E_3,T).......Q(E_n,T)\,, \tag{3}$$

where n is the number of initial random events that participate in the respective "Fault Tree".

The probability of occurrence of the final event as a result of several initial random events that are related to a logical operator "OR" is determined by the following formula:

$$Q(T) = Q(E_1 \cup E_2 \cup E_3 \cup ....E_n) = 1 - Q(\overline{E}_1).Q(\overline{E}_2).Q(\overline{E}_3).....Q(\overline{E}_n) =$$

$$= 1 - [1 - Q(E_1)].[1 - Q(E_2)].....[1 - Q(E_n)]\,. \tag{4}$$

From the diagram of figure 2 can be calculated the absolute probability of a major accident with the tower crane (falling of the entire crane or dropping the load from the hook) due to failure of a mechanical component in operating state z_2 (tower crane in operating state with a load suspended on the hook, during which the tower crane moves and the boom rotates simultaneously) for operating time of the crane in this operating state with a duration – $T = 8\ hours$ [6]:

$$Q_{Z2}(T) = 1 - \exp\left(-E_{12}^{(4)}.T\right)*\exp\left(-E_{13}^{(4)}.T\right)*\exp\left(-E_{17}^{(4)}.T\right)*\exp\left(-E_{3}^{(6)}.T\right)*\exp\left(-E_{1}^{(5)}.T\right)*$$

$$*\exp\left(-E_{2}^{(6)}.T\right)*\exp\left(-E_{1}^{(6)}.T\right)*\exp\left(-E_{1}^{(1)}.T\right)*\exp\left(-E_{14}^{(4)}.T\right)*\exp\left(-E_{8}^{(4)}.T\right)*\exp\left(-E_{9}^{(4)}.T\right)*$$

$$\exp\left(-E_{2}^{(4)}.T\right)*\exp\left(-E_{1}^{(4)}.T\right)*\exp\left(-E_{3}^{(4)}.T\right)*\left[1-\left[1-\exp\left(-E_{11}^{(4)}.T\right)\right]\left[1-\exp\left(-E_{16}^{(4)}.T\right)*\exp\left(-E_{6}^{(4)}.T\right)\right]\right]*$$

$$*\exp\left(-E_{8}^{(3)}.T\right)*\left[1-\left[1-\exp\left(-E_{4}^{(6)}.T\right)\right]\left[1-\exp\left(-E_{9}^{(6)}.T\right)*\exp\left(-E_{8}^{(6)}.T\right)\right]\right].$$

(5)

$$Q_{Z2}(T) = 1 - 0{,}91751 = 0{,}08249. \tag{6}$$

In a similar way, "Fault Trees" can be constructed for other different possible operating states of the tower crane, and based on them to calculate the absolute probability of a major accident due to failure of a mechanical component in the respective operating state. The main purpose of these analyzes should be to look for possible weaknesses in the structure, components and/or kinematic chains of the mechanical drives of the tower cranes.

4. Statistical approach for calculating a total assessment for each organizational and technical cause in relation to its potential for causing a tower crane accident

When assessing the technical and organizational causes of accidents with tower cranes, there is a conditional division of accidents with tower cranes into two main groups: major accidents with tower cranes (risk of serious damage to private property and injury to a large number of people) and general accidents with tower crane (without danger of causing serious damage and injury to a large number of people).

4.1. Assessment of the conditional probability of a specific organizational and technical cause to lead to a major accident with the tower crane

The assessment of the probability (risk), the assumption of a specific technical or organizational cause (so-called "causative agents") to lead to a major accident with the tower crane is the most important information related to the further identification of preventive measures and recommendations to reduce the risks of work with the tower crane and increase the safety of the construction site. The assessment of each technical or organizational cause related to the crane and its equipment according to this criterion is determined by the following dependence:

$$E_{Si} = W_{Si} * P_{Si}, \tag{7}$$

where:

W_S is a weighting factor indicating the importance of the criteria being assessed. Due to the great importance of preventing a major accident with the tower crane (with the danger of causing serious damage and injury to a large number of people) for this criterion is conditionally assumed weighting factor: $W_S = 6$;

P_{Si} is a coefficient of the assessment of the conditional probability of a specific technical or organizational cause (with number i) to lead to the occurrence of a major accident with the tower crane. This assessment varies in range $P_{Si} = (0 \div 1)$ and is determined on the basis of accumulated real statistical data (figure 1) and by using the formulas of the law of total probability and the Bayes' rule [9].

$$P_{Si} = P(S / R_i) = \frac{P(R_i / S) * P(S)}{P(R_i)} = (0 \div 1), \quad (8)$$

where:
S is used to indicate the event of a major tower crane accident (with the danger of causing serious damage and injury to a large number of people);
R_i – the technical or organizational cause with a number i;
m – the total number of technical and organizational causes that are considered to be the "causative agents" of tower crane accidents;
$P(R_i)$ – the probability that an accident that has already occurred with the tower crane (regardless of the size of its consequences) is due to the specific technical or organizational cause R_i;
$P(S) = \sum_{i=1}^{m} \left(P(S / R_i) * P(R_i)\right)$ - represents the full probability of a major accident with the tower crane (with the danger of causing serious damage and injury to a large number of people), provided that an event "accident" has already occurred;

$P(R_i / S)$ – conditional probability, indicating the probability that in the event of a major accident with the tower crane (with a risk of serious damage and injury to a large number of people) this is due to a specific technical or organizational cause R_i. This probability can be determined using the statistical data of figure 1. For example:

- the conditional probability of a major accident with the tower crane due to errors during assembly, disassembly or extension of the tower/boom of the crane is: $P(R_1 / S) = 29/85 = 0{,}34$;
- the conditional probability of a major accident with the tower crane due to external natural influences is: $P(R_2 / S) = 15/85 = 0{,}18$;
- the conditional probability of a major accident with the tower crane due to improper use of the crane (crane operator or rigger error) is: $P(R_3 / S) = 6/85 = 0{,}07$;
- the conditional probability of a major accident with the tower crane due to compromise of mechanical/structural components of the tower crane is: $P(R_4 / S) = 4/85 = 0{,}05$;
- the conditional probability of a major accident with the tower crane due to compromise of the foundation/supporting structure under the crane is: $P(R_5 / S) = 2/85 = 0{,}02$;
- the conditional probability of a major accident with the tower crane due to compromise of electrical components or the crane control system is: $P(R_6 / S) = 1/85 = 0{,}01$;
- the conditional probability that the main cause of a major tower crane accident will eventually remain unidentified with sufficient accuracy is: $P(R_7 / S) = 28/85 = 0{,}33$.

4.2. Assessment of the conditional probability of a specific organizational and technical cause to lead to a general accident with the tower crane

Assessing the probability/degree of danger/risk, if a specific technical or organizational cause/error/omission (with number i) is allowed, this will lead to a general accident with the tower crane (without danger of major damage and injury to a large number of people) is performed according to the following dependence:

$$E_{Li} = W_L * P_{Li}, \tag{9}$$

where:

W_L is a weighting factor indicating the importance of the criterion being assessed. Due to the lesser importance of the current criterion "general accident" (compared to "major accident"), for this criterion is conditionally assumed weighting factor: $W_L = 3$;

P_{Li} is a coefficient of assessment of the conditional probability of a specific technical or organizational cause (with number i), to lead to a general accident with the tower crane (without danger of major damage and injury to a large number of people). This assessment varies in the range $P_{Li} = (0 \div 1)$ and is determined in a similar way to that described in item 4.1 above.

4.3. Determination of a total assessment for each organizational and technical cause in relation to its potential for causing accidents

For each individual technical and organizational cause with number i, a total assessment E_{0i} is calculated in relation to its total potential (risk) for causing any type of accidents (major and general accidents):

$$E_{0i} = E_{Si} + E_{Li}. \tag{10}$$

After the calculation of the total assessment E_{0i} in question for each of the individual organizational and technical causes ("causative agents") for the occurrence of a tower crane accident, the so-called "ranking" of the causes according to the total assessments obtained. Those organizational and technical causes ("causative agents") that have received the highest total assessments are precisely those that are key in terms of ensuring safety when working with tower cranes on the construction site.

5. Conclusion and recommendations

A statistical approach is proposed for calculating the conditional probability (weighting factor) of a specific organizational and technical cause for a serious tower crane accident, as well as an approach for determining a total assessment for each organizational and technical cause in terms of its potential for causing a tower crane accident. The most important recommendations related to ensuring the safe operation of tower cranes on construction sites are the following:

- on the basis of the calculated conditional probability on the basis of real statistical data and the obtained total assessment for each separate organizational and technical cause with regard to its potential for causing accidents, the most important organizational and technical causes ("causative agents") for accidents with tower cranes, which are key from a safety point of view, should be identified. It is to them that the main part of the subsequent organizational and technical measures for increasing the safety of work with tower cranes on the construction site should be directed as a priority after that. Of course, eliminating all possible organizational and technical causes of tower crane accidents is the most ideal option that manufacturers, service personnel and construction managers should strive for, but in practice this is very difficult to do due to the extremely large number of influencing factors;

- it is recommended to perform detailed analyzes of the absolute probability of a major accident with the tower crane due to failures of mechanical components using the Fault Tree Analysis (FTA) for different possible operating states of the tower crane. Their main goal should be to look for possible weaknesses in the construction, components and kinematic chains of the mechanical drives of the tower cranes, to which the subsequent measures to improve the system for maintenance and repair of the tower crane should be directed as a priority;
- it is recommended to work continuously to increase the requirements and control over the qualification of the crane operator and riggers of the tower crane, as well as to increase the requirements and control regarding the quality of maintenance and repair of the tower cranes and periodic inspections/checks on the condition of the tower cranes;
- with regard to the age of the tower cranes used on construction sites, the following is recommended: 1.It is recommended that tower cranes over the age of 10 years undergo a thorough inspection before each installation; 2.Tower cranes over 20 years of age are recommended to be operated only after receiving an official confirmation from the manufacturer, by which he explicitly guarantees/confirms that their safe operation is possible for a longer period of time; 3.Tower cranes over 30 years of age are recommended to be subject to expert assessment of the technical condition or to take the necessary measures for their decommissioning. In Bulgaria, suspension of operation of cranes can in principle be done by written order of the inspectors from the General Directorate "Inspection for State Technical Supervision", when it is considered that the facility is not suitable for further safe operation, regardless of the age of the crane [10].

References

[1] Arslan M H and Kaltakci M Y 2008 Analysis of a Tower Crane Accident *The Open Construction and Building Technology Journal* Vol 2 pp 287-293
[2] MacCann M 2010 Understanding Crane Accident Failures: A report on causes of deaths in crane-related accidents *Crane & Rigging Conference* (Houston, Texas)
[3] Wiethorn J D, Gardiner M R, Bond A E, Cox E P and King R A 2015 Tower crane life expectancy. An examination of recent trends to establish age limits *Technical paper* (Haag Engineering Co.)
[4] Health and Safety Laboratory (HSL) 2010 Tower crane incidents worldwide *Research Report RR820* Health and Safety Executive (HSE)
[5] Ali M K and Mohamad M I 2016 Crane Failure and Accident in Construction *Construction Management, Geotechnics and Transportation* Vol 2
[6] Radlov K 2012 Analysis of the risk of accident of a mobile tower crane with a non-rotating tower caused by technical malfunctions *Proceedings of the VII International Scientific Conference - Civil Engineering Design and Construction* (Varna, Bulgaria) (in Bulgarian)
[7] Zavadsky Yu V 1976 Statistical processing of the experiment: Textbook (Moscow: Vysshaya Shkola) (in Russian)
[8] Dimitrov K and Danchev D 1994 Reliability of construction machinery and systems (Sofia: Tehnika) (in Bulgarian)
[9] Kalinov K 2002 Probability Theory and Statistics (Sofia: NBU) (in Bulgarian)
[10] State Agency for Metrological and Technical Surveillance 2010 Ordinance on the Safe Operation and Technical Supervision of Lifting Equipment (Sofia, Bulgaria) (in Bulgarian)

IOP Conf. Series: Materials Science and Engineering **951** (2020) 012026 doi:10.1088/1757-899X/951/1/012026

Smartphone application for structural health monitoring

G Stoilov, D Pashkouleva* and V Kavardzhikov

Institute of Mechanics, Bulgarian Academy of Sciences, Acad. G. Bonchev Str., Bl. 4, 1113 Sofia, Bulgaria

*E-mail: dessip@imbm.bas.bg

Abstract. In recent years, smartphone technologies have developed rapidly. They are increasingly taking on the role of laptops and compact personal computers, as well as digital cameras, through which measurements can be taken in various industries including the building industry. This ability to receive, process and transmit measurement data makes them a tool allowing civil engineers to inspect the operational condition of building structures and their constituent parts (so-called "Structural Health Monitoring"). In this paper, a contemporary review of publications related to the use of the smartphone as a measuring tool in structural health monitoring is made. It is seen as an opportunity to develop a "real-time" smartphone application, to measure and visualize fields of displacement and deformation of structural elements. This possibility is based on digital image correlation analysis. The working principles of the smartphone application using the smartphone camera are described and the capabilities of the smartphone application, developed by the authors, for the measurement of two-dimensional fields of displacement and deformation are illustrated.

1. Introduction

Over the past 15 years, smartphones have developed rapidly together with the advancement of technology. They have increasingly taken on the role of both portable and compact personal computers, as well as digital cameras, through which measurements can be made in various fields such as sport, medicine, defence, transport, energy, industry and construction.

Modern smartphones interact with the environment through a number of built-in and external sensors used in various software applications. Utilizing these sensors the smartphones are able to collect data that is used by specially developed applications for storage, processing and analysis. Due to their mobility, large data storage capacity, significant computing power and, last but not least, low cost smartphones, in addition to being a means of telecommunications, may with a suitable application, take on the role as a measuring system.

An example of using a smartphone as a measuring system is when inspecting the operational condition of building structures and their constituent parts (so-called Structural Health Monitoring (SHM)). In the monitoring of engineering facilities, there are tasks for measuring the displacement and deformation fields. They cannot be solved with traditional contact methods such as the use of strain gauges, fibre optic sensors, ultrasonic sensors, etc., despite the fact that these technologies are also undergoing development and progress. In the last two decades, innovative sensors and displacement measurement systems [1, 2] have been developed in the field of SHM, based on displacement and strain sensor, laser Doppler vibrometers, GPS satellite research, terrestrial laser scanning.

Initially, Morgenthal and Hopfner [3] measured displacements using the speaker and microphone of an Android-based smartphone. They concluded that significant inaccuracies might arise due to component calibration and fluctuations in temperature.

In [4], Zhao and team have proposed the use of a smartphone with an application based on the laser projection sensing method for real-time displacement measurement in bridges. A laser device emitting in visible region of spectrum is fixed on a structural element of the observed bridge and the laser spot is projected on a projection plate located in appropriate position. The iPhone recognized the movement of the laser spot on the projection plate, and structural displacement was calculated using image processing methods. Two mobile applications are used - D-viewer and Orion-CC for the purpose. In [5], the same author continues the development this technology by comparing the possibilities of two different smartphones for monitoring a series of static and dynamic displacements. The experiments were conducted to verify the stability of the newly developed D-Viewer mobile application. The smartphones used in this study had a short monitoring range and low acquisition frame rate. The future will see the improvement of these parameters with better data processing and camera units in the smartphones. In a 'follow-up publication' [6], the same authors used the D-Viewer mobile application again to measure bridge displacements.

A more recent work [7], sees, for the first time, a smartphone camera used for recording images of a speckle pattern on the surface of engineering objects while undergoing deformation. These speckle images are transmitted to a computer that implements the Digital Image Correlation algorithm (DIC) to determine the deformation fields. Other authors are exploring the possibility of using a smartphone camera, DIC equipment, D-Viewer mobile application and the smartphone processor to evaluate in-plane and out-of-plane displacements of bridges by tracking the movement of two circle-shaped markers on the observed surface [8].

In [9], the deformations of compressive concrete blocks are visualized, measured and tracked, also using a smartphone and DIC technique. The changes in displacements fields of monitored surfaces at different loads are analyzed. In the last three cited works [7-9], the smartphone works in combination with the DIC method, but for the realization of this algorithm, not only the potential of this device is used.

Analyzing the cited publications it can be concluded that the creation of smartphone applications to introduce and validate these devices as measuring tools is a current task. The aim of this work is to develop a smartphone application, by which the displacement/deformation fields of a monitored surface can be visualized and measured using the DIC algorithm implemented in real time from the smartphone processor.

2. Working principles of App

The main idea is to create an application which implements the DIC method to obtain displacement/deformation fields using of a smartphone as standalone unit. DIC is an optical method [10, 11], which tracks the displacements and deformations of the so-called speckle structures on the object surface. It is based on comparison of consecutive images pairs. The first (reference) image is taken before deformation, and the second - after deformation of the observed surface. In some publications, this surface is called a "zone of interest" (ZOI). The reference image is divided into equal-sized "correlation subsets" - most often square matrices of pixels. By numerically solving a correlation problem to find the maximum match of the pixel intensity of each subset belonging to first image with one of the subsets belonging to second image, we find the location where the considered subset has moved within the time interval, between the capture of that first image and the second image capture.

The operation algorithm of the application is as follows: 1. The original colour image is captured and is converted into a greyscale image on the smartphone; 2. Quick assessment of the image quality is carried out, taking into account the contrast and distribution of pixels intensity in ZOI; 3. A switch to measurement mode and save next frame as a reference; 4. Filtering and binarization of the frames pair is applied; 5. Performing correlation calculations to obtain the displacements of the centres of the matched subsets for which speckle pattern quality exceeds a predetermined threshold value (see step 2); 6. The application analyzes the images frame by frame to calculate in-plane displacements of the subsets centres. Then, by numerical differentiation, the field of deformations is determined.

3. App description

This application is designed for Android 8.0 and higher operating systems versions. The graphical interface is implemented using Java programming language. Functions for processing information from the camera and performing correlation analysis are written in C++ using parallel computing multiprocessor architecture of modern smartphones and OpenCV library. The authors intend to develop this application as "Open Source" to facilitate its use, development and improvement by other researchers that are working in this field.

The application aims to provide the user with an easy and accessible tool for gathering quantitative information on displacements and deformations of the inspected objects. It allows the user, equipped only with a modern smartphone to acquire data applying DIC measurement technology, by setting all the necessary parameters for optimal running the algorithm in each particular case and to visualize the results.

The user must perform the following procedures:

1. Open the start screen (figure 1).
2. Allow autofocus and capture images.
3. Selection of the correlation parameters (figure 2).
4. Selection of other parameters for analysis (figure 2).
5. Selection of measurement mode (displacement or deformation).
6. Start the measurement.
7. Acquisition and visualization of data.
8. Stop the measurement.
9. Extract output data files from the smartphone, if necessary.

After opening the initial screen (figure 1(a)), a rectangle in the middle of the frame is initially displayed, which gives an idea of the size of a correlation subset. The focus is adjusted automatically when the distance between the smartphone camera and the monitored object varies. Colour of this marker is used to indicate that the central frame area is in focus. Green denotes a focused frame and Red indicates an unfocused frame. Visualizing the marker with thick red lines shows the operator that the current distance from camera to the monitored surface is less than the minimum focal distance of the smartphone being used. The interface with its buttons occupies a large part of the screen and therefore a full screen mode is provided (figure 1(b)). This feature can be activated by shaking the smartphone.

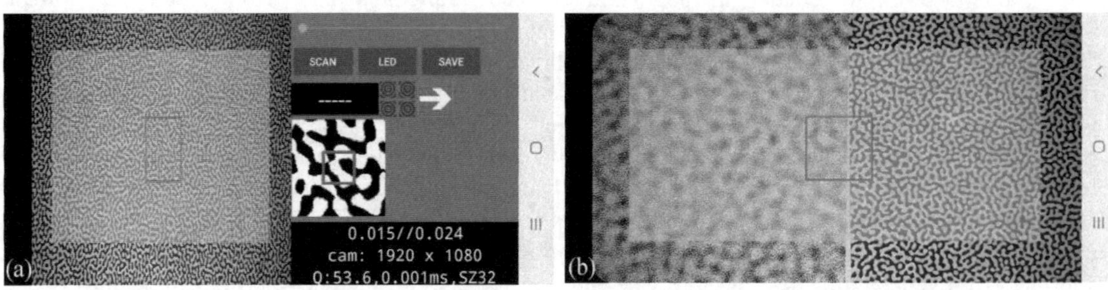

Figure 1. Screenshot of initial screen (a) and full screen modes (b).

From the start screen, the user can go to settings screen, through which he can set and adjust the values of parameters. Due to the necessity of large number parameters during the measurement and monitoring process, a set of 4 buttons for setup pre-selected values are also provided on the respective display screen (figure 2). They define a set of suitable parameters for operation in measurement mode with low or high resolution camera, respectively, and with a size of the correlation subsets of 32 or 64 pixels.

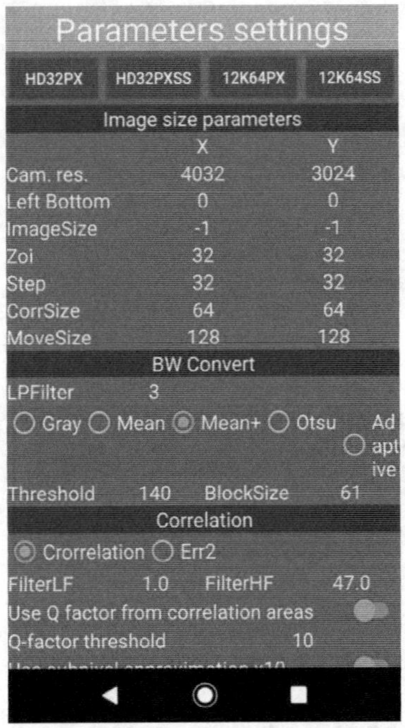

Figure 2. Screenshots of set up data window.

Capturing of the first (reference) image in the series of consecutive images is initiated by pressing the Scan button (figure 1(a)) or by tapping (clicking) on the screen in "full screen" mode. In this action, touching the smartphone, the operator can hardly avoid "shaking" the whole device including the built-in camera. Therefore, the software provides a delay for the moment of shooting the initial frame, consistent with the time required for relaxation of such an event. As an alternative measure to avoid this inconvenience, a RemoteShutter application has been developed, which is launched on a second smartphone. These two devices are connected via Bluetooth. Through the second smartphone, placed at a distance from the main smartphone, an operator can visually monitor the development of the process. The visual information received on this channel has a reduced quality, but it can be used to guide when to start or stop the measurement.

4. Preliminary test

To illustrate the workings of the DIC application for a smartphone, we conducted a modelling laboratory experiment by simulation of a mosaic plate displacement. The obtained real-time displacement field is presented in figure 3 (superposition of translation and rotation). The mosaic pattern is a type of 'speckle pattern' suitable for the implementation of the DIC algorithm. In cases when the studied surface does not have a good quality speckle like the mosaic pattern, different ways have been developed for computer generation of such patterns and their application on the surface to be monitored [10,11].

The result of a physical experiment simulating deformation field of a computer-generated speckle pattern is shown in figure 4.

The experiment was performed with a Samsung Galaxy S10e smartphone. This device has two cameras with a resolution of 12 MP and 16MP and a focal length of their lenses of 26 mm and 12 mm, respectively.

Figure 3. Example results of the translation and rotation measurements.

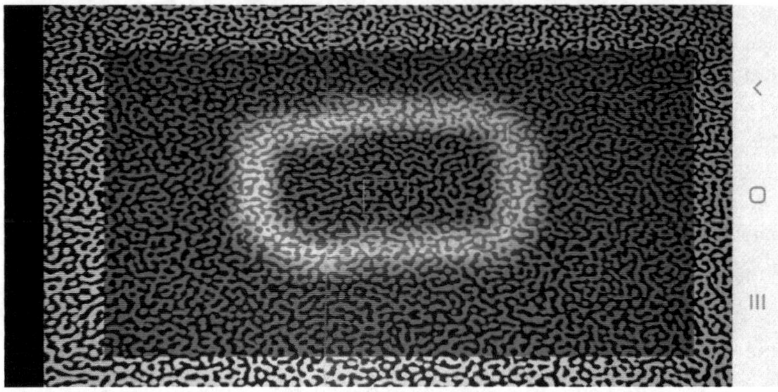

Figure 4. Example results presenting a deformation map.

5. Conclusions and future works

In this paper, we present a mobile application for real-time measurement and monitoring of displacement/deformation fields based on Digital Image Correlation algorithm. It is shown that technology, based on the smartphone built-in camera and the proposed smartphone application, allows 2D measurements to be realized quickly and with satisfactory accuracy. This equipment is relatively easily accessible, works remotely, and is non-contact, non-destructive and usable in both the laboratory and out of door conditions. The preliminary experiments suggest that a smartphone supplied with this software application can be utilized as a measuring tool in SHM.

In the future, this line of study will be focused on systematic and more detailed research of the effectiveness and the accuracy of measurements that could achieved with the proposed technology. We anticipate that in the near future this relatively inexpensive equipment will be applied for express and/or long time monitoring of real world engineering structures.

Acknowledgement

This work was partially supported by the „Mechatrol" group – a Notified Body to the Construction Products Regulation (EU) No 305/2011 at the Institute of Mechanics - BAS (Body: NB1938).

References

[1] Ye X, Dong C and Liu T 2016 A review of machine vision-based structural health monitoring: methodologies and applications *J. of Sensors* **2016** 7103039

[2] Sony S, Laventure S and Sadhu A 2019 A literature review of next-generation smart sensing technology in structural health monitoring *Struct Control Health Monit.* e2321

[3] Morgenthal G and Höpfner H 2012 The application of smartphones to measuring transient structural displacements. *J Civil Struct Health Monit.* **2** 149-61
[4] Zhao X, Ri K, Han R, Yu Y, Li M and Ou J 2016 Experimental research on quick structural health monitoring technique for bridges using a smartphone *Adv Mater Sci Eng.* **2016** 1871230
[5] Zhao X, Liu H, Yu Y, Zhu Q, Hu W, Li M and Ou J 2016 Displacement monitoring technique using a smartphone based on the laser projection-sensing method *Sensors and Actuators A: Phys.* **246** 35-47
[6] Zhao X, Zhao Q, Yu Y, Chen Y, Liu H, Li M Ou J 2017 Distributed displacement response investigation technique for bridge structures using smartphones *J Perform Constr Facil.* **31** 04017029.
[7] Malesa M, Naploszek D, Kuczynski K and Skrzypczak P 2016 Mobile version of Digital Image Correlation for deformation measurement of engineering objects *Proc. IS&T International Symposium on Electronic Imaging* 3DIPM-409
[8] Wang N, Ri K, Liu H and Zhao X 2018 Structural displacement monitoring using smartphone camera and digital image correlation *IEEE Sens J.* **18** 4664-72.
[9] Xie B, Zhao X, Li J, Yan X and Lv X 2019 Research on 2D digital image correlation measurement based on smartphones *Proc. SPIE Health Monitoring of Structural and Biological Systems XIII* **10972** 109722F
[10] Stoilov G, Kavardzhikov V and Pashkouleva D 2012 A Comparative Study of Random Patterns for Digital Image Correlation *J of Theor and Appl Mech* **42** 55-66
[11] Stoilov G, Kavardzhikov V and Pashkouleva D 2018 Multiscale monitoring of deformation fields by digital image correlation method *J of Theor and Appl Mech.* **48** 23-40

Analysis of the construction site waste management plans in Bulgaria, adopted in the period 2015-2017

V A Stoyanov[1,2]*, V Petkova[3] and V Andonova[4]

[1] Department of Technology and Construction Management, Faculty of Construction, USEA (VSU) "L. Karavelov", 175 Suhodolska St., 1373 Sofia, Bulgaria;
[2] Academy of the Ministry of Interior, Department of Safety and Prevention, 17 Pirotska Str., 1309 Sofia, Bulgaria
[3] Institute of Mineralogy and Crystallography, Bulgarian Academy of Science, Acad. Georgi Bonchev Str., Block 107, 1113 Sofia, Bulgaria;
[4] Euro Perspectives Foundation, 1-9 Chepino Str., 1680 Sofia, Bulgaria

* E-mail: vensy.stoyanov@vsu.bg

Abstract: At the end of 2017, a new "Ordinance on construction and demolition waste management and use of recycled building materials" was adopted in Bulgaria without assessing the application of the existing one. Then and until now, there is no reliable public data available on construction and demolition waste, based on which to make a statistical assessment and make management decisions. To clarify the manner of preparation and implementation of construction site waste management plans, a study of existing accepted projects was conducted. They were evaluated based on their content, chosen measures, and recognition of the application of priority order (hierarchy) in the management of the generated construction and demolition waste. Based on the analysis, it is argued that to make the Ordinance on construction and demolition waste management more effective, it should be accompanied by increasing the reliability and quality of waste data, implementing specific management decisions based on knowledge of material flows, development of a market for reused construction products and specialization of the designers of these plans.

1. Introduction

Legislation on waste management in Bulgaria was introduced in 2012, transposing the Directive 2008/98/EC [1] on waste (Waste Framework Directive, thereafter the FWD) into national law. A special "Ordinance on construction and demolition waste management and use of recycled building materials" was introduced in the same year [2]. In 2016, the Council of Ministers, without examining the impact of the existing Ordinance, proposed a draft for the adoption of a new Ordinance on construction and demolition waste management (thereafter the CDWM), which was issued at the end of 2017, and has been used without being updated for almost 3 years [3, 4]. According to the report of the petitioner (Minister of Environment and Water), the current regulation does not transpose European legislation [3], but the review of the document shows that the major principles of the FWD are included, such as an obligation to handle waste in a way that does not hurt the environment or human health and an encouragement to apply the waste hierarchy.

The Ordinance of CDWM stipulates that the main driver for achieving the national target for 70 wt% recovery of materials by 2020 of this waste streams is the development of construction site waste management plans (thereafter the SWMPs). These plans must be designed for all new projects for construction, reconstruction, overhaul, and deconstruction of construction works, which have a relatively large total built-up area (for example for reconstruction and overhaul over 500 m^2 in [2] and over 700 m^2 in [4]) or are relatively long linear objects. The SWMPs shall ensure that enough resources and time are allocated to ensure [2]:

- targets for material recovery, including preparation for re-use, use of recycled construction materials and/or recovery of construction waste in backfills;
- measures for the selective demolition, where applicable;
- measures for separate collection, utilization and disposal of construction and demolition waste (thereafter the CDW);
- measures for prevention and minimization of the generated CDW on the construction or demolition site;
- instructions for keeping hazardous waste records;
- measures for the application of priority order (hierarchy) in the management of the generated CDW.

The only measurable targets are for material recovery of non-hazardous CDW, which apply to the year of opening the construction site, gradually increasing during the period 2014-2020. For example, for tiles and ceramics (code 17 01 03) this target increases from 30 wt% (2014) to 70 wt% (for 2020). The SWMP includes forecasts for the amounts of some dense CDW, which must be larger than the targets. An additional target is the use of recycled construction materials or treated CDW for recovery in backfills in amounts, depending on the type of construction and the scope of the permit for construction. According to the Waste Management Act [5], the contracting authority is responsible for the creation and implementation of SWMP and the specialists who prepare this plan are designers with full or limited design qualification. For this reason, following the introduction of the Ordinance of CDWM, several courses have been held for designers on how to design a SWMP and the Ministry of Environment and Water issued several national guidelines for the implementation of the Ordinance of CDWM. In 2010, the National Statistical Institute announced that they would start collecting data on CDW, and in 2017 the Bulgarian Universities included in their programs or curricula a discipline "Construction and Demolition Waste Management".

These activities only partially cover the activities set for implementation in the strategic waste management plans [6, 7]. At the moment there is no information available on the implementation of the main indicators set for achieving the targets, which can be established and proven by the lack of vision and strategies after 2020, lack of detailed and summarized statistics on CDW, lack of monitoring mechanisms and reports to national authorities, lack of data on the implementation of the SWMP, etc. Upon request for access to such information, the responsible institutions are denied. The data on CDW that Bulgaria submits to Eurostat is not reliable, as the misunderstanding of CDWM policies is influenced by national economic, social and technological factors [8, 9]. Research and popular publications in the field of CDWM are isolated [10, 11, 12].

On the one hand, Bulgaria, as part of the European Union, must be involved in solving some tasks related to CDWM, which is seen as a cornerstone of the circular economy [13, 14, 15]. On the other hand, the presented situation shows a great degree of uncertainty about the application of CDWM in Bulgaria. The present study attempts to understand the management problems in the field of CDW. For this purpose, information from publicly available real projects in the field of CDWM is examined. Based on the requirements, these SWMPs are analysed in terms of managerial approaches such as:

- determination of the resources, time and time plan for the activities;
- effective coordination, planning and organisation of the activities;
- adequate management and coordination structure on the construction/demolition site;
- proper procedures for monitoring and communication on the construction/demolition site;
- provided appropriate training necessary for all operators.

IOP Conf. Series: Materials Science and Engineering **951** (2020) 012027 doi:10.1088/1757-899X/951/1/012027

2. Materials and method

2.1. Data collection
The data for analysis were taken from sites up to date as of 30.06.2020, the information from which can be verified. Due to the extremely small number of SWMP available after 2018, it was agreed to consider plans adopted in the period 2015-2017, preceding the acceptance of the current CDWM Ordinance. To analyse comparable data, the choice of SWMPs was limited to:

- investment projects financed with public funds;
- plans, drawn up by designers with full design qualification;
- plans that include both parts: construction and demolition;

These data were confirmed by reviewing the name and content of the investment project, the manner of its approval, including download from the public procurement site (table 1). The design qualification was proved by reviewing the register of designers with full design qualification [16]. To understand the profile of the designer, information was taken about their speciality, which is derived from the section to which belong.

2.2. Data preparation
The data from the projects were processed by reviewing the projects themselves, from which several indicators were derived that can be quantified. When construction or deconstruction activities were mentioned in more than one page, they are marked as "partially described" (table 2). When there was a detailed description of the individual activities and processes, they were marked as "available", although most of the projects copy "technological part", which is a mandatory part of the content of investment projects [17]. The investigated plans have a part of the removal of construction products, none of them was audited before the demolition activities, which is a mandatory norm in most EU countries [18].

Although a reference to the Ordinance of CDWM is obligatory when SWMPs are designed, during the review of the projects it was noticed that a significant part of the explanatory note was a text that has been copied from the Ordinance. "Available" indicates cases when the copied texts are over one page. Another indicator that can be easily determined in the SWMPs is whether measures for separate collection of construction or demolition waste are considered (table 2). Here the assessment is "available", "unavailable" and "partial described", as the latter means that the designer has given only recommendations, without considering the way of organizing the activities and the processes of their implementation.

The main requirement for the approval of a SWMP by the mayor of the municipality or an official authorized by him is to set forecasts of mass percentages for material recovery of CDW higher than those established in the Ordinance of CDWM. The decision for approval can be taken, but whether real actions are set for its implementation can be understood from the way of calculating the amounts of CDW. The assessment is considered to be "available", "unavailable" and "unclear described", the latter means that the designer has accepted an amount of waste with unclear justification (table 2). The assessment "unclear" includes the case when the designer has accepted the amount of waste as a percentage of the mass of all construction products used in construction work (often this is 1 wt% for concrete and steel and 3 wt% for other construction products).

An additional indicator for assessing the accuracy of the prepared and adopted SWMP is the monitoring of the method of calculating the amount of CDW. It is generally accepted in construction that the used solid materials are determined in volume, while the Ordinance of CDWM requires the amount of generated waste to be forecasted in mass with an accuracy of 50 kg. In case of the wrong conversion, it may happen to claim the creation of amount of CDW that exceeds the capacity of the contractor or leading to a violation of the principle of reducing the amount of waste. As the CDWM Ordinance does not specify a reference document for the densities of materials and waste generated by them, it was decided to find whether a specific document for this transformation is indicated and in which the individual types of waste are converted from volume to mass. The accepted assessment is "available", "unavailable" and "unclear described" (table 2).

Table 1. Summary for the studied 20 site waste management plans (SWMPs) (type of projects: B – building, C – low-rise construction, P – plumbing) (underlined letters must be entered in Cyrillic).

№	Project/Link	Type	Speciality	Year	Pages
1	2	3	4	5	6
1	Extension and expansion of an existing kindergarten in Regulated Land VI - municipality, quarter 36, village of Orizare, Municipality of Nessebar. www.cdg-radost.na4alobg.com/puso.pdf	B	StrEng	2015	12
2	Reduction of greenhouse gas emissions through implementation of energy efficiency measures in the administrative building of Ardino municipality. www.ardino.bg/docs/profil_na_kupuvaca_docs/Cast PUSO.pdf	B	StrEng	2015	15
3	Multi-family residential building in Sofia, zh.k. "Nadezhda", bl.403, ent. A, B and C, with Reg. №432-31PM-117-211 of the Agreement for financial assistance and implementation of energy efficiency renovation under project BG161PO001-1.2.01-0001. www.zop1.bg/download/1001969	B	StrEng	2015	18
4	Energy efficiency measures and accompanying activities of the building of "Otez Paisii" Secondary School, "Nikola Velchev" Secondary School and their dining rooms, Samokov. so-slatina.org/wp-content/uploads/2018/04/puso-Hubavka5.pdf	B	WatEng, MinEng	2015	23
5	Internal renovation and improvement of the energy efficiency of the building of the exemplary community centre "Prosveta - 1908" Zlatograd – Second stage. www.prosveta.info/documents/Chitalishte_etap2_p3_Chast_PYSO.pdf	B	Arch	2016	26
6	Repair and restoration of an administrative building of the Black Sea Region Basin Directorate with identifier 10135.1506.802.1 according to the cadastral map of the city of Varna. www.bsbd.org/UserFiles/File/AOP/00889-2016-0001/11 Upr na stroit otpadatsi.pdf	B	TechnEng	2016	26
7	Thermal insulation and accompanying measures of the building of the IOMT-BAS in connection with the project "Investment Program for Climate" of the National Trust EcoFund. www.iomt.bas.bg/zop/project_2016/inv_project/Chast-PUSO.pdf	B	WatEng, MinEng	2016	18
8	Replacement and reconstruction of boiler and heating installation of Belogradchik Municipality. belogradchik.bg/wp-content/uploads/2016/08/Chast-PUSO.pdf	B	HVAC	2016	12
9	Overhaul of hospital rooms and offices at the First Clinic of Nervous Diseases, located on the 14th floor in a high body of MHAT "St. Marina" EAD – Varna. www.svetamarina.com/uploaded/system_documents/3577/PUSO_-_14_etaj.pdf	B	StrEng	2016	15
10	Repair of detention facilities 01 RU SOFIA - SDVR, located in Sofia, 8 Shipchenski Prohod Str. www.mvr.bg/docs/librariesprovider8/dokumenti-ot-profil-na-kupuvacha /i_raionno_puso.pdf	B	StrEng	2016	21

1	2	3	4	5	6
11	Renovation of the assembly hall and lobby to the assembly hall of 49 Benito Juarez Primary School for the implementation of order №05-02-160 of the Sofia Directorate for Fire Safety and Protection of the Population. https://www.49ousofia.com/sites/default/files/profilk/PUSO.pdf	B	Arch	2017	8
12	Construction, reconstruction and rehabilitation of pedestrian alleys and sidewalks, Bankya – Second stage. https://www.bankya.bg/downloads/A4_5.pdf	C	StrEng	2015	8
13	Reconstruction and improvement of the adjacent yard to the kindergarten "Priateli", located in Regulated Land - I, for nurseries, district 62 according to the zoning plan of the town of Kuklen, Kuklen Municipality. http://buyers-rofile.kuklen.org/uploads/files/20180516dpkq596321/ 29232018161605_Plan za upravlenie na stroitelni otpadatsi.pdf	C	StrEng	2016	17
14	Road PDV-2055 / Road 11-64 railway station Kaloyanovo - Razhevo Konare - Glavatar from km 0+000 to km 4+981. new.kaloianovo.org/attachments/article/63/ПУСО.pdf	C	StrEng	2016	15
15	Reconstruction and rehabilitation of the street network in the town of Madan, Madan municipality - 3966.90 m. www.madan.bg/assets/migrate_files/-REKONSTRUKTSIYa-NA-VODOPROVODNA-KANALIZATSIONNA-M/1obzapiska_puso_mada.pdf	C	WatEng	2016	11
16	Improvement of the central square in the General Nikolaevo neighbourhood, Rakovski, municipality of Rakovski. rakovski.bg/profile/uploads/07.pdf	C	WatEng, SurvEng	2017	14
17	Reconstruction, repair and furnishing of a kindergarten, Ugarchin. https://e-obp.eu/bp/Document/ 988eaed5-8aa3-4121-b44f-2b35c9335c3b (Pregled) PUSO.pdf	C	StrEng	2016	8
18	Reconstruction and completion of the water supply network of the village of Gospodinovo, Byala municipality. https://e-obp.eu/bp/Document/ 6c7f7802-5c1c-4b0d-a09c-281db7fdc354Proekt s. Gospodinovo - chast 2.zip	P	StrEng	2016	14
19	Reconstruction of an open drainage canal and the adjacent sidewalks on „28-ma“ Street in the village of Dalgo Pole, Kaloyanovo Municipality. new.kaloianovo.org/attachments/article/55/Plan za upravlenie na stroitelnite otpadatsi.pdf	P	Road Eng	2017	16
20	Reconstruction and replacement of part of the internal water supply network in the village of Opan, Opan municipality - Second stage. www.opan.bg/uploads/procurements/investitsionen-proekt-chast-upravlenie-na-stroitelnite-otpadatsi.pdf	P	WatEng	2017	29

Remark (abbreviations for designer's speciality): Arch – architect, StrEng – structural engineer, SurvEng – Engineer of Geodesy and Applied Geodesy; WatEng – engineer of hydrotechnics and hydro constructions, MinEng – engineer of mining, geology and ecology, HVACEng – engineer of heating, ventilating and air conditioning, TechnEng – Engineer of Construction technology, RoadEng – Engineer of transport construction and facilities

Table 2. Evaluating the manner of development of site waste management plans (SWMPs)
(● – available; ○ – unavailable; ◒ – partial described; ◍ – unclear described)

Project №	Description of construction activities	Copying large texts of the ordinance	Measures for separate waste collection	Calculation of amounts of demolition waste	Calculation of amounts of construction waste	Reference to volume to mass calculations	Satisfaction of the requirements of the ordinance
1	●	○	◍	○	◍	○	●
2	●	●	○	○	○	○	●
3	●	●	○	○	○	○	●
4	●	●	○	◍	◍	○	●
5	●	●	○	◍	◍	○	●
6	●	●	○	○	◍	○	●
7	●	●	○	○	○	○	●
8	◒	●	○	○	○	○	●
9	●	●	◍	○	○	○	●
10	◒	●	○	◍	◍	○	●
11	◒	○	○	◍	◍	○	●
12	◒	○	○	◍	◍	◍	●
13	◒	●	◌	○	◍	◍	●
14	◒	●	○	○	○	○	●
15	●	●	○	○	○	○	●
16	◒	●	○	○	○	○	●
17	○	○	○	○	○	○	●
18	●	●	◍	○	◍	○	●
19	●	●	◍	○	◍	○	●
20	●	●	○	◍	◍	●	●

One of the main principles for understanding and implementing pan-European waste management policies is adherence to the policies adopted by the five-level hierarchy. This hierarchy is enshrined both in the Law on waste management [5] and in the Ordinance of CDWM [2, 4]. Initially, it is assessed whether activities followed the waste hierarchy are planned during the development of the project. Then it is considered whether the project indicates or applies specific activities to prevent or reduce the hazard and amount of construction waste, preparation for re-use, material recovery, incl. transfer for recycling, as well as the use of construction waste for backfills and the use of recycled materials from CDW. The evaluation is "available", "unavailable" and "unclear described", the latter referring to the cases where the project does not contain information on implementation (table 3).

3. Data analysis and discussion

The examined SWMPs contain an average of 16 pages and there is no significant difference from the type of project. Most often, the plans are designed by a structural engineer, which is explained by the fact that these specialists represent the largest section of Chamber of Engineers in the Investment Design (CEID), which maintain and update registers of the engineering designers. Architects relatively rarely design SWMPs, which is due to the specifics of training of architects in Bulgaria (it is art-oriented, not technological) and the fact that specialized courses for the training for the Ordinance of CDWM were conducted with the initiative of CEID and not of Chamber of Architects in Bulgaria (the professional organisation of architects, landscape architects and urban planners), who maintains the register of practising architects.

Table 3. Evaluating the application of hierarchy principle in the site waste management plans (SWMPs) (● – available; ○ – unavailable; ◍ – unclear described)

Project №	planning activities followed the waste hierarchy	prevention or reduction of waste generation		preparation for reuse activities		recycling activities, incl. back-filling operations		use of recycled materials from CDW	
		indication	applied activities	indication	applied activities	indication	applied activities	indication	applied activities
1	◍	○	○	○	○	○	○	●	○
2	◍	●	○	◍	○	◍	○	●	○
3	◍	●	○	◍	○	○	○	○	○
4	◍	●	○	◍	○	●	○	○	○
5	◍	●	○	◍	○	●	○	●	○
6	○	○	○	●	○	○	○	●	○
7	○	●	○	●	○	○	○	○	○
8	○	○	○	●	○	○	○	○	○
9	◍	○	○	●	○	○	○	○	○
10	○	●	○	○	○	○	○	○	○
11	○	○	○	○	○	○	○	●	○
12	○	○	○	○	○	○	○	●	○
13	◍	○	●	●	○	○	○	○	○
14	○	○	●	○	○	●	○	●	○
15	◍	○	●	●	○	●	○	●	○
16	○	○	●	○	○	●	○	●	○
17	○	○	○	○	○	●	○	○	○
18	○	○	●	○	○	●	○	●	○
19	○	○	○	○	○	○	○	●	○
20	◍	○	○	○	○	●	○	○	○

Almost all projects copy content from the "technological part" part of investment projects. This regularity is because the SWMP, which together with the safety and health plan at the construction site, is designed based on the remaining parts of the investment project (architectural, structural, installations, etc.) and not simultaneously with them. Large parts of the texts and tabular material of the Ordinance of CDWM are also included in the projects.

In this way, the designer informs the contractor of construction/demolition about the commitments for the implementation of the Ordinance of CDWM, which leads to an increase in the number of pages of the SWMP. Thus, the designer of SWMP, who is a structural engineer, is forced through his project to meet state requirements, to comply with the techniques and technologies used by the contractor of construction or demolition and to notify the contracting authority of his commitment to the generated waste. Insufficient specialized training, lack of previous experience, as well as the incomplete infrastructure for the execution of projects in this area, leads to the formalization of design in the field of CDW. The presence of common texts in the examined SWMP, often irrelevant to the project activities, shows that the designers learn from each other by accepting the approved projects SWMP on the Internet as examples of good practices in the field of construction and demolition waste management.

In SWMP, the measures for separate collection of CDW streams are either not described or are described unclearly. The main reason for this situation is that in the Ordinance of CDWM does not require planning based on material flows [19]. Despite the information provided by the European Commission and good examples from leading European countries, this approach is not recognized by the designers of the SWMPs. Only three of the analysed SWMPs (No 6, 7 and 18 in table 1) indicate that there is hazardous waste, but the designers do not indicate the quantities or prescribe specific activities with them. Calculations of the amount of construction and demolition waste are not made for the majority of the projects. There is no reference to the method of converting the amount of waste from volume to mass. On the one hand, this incorrectness can be considered as a gap in the requirements of the Ordinance of CDWM, but on the other hand, designers have the opportunity to use other regulations and standards in their calculations.

One of the main principles adopted in the legislation and policies for waste prevention and management is the implementation of a priority order of activities to achieve the most environmentally friendly results. This policy must be set by the Ordinance of CDWM and taken into account through the approved SWMPs. The Ordinance of CDWM specifies a text for compliance with the hierarchy, but half of the analysed projects are not based on its application. The other half of the SWMPs plan activities follow the waste hierarchy, but this planning is unjustified and unclear. The projects mention the prevention or reduction of waste generation, preparation for reuse, recycling activities and use of recycled materials from CDW, but this is not explained with specific activities that allow calculations of quantities of waste or materials. The hierarchical approach in CDWM is most recognizable in plumbing replacement projects, but they do not prescribe specific activities. The unclear indications of the activities in the hierarchy are mainly for activities on construction and demolition of buildings.

These circumstances do not lead to difficulties in conformity assessment the quantitative requirements of the Ordinance of CDWM. This means a formalism in the planning of CDWM, which in turn does not allow the accumulation of experience and expertise.

4. Conclusions

Based on the analysis of the content of construction site waste management plans in Bulgaria, adopted in the period 2015-2017, it can be argued that to make the "Ordinance on construction and demolition waste management and use of recycled building materials" more effective, it must be accompanied by increasing the reliability and quality of data for construction and demolition wastes. The requirement of the European Directive to achieve 70 wt% materials recovery by 2020 of this waste streams formalizes the design and to avoid this construction site waste management plans must describe specific management decisions for implementation based on knowledge of material flows, application of different management approaches to construction waste and demolition waste and development of the market for reused construction products. It is mandatory to apply good management practices, such as auditing before demolition activities. To meet the described requirements requires specialization of professionals in this field of the construction industry.

References

[1] Directive 2008/98/EC of the European Parliament and of the Council of 19 November 2008 on waste and repealing certain Directives, OJ L 312, 22.11.2008, pp. 3-30

[2] Ministry of Environment and Water 2012 *Ordinance on construction and demolition waste management and use of recycled building materials*, State Gazette № 89, 13.11.2012, pp. 36-89; withdrawn by State Gazette № 98 on 8.12.2017 (in Bulgarian)

[3] Ministry of Environment and Water 2016 *Draft ordinance on construction and demolition waste management and use of recycled building materials*, p 33, 19.10.2016, https://www.strategy.bg/FileHandler.ashx?fileId=7972, accessed on 30.06.2020 (in Bulgarian)

[4] Ministry of Environment and Water 2017 *Ordinance on construction and demolition waste management and use of recycled building materials*, State Gazette № 98, 8.12.2017, pp. 18-32; (in Bulgarian)

[5] Ministry of Environment and Water 2012 *Waste management act*, State Gazette № 53, 13.07.2012, pp. 2-53, as amended by State Gazette № 81, 2019; (in Bulgarian)

[6] Ministry of Environment and Water 2014 *National waste management plan* 2014-2020, https://www.moew.government.bg/static/media/ups/tiny/filebase/Waste/NACIONALEN_PLAN/_/NPUO_2014-2020.pdf, 228 p, accessed on 30.06.2020 (in Bulgarian)

[7] Ministry of Environment and Water 2011 *National Construction and demolition Waste*

Management plan for the territory of the Republic of Bulgaria for the period 2011-2020, https://www.moew.government.bg/static/media/ups/tiny/file/Waste/cdw/NSPUOSR-final.pdf, 102 p, accessed on 30.06.2020 (in Bulgarian)
[8] Saez P V and Osmani M 2019 A diagnosis of construction and demolition waste generation and recovery practice in the European Union *Journal of Cleaner Production* **241** 118400
[9] Malinauskaite J, Jouhara H and Spencer N 2017 Waste prevention and technologies in the context of the EU Waste Framework Directive: lost in translation? *European Energy and Environmental Law Review* **26** 66
[10] Zaharieva R 2014 *Proc. XIV Int. Sci. Conf. VSU'2014 (Sofia)* vol. 3 (Sofia: USEA "L. Karavelov") p 356 (in Bulgarian)
[11] Zaharieva R 2014 *Proc. XIV Int. Sci. Conf. VSU'2014 (Sofia)* vol. 3 (Sofia: USEA "L. Karavelov") p 361 (in Bulgarian)
[12] Stoyanov V, Petkova V and Andonova V 2018 *X Jub. Int. Sci. Conf. "Civil Engineering Design and Construction"* (Varna) (Sofia: NTSSB) p 305 (in Bulgarian)
[13] OECD 2019 *Waste Management and the Circular Economy in Selected OECD Countries: Evidence from Environmental Performance Reviews* (Paris: OECD Publishing) p 118
[14] Wahlström M, Bergmans J, Teittinen T, Bachér J, Smeets A and Paduart A 2020 *Eionet Report - ETC/WMGE 2020/1 (Construction and Demolition Waste: challenges and opportunities in a circular economy)*, January 2020 p 59
[15] Ruiz L A L, Ramón X R and Domingo S G 2020 The circular economy in the construction and demolition waste sector – a review and an integrative model approach *J. Clean. Prod.* **248** 119238
[16] Register of designers with full design qualification as of 30.09.2017, State Gazette № 21, 9.03.2018, pp 42-155 (in Bulgarian)
[17] Ordinance № 4 of 21 May 2001 on the scope and content of investment projects, State Gazette № 51, 2001, as amended by State Gazette № 44, 2017 (in Bulgarian)
[18] European Commission, 2018, Guidelines for the waste audits before demolition and renovation works of buildings https://ec.europa.eu/docsroom/documents/31521/attachments/1/translations/en/renditions/native, May 2010 p 37, accessed on 30.06.2020
[19] Augiseau V and Barles S 2017 Studying construction materials flows and stock: A review. *Resources. Conservation and Recycling* **123** 153

Assessment of the impact of destabilizing factors on implementation of investment and construction projects

A A Lapidus, I L Abramov, Z A K Al-Zaidi*

Moscow State University of Civil Engineering, Moscow, Yaroslavskoe shosse 26.

* E-mail: zaidalzaidi20142016@gmail.com

Abstract. Among the distinctive features of construction projects, it should be noted the complexity and specificity of the final product in the form of buildings and structures for various purposes. Investment construction projects are affected by a large number of destabilizing factors at all stages of implementation. These factors negatively affect the final indicators of the construction investment project: failure to meet deadlines, the likelihood of construction in progress, increased cost, low quality. In this regard, conducting research on the study of destabilizing factors affecting the production activities of construction company during the implementation of investment projects is relevant in difficult times for the construction industry. Construction industry, as a complex, dynamically changing system, susceptibility to various destabilizing factors, requires a constant readiness to achieve the set goals and the final result at each stage of their life cycle. This article discusses the destabilizing factors that arise during the implementation phase of an investment construction project. In order to identify and classify the negative factors that may be encountered participants construction activity, it was prepared a questionnaire for the experts who have the necessary experience in the construction industry. Based on the polls conducted, destabilizing factors were identified and ranked. Statistical methods were used to evaluate the factors studied and determine their significance and impact on the project activities. The results of the study showed that financial, technical and labor factors have a significant impact on the construction process, therefore, special attention should be paid to the development of organizational and technological measures to prevent them. This will eliminate or minimize the influence of these factors on the production activities of construction company.

1. Introduction

The efficiency of a construction company at all stages of production depends on the impact of internal and external destabilizing factors. The quality of construction and assembly works, the duration of construction operations, the company's profit, the investment payback period - all these key indicators are prone to change in the course of implementation of an investment construction project affected by the above factors. To stabilize its performance, the company seeks to reduce the negative impact of external and internal factors using additional resources thus increasing its costs and ensuring its stable competitive operations.

The existing regulatory mechanisms aimed at improving the efficiency of operations of construction companies, i.e. raising the quality level of implementation of investment and construction projects, as we can see recently, are not effective [1].

Stable functioning, sound management and correct forecasting of the future production operations of a construction company can ensure the timely identification of destabilizing factors thus preventing the risk of unfinished construction and degradation of the technical and economic indicators of investment and construction activities. It must be borne in mind that the negative factors that affect the company from outside are difficult to identify, predict, and control. For this reason, the problem of the negative impact of external factors, including geopolitical factors, has been discussed at the international level for several decades. Experts agree that measures should be taken to prevent and eliminate the effects of negative factors in the interests of entrepreneurs. In addition, the liability for damage caused by effects of external destabilizing factors should not completely lie on general contractors and (or) subcontractors [2].

A construction company can exert a great influence on its operations and adjust the production capacities and efficiency of its work in advance, inter alia, by managing external destabilizing factors and minimizing risks of uncertainty at the stage of construction operations [3]. At the same time, experience shows that management of destabilizing factors associated with adverse external environment effects or incorrect behavior of third-party contractors related in one way or another to the activities of the construction company (suppliers, machinery operators, investors, banks, public services, etc.) is in most cases ineffective. As for internal destabilizing factors that affect deviations in the performance indicators of the company, they directly depend on the ability of the executive staff of these companies to manage and, in some cases, prevent these factors.

Forecasting of the probability of destabilizing factors requires consideration of the following circumstances [4]:

- Contractors are often indifferent to negative factors and their consequences with a resulting lack of necessary experience in dealing with them. For this reason it is vital for all parties concerned to be aware of the importance of the problem under review;
- Lack of measures to prevent and eliminate the possible consequences of destabilizing factors causes damage not only to the construction company, but also to the national economy as a whole;
- In order to create a management structure for destabilizing factors that may affect the implementation of a particular project, general contractors and subcontractors should classify and rank the most significant factors.

1.1. Types of destabilizing factors

Despite the diversity of destabilizing factors, construction companies focused on solving tactical problems classify, rank and study such factors using various research methods and methodological tools for statistical processing of information [5-20].

Destabilizing factors affecting construction companies can be classified by the degree and nature of their impact:

- Design factors (effect on the achievement of project objectives);
- Business factors (effect on operations);
- Environmental factors (external in terms of relation to the project; may have an impact on the project objectives);
- Internal factors of each participant in the construction chain.

There are several types of destabilizing factors that have been well studied and analyzed [6,7]:

1. Organizational factors arising from organizational plans for implementation of the project;
2. Labor and technical factors related to human resources and machinery;
3. Political destabilizing factors and security factors. Factors arising from political changes and the security situation;
4. Economic destabilizing factors - factors related to financial allocations and obstacles that may arise in this area;
5. Legal destabilizing factors - factors that arise as a result of violation of contractual obligations and laws.

Research sources [8-19] also mention a number of other important factors.

Further research should be focused on the development of an appropriate methodology that improves the stability of a construction company amidst destabilizing factors, with due regard for system engineering principles. The authors of this article propose to divide the destabilizing factors into categories. This will allow developing an algorithm for maintaining the efficiency of construction operations in response to destabilizing factors. Using the categories of destabilizing factors (table 1), statistical analysis techniques and the expert evaluation method, it is possible to rank the factors that affect the efficiency of construction operations and, therefore, investment construction projects in general, and to determine the degree of their impact.

Table 1. Categories of destabilizing factors (compiled by the authors)

Destabilizing factor category	Causes of destabilization
Organizational	Acquisition of inappropriate land for the required type of construction
	Lack of necessary equipment and machinery for construction and assembly works
	Lack of material resources (territorial)
	Material supply failures
	Low-skilled labor resources
	Non-compliance with safety regulations
	Bankruptcy of subcontractors
	Poor planning of production and financial activities
Political, contractual support	Changes in legal regulations
	Delays in obtaining expert assessment for the project, denial of construction permits
	Unstable political situation
	Major changes in contract terms
	Inadequate engineering supervision
	Bureaucratic costs
	Non-compliance with RF government regulations
	Terrorism and crime
Economic	Delayed payment of wages
	Technical failures when transferring funds
	Lack of own financial resources
	Insolvency of consumers
Design	Errors in design documentation
	Major changes in design documentation
	Untimely modification of design documentation
Environment	Environmental pollution, damage to the environment
	Non-compliance with environmental laws and regulations
	Climatic conditions

2. Method

The expert assessment method is known as an independent research tool with a wide scope of application:

- Identification and prioritization of management goals and objectives;
- Identification and prioritization of alternative solutions to problems;
- Alternative distribution and prioritization of resources for solving problems.

There is a variety of expert assessment methods currently used in tackling these typical problems: questionnaires, interviews, discussions, meetings, business games, case analyses, etc. Questionnaires are the most efficient and widely-used way to obtain information.

In this connection, the study performed by the authors is based on questioning designed to identify the most important destabilizing factors in the construction industry. The questioning was conducted by correspondence to obtain answers of the greatest possible number of participants. Questionnaires were sent to state and private construction companies.

The collected information was systematized and processed in Excel to calculate the arithmetic average of each destabilizing factor, where the average represents the degree of importance (or degree of impact) of the factor affecting the construction project as a whole.

The questionnaire is in the form of a table, which lists destabilizing factors. Survey participants were asked to evaluate the importance and impact of each factor on an investment construction project on a scale from 0 to 100.

2.1. Determining the number of experts

The minimum number of experts to be involved in the study for ensuring a representative set of statistical data based on the confidence coefficient of the result (A = 0,95) is determined by the formula (1) [20-22].

$$E = \frac{c^2 r_a r_o}{\Delta^2}, \tag{1}$$

where:

E – the minimum required number of experts;
c – confidence coefficient (0,95);
r_a – the proportion of sample elements that have this attribute (0,95);
r_o – the proportion of sample elements that don't have this attribute (0,05);
Δ – representativity error (0,05)

$$E = \frac{(0,95)^2.(0,95).(0,05)}{(0,05)^2} = 17,2 \sim 18 \text{ Per.}$$

The information obtained during the survey is compared with other studies in this area. Subsequently, the collected questionnaires can be used by other researchers, who can make any additions as they think fit.

100 questionnaires (see table 2) were distributed among state-owned and private construction organizations. There were 73 responses received. This number exceeds the required number of samples and makes the results highly accurate.

Interviewees were divided into groups depending on experience and length of service. Figure 1 shows the number of responses given by each group of participants. As you can see, the vast majority of respondents had more than 15 years of work experience.

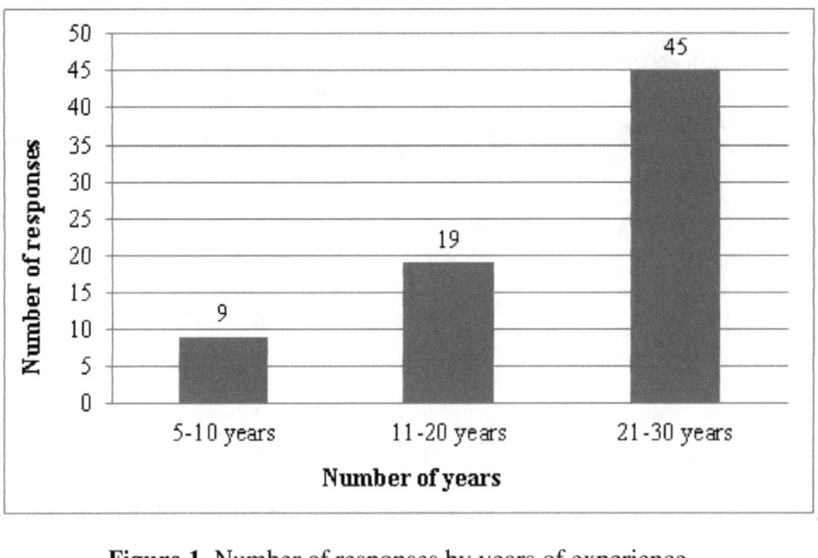

Figure 1. Number of responses by years of experience

Table 2 shows the weight of significance of destabilizing factors that were calculated on the basis of the results of the questioning. The significance range varies from 0% (no impact) to 100% (very high impact).

Table 2. Weights of the significance of destabilizing factors

Destabilizing factors	Symbol	Weight of factor %
Unstable dynamics of labor productivity of employees	S_1	69
Non-compliance with safety measures	S_2	58
Incompetence of work performers (workers)	S_3	75
Lack of qualified engineers and technicians, quality materials and up-to-date equipment	S_4	52
Use of up-to-date equipment and technology without required experience or proper training	S_5	56
Political and social pressure	S_6	44
Supply of substandard materials	S_7	71
Errors in the design documentation	S_8	63
Administrative barriers erected by the state	S_9	41
Legal disputes between participants in the project at the construction stage	S_{10}	34
Sudden discontinuance of funding	S_{11}	79
Inflation and price fluctuations	S_{12}	73
Corruption	S_{13}	62
Discrepancy between specifications and implementation due to misunderstanding of diagrams and specifications	S_{14}	49
Inadequate engineering supervision	S_{15}	33
Adverse climatic conditions	S_{16}	40
Changes in legal regulations	S_{17}	29
Inaccuracies in work scheduling	S_{18}	37
Delays in obtaining expert assessment for the project, denial of construction permits	S_{19}	35
Terrorism and crime	S_{20}	45
Cooperation between the general contractor and subcontractors	S_{21}	39
Equipment breakdown and emergency situations	S_{22}	76

The destabilizing factors under review were ranked according to the degree of significance in Table 2; the results are presented in figure 2.

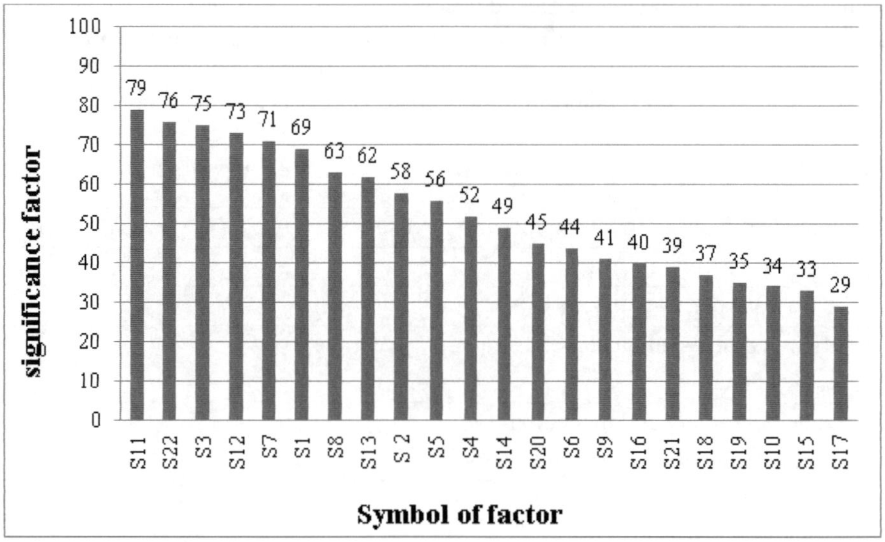

Figure 2. Ranking of destabilizing factors by degree of significance (compiled by the authors)

Figure 3 shows an example of determining the degree of significance of sudden discontinuance of funding as a destabilizing factor. This factor is indicated by S_{11}. The example shows that 40 out of the 73 respondents attribute 100% significance to this factor, while those who attach 75% significance are 15 out of the 73, etc.

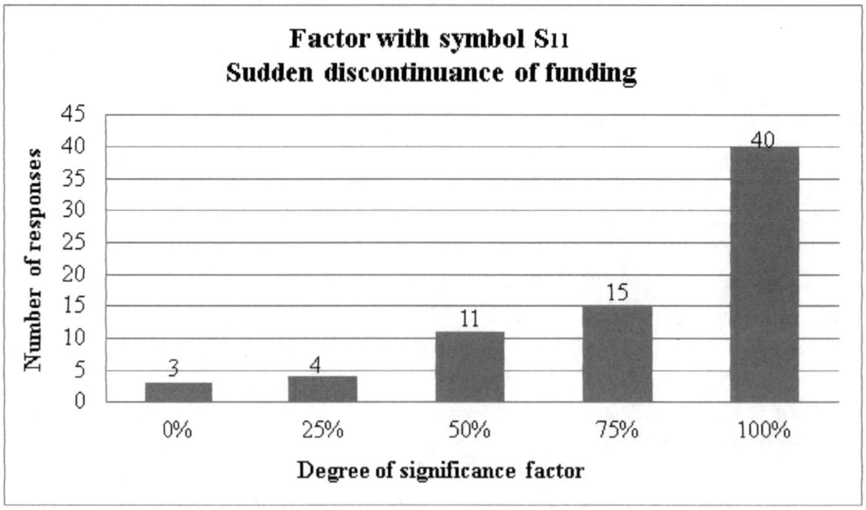

Figure 3. Factor weight (compiled by the authors)

Conclusion

This study was focused on determining the significance of destabilizing factors and their impact on the main indicators of investment construction projects (duration, costs, and quality of construction). The analysis covered some projects implemented by state-owned and private companies with engineering and technical personnel, project managers, and design and production research consultants involved as experts.

The results showed that there are a large number of destabilizing factors (financial, technical, labor, legal, etc.) that have a significant impact on the activities of construction companies.

Depending on the area of their occurrence, destabilizing factors can be classified into internal factors dependent on the operations of the company and external factors determined by the environment.

The information obtained during the analysis will be used in further studies to develop a methodology that includes measures to prevent and eliminate the consequences of destabilizing factors that arise in the implementation of investment construction projects. Such measures will ensure the timely adoption of well-considered and developed organizational decisions and identification of specific responses to destabilizing situations.

Destabilizing factor management plans should include a number of important points providing for:

1. Training of participants in a construction project in responding to destabilizing factors, reducing their level of impact, and methods of maintaining a given stable state of the construction company at all stages of the project.
2. Elimination of bureaucratic barriers that make it difficult to obtain the appropriate licenses and permits for construction and assembly works.
3. Refusal to implement construction projects in politically unstable regions (to avoid financial losses).
4. Checking basic materials and assessing their compliance with specifications at each stage of the project, using high-tech and highly efficient equipment.

The development of a decision-making mechanism aimed at reducing the level of impact of destabilizing factors allows preventive measures and responses to deviations at the initial stage rather than post factum when it is difficult or impossible to exclude losses.

References

[1] Shekalin A N, Gorbunov V N and Sinitsyn M A 2015 Identification of risk factors taking into account particular features of investment and construction operations, Modern Scientific Research and Innovation, **45** (No. 1-2), 67–76.

[2] Gabrielyan M O and Tretyakov O B 2016. Classification of risks in investment and construction activities // University Herald (No. **5**) 60-67.

[3] Koshelev V A and Sosunova L A 2014. Risk analysis in housing construction: methods and tools Russian Entrepreneurship (No. **3**) 34-41.

[4] Entisar Kadhim Rasheed 2015 Valuation the Impact of Risks on the Goals and the Safety of Construction Projects in Iraq *Journal of Engineering/Iraqi academic scientific journals*. **21** (No. 4) 1-19.

[5] Abramov I L, Stepanov A and Ibrahim I F 2017 Advantages of pre-fabricated reinforced concrete construction in Iraq. *Matec web of conferences* **26**. Rsp 2017 - 26th r-s-p seminar 2017 theoretical foundation of civil engineering p. 00001.

[6] Ibrahim Badi 2017 Risk Assessment in Misurata Construction Projects *First Conference on Industrial Technology CIT 2017* (Libya: University of Misurata) 1-6.

[7] Souad Nasser Akkab 2009 Study of Design and Implementation Factors That Leads to the Increase of Project Cost Risks During Construction Stage *Engineering and Technology Journal* **27** (No. 12).

[8] Abramov I L 2019 Quantitative assessment method for sustainability of a construction company *Vestnik MGSU* **14** (No. 12) 1619-1627.

[9] Nayfa M, Amoudi O and Hamada M. 2012. Management the risks of the construction phase of the project construction in Syria *Damascus University Journal for Engineering Sciences* **28** (No. 1) 129-150.

[10] Lapidus A A 2011 Problems of the introduction of innovative solutions in technology and organization of construction *Technology and organization of construction production*, MGSU (No. 4) p. 1.

[11] Lapidus A A and Abramov I L 2018 Implementing large-scale construction projects through

application of the systematic and integrated method, *IOP Conference Series: Materials Science and Engineering* **365** (Issue 6)
[12] Koshelev V A 2015 Sources of risks in construction *Naukovedenie* **7** (No. 1) 1-13.
[13] Jayasudha K and Vidivelli B 2016. Analysis of major risks in construction projects *ARPN journal of engineering and applied sciences* **11** (No. 11) 6943-6950.
[14] Dziadosz A and Rejment M 2015 Risk Analysis in Construction Project - Chosen Methods *Procedia Engineering* **122** 258-265.
[15] Zhadanovsky B and Sinenko S 2018. The methodic of calculation for the need of basic construction machines on construction site when developing organizational and technological documentation *E3S Web of Conferences High-Rise Construction 2017(HRC 2017)* – Art. No. 03077 **33** p. 6.
[16] Shatrova A I 2018 Organizational and technological solutions for improving the strategic planning efficiency in construction operations//Science and Business: Ways of development. No. 12 (**90**). - P. 29-32.
[17] Kazaryan R, Pogodin D, and Shatrova A 2019 Aspects of scheduling processes and results of the reorganization of projects in high-rise construction objects *E3S Web of Conferences* **97** 04002.
[18] Jurgaitis A, Topchiy D, Kravchuk A and Shevchuk D 2019 Controlling methods of buildings' energy performance characteristics *E3S Web of Conferences* **91** 02026.
[19] Ushenin D V 2018 System-integrated application of innovative technologies in construction. Days of student science: a collection of reports of a scientific and technical conference based on the results of research work by students of the Institute of Civil Engineering and Architecture (Moscow: NIU MGSU) 1290-1292.
[20] Musatova T E 2015 The methodology for predicting the efficiency of an innovative project based on expert assessments *Current problems of science and education* (No. 1) p. 634.
[21] Ruposov V L 2015 Methods for determining the number of experts *Herald of Irkutsk State Technical University* **98** (No. 3) 286-292.
[22] Florez L and Cortissoz J C 2016 Defining a mathematical function for labor productivity in masonry construction: A case study *Procedia Engineering* **164** 42-48

IOP Conf. Series: Materials Science and Engineering **951** (2020) 012029 doi:10.1088/1757-899X/951/1/012029

Construction contracts expenses and revenues reporting

Sv Stefanov

University of Economics - Varna, 77, Knyaz Boris I Blvd., Varna, 9002, Bulgaria

E-mail: svetstefanov@ue-varna.bg

Abstract. According to the Conceptual Framework of the International Accounting Standards, expenses are recognized when a decrease in future economic benefits related to a decrease in an asset or an increase of a liability has arisen that can be measured reliably. Under the matching principle between revenue and expenses, the latter are recognized based on the direct relation between the actually paid expenses and the earned revenue during the reporting period. In this context, expenses forming the value of finished goods are recognized when the revenue received from the sale of such goods is earned. In cases when the economic benefits are expected to be acquired during several reporting periods, expenses should be deferred and recognized during different reporting periods within which the acquisition of economic benefits is converted to cash. To defer them, a certain allocation base which is an item of the accounting policy followed by the entity should be applied. The key issue in the accounting of expenses and revenues under construction contracts is related to the allocation of the contract revenues and expenses between the reporting periods during which such works have been executed. Accounting standards outline the criteria for when to recognize expenses and revenue so that to identify the correct time when they should be recognized in the financial statements of the entity carrying out the works. The main concept underlying the accounting standards is that contract revenues and expenses associated with a construction contract are to be recognized as revenues and expenses by reference to the stage of completion of the contract activity at the end of the reporting period. Construction contracts applied in practice are classified as fixed-price contracts and cost-plus contracts. A fixed-price contract is a construction contract in which the contractor agrees to a fixed contract price, or a fixed rate per unit of output, which in some cases is subject to cost escalation clauses. A cost-plus contract is a construction contract in which the contractor is reimbursed for allowable or otherwise defined costs, plus a percentage of these costs or a fixed fee.

1. Introduction

Organizing financial reporting in each civil engineering company is of major significance for its optimal management and governance. Financial reporting is usually understood as a purposeful arrangement and improvement of reporting process methods and techniques, as well as collecting, processing, summarizing, and analyzing information about the facts, processes, and relationships in the business of an entity that is necessary for its management.

2. Features of construction industry and its impact on the reporting of revenues and expenses of construction entities

A major characteristic of a constructed product is its immobility. On the other side, the production process is mobile. Workers and machinery move from one place to another within the construction site depending on the performance of individual types of work.

Constructed production may be characterized as individual and small-scale. Each new project is executed in different conditions compared to the previous one. The organization of production, and hence the scale and structure of production costs are influenced by many factors that must be considered when constructing a project – soil category, seasons, sources of raw materials supply, etc.

There are also many other features characteristic to a constructed products: relatively long production cycle; the significant size of each physical unit of output; a large number of types of works and abrupt change of the structure of works on the same site; numerous relations of the construction entity with natural and legal entities (contracting authorities, suppliers, designers, subcontractors, etc.); production processes are running mainly outdoor, etc.

In entities engaged in construction activity, movable fixed assets prevail – various types of construction machinery and transportation vehicles. Their relocation from one construction site to another is related to the costs of their transportation, installation, and dismantling.

A significant part of the materials (such as concrete, sand, gravel, bricks, etc.) is very often supplied directly to the construction site. This requires accurate recording of all delivery costs and precise reporting of their use. Also, specific assets such as timber for formwork, wooden and metal formwork panels, and metal scaffolding are used on the sites. What is common for them is that they are used repeatedly on one and sometimes on several sites, which requires to adopt a certain allocation base to distribute their value when referring them as an expense. Another part of the materials (e.g. fittings, joinery, solutions, etc.) can be further processed and produced by the ancillary activities of the entity. The costs of this activity are complex and since they are subject to allocation to respective construction sites they also need the adoption of a special allocation base.

Prior to commencing the main construction project, various temporary buildings and facilities (such as warehouses, sheds, water pipes, power networks) are to be constructed, i.e. to incur costs for temporary construction. These costs and constructed temporary buildings and facilities are specific objects of reporting in the construction entities. [1]

These and other features have a significant impact on the organization of reporting production costs and calculating the cost of production. In this regard, an important role plays the selection of cost objects and account reporting objects. The cost object is understood as production having a certain consumer value and various degrees of completion, the prime cost of which is determined to measure the expenses incurred for its production. Depending on the characteristics of the technological process and the form of business relationship with contracting authorities, respectively with the main contractor, cost objects in construction entities are: all items of the works performed on a project (sub-project), a certain portion of it (stage, structural element) or individual types of work. Or, to put it otherwise, a separate project is any individual construction with a different location. A stage is part of the project having a stand-alone significance and may be put into operation before completing the entire project. A subproject is an integral part of the project having a separate functional purpose. Basic types of construction and building work – a quantity of a certain type of work that after completion creates conditions for another type of construction works.

3. Features of expense reporting

The organization of cost reporting in a construction entity is closely related to the methodology of calculating the value of constructed production. When the reporting objects coincide with the cost objects or part thereof, the most accurate determination of their prime cost is provided. In this case, the conditionality in the allocation of costs drops out. In addition, verifying performance results when handing over the works according to their place of execution, as well as identifying the specific factors influencing the dynamics and the level of costs becomes easier.

Accordingly, the object of reporting of construction expenses is part of the production process (stages, structural elements, and types of works), within which costs are grouped as per the requirements of calculation. When the reporting objects coincide with the cost objects or part thereof, the most accurate determination of their prime cost is provided. In this case, the conditionality in the allocation of costs drops out. In addition, verifying performance results when handing over the works according to their place of execution, as well as identifying the specific factors influencing the dynamics and the level of costs become easier. [2]

In civil engineering, although production has a pronounced individual nature, matching reporting objects to cost objects can be only relative. Moreover, if several production units are involved in a certain project, the reporting is also made by contractors, i.e. in this case cost objects are more consolidated than the reporting objects. However, after summarizing the cost estimates per individual units, reporting objects match the cost objects.

The entity's operating expenses are reported with Account 611– Operating Expenses. For construction entities, it may be titled Civil Engineering Expenses or General Construction Works Expenses. Detailed accounts by individual projects (contracts) and by calculation items are added to it. Individual project expenses have been grouped by the following main items of calculation – raw materials, salaries and wages, social security contributions, hired services, construction machinery operating costs, etc. The Item of Raw Materials includes also: costs of reusable materials (metal scaffolding). These materials are used repeatedly in construction production. This implies that their value is included in the costs of producing parts, usually depending on their actual use. Therefore, the construction entity should determine wear rates per single-use.

The Item of Salaries and Wages reports the basic salary determined by quotas and rates as per the implemented payment system – to employees engaged to perform the construction works. These include agreed additional labor remuneration, out-employment remunerations, remuneration of persons engaged under management, and control agreements. For civil engineering, an object of reporting through this account are the costs for basic salaries under the applied payment systems: piece-rate pay, piece-rate bonus pay, standard wage rates, etc.

The Item of *Social Security Contribution* includes state social security contributions, health insurance contributions, supplementary mandatory social security contributions that are paid by the employer. In addition, the following are reported through the account: social security contributions on special grounds (benefits, seniority bonuses, allowances and staff benefits). Contributions to the Unemployment Fund are also reported there.

Hired Services Expenses – this item reports expenses for third parties' services, excluding services for acquiring fixed assets and current tangible assets. These are expenses related to the entity's business such as repair of own and leased fixed assets, rents, transportation services, insurances, postal services, advertising and ads, business expenses. Expenses for courier services, subscription fees, notarial and other similar fees, security, commissions, training and qualification, and other expenses can also be included in the Item of Hired Services.

The Item of Construction Machinery Operating Costs includes all costs associated with the maintenance and operation of construction materials: loading and unloading, transportation to sites, installation and dismantling of facilities, temporary storage buildings, costs for electricity, fuels, lubricants, water, air, auxiliary materials, salaries of service personnel, depreciation and rental of machinery, maintenance and repair costs.

The Item of Other Expenses should be reflected in the following types of expenses: mission expenses in the country and abroad; exhibitions and fairs expenses; VAT paid on supplies not entitled to tax credit; scholarships for students; penalties and other financial payments subject to business contracts of the entity; fines and sanctions for violated regulatory acts; lack of assets reported at the expense of the entity; expenses for the conservation of assets upon their temporary decommissioning; and others. [3]

The main concept underlying the Accounting Standard 11 [4] and the International Accounting Standard 11 (Construction Contracts) [5] is that where the outcome of the construction contract can be estimated reliably, contract revenues and expenses associated with a construction contract are to be recognized as revenues and expenses by reference to the stage of completion of the contracting activity at the end of the reporting period. Construction contracts applied in practice are classified as fixed-price contracts and cost-plus contracts.

A fixed-price contract is a construction contract in which the contractor agrees to a fixed contract price, or a fixed rate per unit of output, which in some cases is subject to cost escalation clauses.

A cost-plus contract is a construction contract in which the contractor is reimbursed for allowable or otherwise defined costs, plus a percentage of these costs or a fixed fee.

In the case of a fixed-price contract, the outcome of a construction contract can be estimated reliably when all of the following conditions are satisfied:

1. Total contract revenue can be measured reliably;
2. It is probable that the economic benefits associated with the contract will flow to the entity;
3. Both the contract costs to complete the contract and the stage of contract completion at the end of the reporting period can be measured reliably;
4. The contract costs attributable to the contract can be clearly identified and measured reliably so that actual contract costs incurred can be compared with prior estimates.

In the case of a cost-plus contract, the outcome of a construction contract can be estimated reliably when all of the following conditions are satisfied:

1. It is probable that the economic benefits associated with the contract will flow to the entity;
2. The contract costs attributable to the contract, whether or not specifically reimbursable, can be clearly identified and measured reliably.

The stage of completion may be measured in many ways. The entity may use methods ensuring a reliable measurement of the extent of work that has been completed, such as:

1. The ratio of contract costs incurred to date to the total estimated contract costs;
2. Conducting a survey of work performed;
3. Physical units of work completed under the contract.

Under the meaning of the accounting standards, if the outcome of a construction contract cannot be estimated reliably, contract revenue should be recognized only to the extent that contract costs incurred are expected to be recoverable and contract costs should be expensed as incurred for the period they have been made.

4. Features of revenue reporting

Revenue of completed and handed over for use construction projects completed in full and to a certain degree of completion, as per the contract.

It should be noted that in practice there has been adopted a conditional division of revenues from:

- Finished construction product – where the projected has been constructed by the construction entity with own or borrowed funds, including advance payments from prospective buyers under preliminary contracts for subsequent sale as a finished construction product in a certain stage of completion.
- Construction services – when sites, improvements, repairs, restoration works are performed at the request of an investor or a contracting authority and are financed by him.

Account 701 *Cash Inflows from Primary Activity*, detailed account 1 – *Revenues from Construction Activity* is used for finished construction product revenues, as in addition to it, detailed accounts for individual construction sites have been kept.

Account 703 *Service Sale Revenues* is used for the revenues of other construction services and sales. Where a construction entity has revenues from other types of services, four-digit accounts for services such as design, ancillary activity, transport, etc. are kept to the account.

For account reporting, allocating and reporting revenues as such derived from finished construction products and from construction services is conditional. It is possible for the construction entity to report all revenues by account 701, and when there is more than one primary activity – to use four-digit accounts, one of them being for finished and delivered projects and completed construction and installation works.

Writing-off the book value of finished projects and completed construction and installation works is carried out in compliance with the revenues and expenses matching principle.

A construction entity may realize construction activity revenues in the following cases:

- Constructing on-site temporary buildings and facilities for the needs of the investor. They are reported and paid as per the adopted manner of reporting and payment of general construction works;
- Constructing on-site temporary buildings and facilities for own needs with funds provided by a contract for temporary construction. Funds obtained thereof are reported as fixed assets financing or running costs, and financing revenue is formed up to the amount of the fixed assets depreciation and the running costs incurred;
- The revenues from the respective projects are increased by the penalties received by the investor under the contract;
- The project price, respectively the value of the completed construction and installation works is reduced with the final deductions made;
- Other revenues have been formed as per the provisions of the contract or under an additional agreement, for early commissioning of the project, high quality, etc.

5. Disclosure in financial reports
The appendix to the annual financial statements should disclose information on:
a) The amount of contract revenue during the reporting period;
b) The methods used to determine contract revenues recognized during the reporting period;
c) The methods used to determine the stage of completion of contracts in progress;
d) Contracts being in progress at the end of the reporting period, where the entity should disclose: a) the sum of expenses incurred and revenues recognized at the end of the year; b) the sum of advance payments received; c) the sum of deductions made under the contracts; d) the amount of the receivables from customers for the work completed under the contract; e) the provision accrued.

6. Conclusion
In the conditions of increasing competition between economic subjects, the prime cost of constructed production becomes a key factor for construction organizations to achieve market success. To be properly accounted and included in the cost of a particular construction project or service, the costs incurred should be properly classified. When at the time expenses are incurred, they cannot be attributed immediately (directly) to the prime cost of a specific construction project or service, the latter should be allocated among individual items of calculation on a certain basis, which is an element of the accounting policy of the entity.

Through the application of the Calculation Method each specific construction project receives a value assessment. To achieve such an assessment, the expenses incurred in the entity are to be covered and classified, as well as certain calculations for their allocation between individual construction projects are to be made. Thus, these calculations are facilitated, and the more accurate and detailed the expenses reported in the accounts are, the more accurate the assessments become.

References
[1] Rangelov Ev 2008 *Accounting of Construction Entities* SEK 5-10
[2] Ivanova St 2018 *Cost price of the construction production. Cost of Constructed Production. Socio-economic Analyzes* Book 1. St.Cyril and St.Methodius University of Veliko Turnovo, Faculty of Economics, 75
[3] Rangelov Ev 2017 *Steadfast Practice. Handbook for Accountants of Construction Companies*, Straton, 55-56
[4] National Accounting Standard 11 Contractors for construction, 2016, https://kik-info.com/schovodstvo/nss-2016.php
[5] International Accounting Standard 11 Construction Contracts, 2018, https://www.iasplus.com/en/standards/ias/ias11

IOP Conf. Series: Materials Science and Engineering **951** (2020) 012030 doi:10.1088/1757-899X/951/1/012030

Institutional risks impact on the growth potential of construction entrepreneurship organizations

B Chaparov[1], A Gospodinova[1,*], I Fedoseev[2] and M Yudenko[2]

[1] University of Economics - Varna, 77, Knyaz Boris I Blvd., 9002 Varna, Bulgaria
[2] Saint Petersburg State University of Economics, 21, Sadovaya street, 191023 Saint Petersburg, Russia

*E-mail: annagospodinova@ue-varna.bg

Abstract. Theoretical approaches to the concept of "growth potential" of entrepreneurial organizations were researched. A hypothesis is put forward about the objective presence of institutional risks in the organizations activities in the form of ineffective norms and rules that reduce the growth potential of organizations in the construction industry. The institutional norm of "contribution to the infrastructure development" in construction entrepreneurship and its impact on the activities of developers in Russia and Bulgaria are studied. The conclusion is made about the instability of this norm and the lack of unification of its application in the regions, which creates institutional risks for the construction business and consumers of construction facilities. Institutional risks are directly related to administrative barriers in the entrepreneurs' activities, and their interaction reduces the investment attractiveness of facilities, increases transaction costs and reduces the growth potential of organizations.

1. Introduction

The problem of institutional risks impact on the growth potential of construction business organizations is extremely urgent in the medium and long term. The purpose of the article is to determine the role of institutions in the social and economic practice of entrepreneurs, as well as the impact of institutional risks on the economic potential of construction business organizations. Social institutions play an important role in society: they are fundamental factors in the economic systems functioning in the long term. But the opposite effect may arise when a large number of formal norms will impede the development of economic entities and lead to a decrease in the growth potential of such entities. A successful solution to this problem is a necessary prerequisite and a guarantee that, through the application of effective norms and rules, the construction complex of Russia and Bulgaria can become a locomotive of growth and development of the economy as a whole. Achieving this urgent strategic target is possible only as a result of the formation and functioning of such social institutions that would be effective for the development and functioning of entrepreneurs in construction. The more effective is the norm governing the activities of the construction process participants, the less institutional risks it bears.

2. Literature Review

Before researching the issue of institutional risk impact on the activities of entrepreneurial organizations, we will find out which approaches in the scientific and educational literature are applied to the concept of "growth potential". There is currently no single theoretical approach to the interpretation of this concept. The authors of the monograph Fedoseev et al. (2018) hold a view that the initial process of economic growth of organizations in the construction industry and the condition for the formation of their potential is the process of construction production concentration, which is caused by the impact of factors of the external and internal environment. At the same time, the concentration process, reflecting the potential of organizations enlargement, leads to an increase in the volume of work on the market, thereby exerting an influence on the change in its capacity.

The article by Yakovleva and et al. (2019) researched a methodological solution for applying semantic modeling to the enterprise potential research, which includes the systematization of existing indicators used in traditional financial analysis.

The point of view of Chepachenko, Yudenko and Palicki (2018) is correct, according to which the growth potential is associated with the economic capabilities of the organization. Economic potential represents the opportunities that an organization can show in its field of activity, and it does not matter if they were actually realized or only hypothetically possible. Not always all possible resources are involved simultaneously and with the necessary efficiency. This is why potential is determined by equally important factors: the availability of resources; the ability to apply them effectively; the degree of available resources involvement in production.

Scientists from India, Nandeeswaraiah and Ramana (2018) analyze labor potential, as an independent important direction in the development of small and medium-sized businesses. Scientists write that moreover, Shri. Manmohan Singh ji, former prime minister of India, once announced that the success of the enterprise depends on how able the enterprise utilizes the potential of skilled force for them. And it should be noted that the problems of the development of small and medium-sized businesses in India have much in common with problems in Russia.

A number of researchers consider the concepts of growth potential and development potential to be equivalent. But, taking into account that the concepts of growth and development of economic entities are not identical economic categories, it is necessary from a methodological point of view to more clearly distinguish between these concepts in order to productively research, measure and justify the method of assessing the growth potential and development potential of entrepreneurial organizations.

The concepts of “growth” and “development” of economic entities, including construction organizations, being important economic categories, have long received great importance and attention in economic theory. The role and significance of these processes is evidenced by the decisions of economists, owners, who hold the opinion of an acceptable annual increase in sales at the level of the annual growth rate of the real estate market, and in other cases not less than 2-3%. How acceptable is this economic growth potential to ensure the livelihoods, growth and development of any business? The answer is due to taking into account the key factor of time affecting the processes of growth and development. So, in Russia, at a constant annual growth rate of product sales of 2.5%, the total volume of product supply will double in 28 years and 8 months (by 2046), and at 5% per year, doubling will be in 14 years (by 2032). Such is the price of the time factor and the potential annual growth rate of product sales. Therefore, for each production organization, a time factor is important, affecting the implementation of its economic growth potential, which is largely derived from the process of building up and implementing the potential of its economic development, which allows not only to ensure higher production rates, but also to increase its competitiveness potential, increasing market share.

Chepachenko, Yudenko and Palicki (2018) consider that the need to build up capacity through competitiveness encourages owners and senior management of organizations to increase the production capacities of organizations in an effort to obtain higher performance results, which is impossible outside of building the potential for the development of their construction business in a competitive environment and in the face of risk.

A scientist from Bulgaria Chaparov (2012) connects the growth potential of a construction enterprise with its sustainability, noting that considering the program of a construction enterprise from the point of view of its financial sustainability, four key aspects of activity can be identified, namely: financial activities, relations with consumers of the manufactured construction product, organization of internal business processes, growth and development.

IOP Conf. Series: Materials Science and Engineering **951** (2020) 012030 doi:10.1088/1757-899X/951/1/012030

3. Objectives

The objectives of the study were listed below:

- To research the concepts of "growth potential" and "development potential" of entrepreneurial organizations.
- To confirm the hypothesis about the institutional risks impact on the activities of organizations in construction towards reducing the growth of the business's potential.
- To substantiate the assumption about the impact of institutional risks on the appearance of excessive administrative barriers in entrepreneurial activity.

4. Methodology for determining the enterprise sustainability

The article considers the generally accepted macroeconomic approach in that the economic growth of the national economy is the goal of development of any economy, whatever its nature, and development management is a key factor in sustainable economic growth Oparin, Chepachenko and Yudenko (2017) are you sure that to study institutional risks from the point of view of their impact through social institutions (norms, rules) on the growth of the economic potential of organizations in the field of construction.

It should be observed that institutional risks closely interact with administrative barriers and transaction costs. For a more complete cover of the stated topic, it is necessary to identify the relationship between factors affecting the economic growth of the organization and indicators that determine this economic growth. This is, first of all, the coefficient of sustainability of organization economic growth. This coefficient is calculated as a relative indicator, in the numerator of which is net profit reduced by the size of the dividend fund, and in the denominator is the equity of the organization. Given the savings in transaction costs from the reduction of institutional risks, the coefficient will be calculated as follows.

Coefficient of sustainability of economic growth is defined as:

$$K = (NP - DF + \Delta TA)/AE \times 100, \quad (1)$$

where: K is the coefficient of sustainability of economic growth, %;
NP – net profit of the organization, rubles;
DF – dividend fund of the organization, rubles;
AE – the average annual amount of equity, rubles;
ΔTA – savings in transaction costs from the reduction of institutional risks, rubles.

Table 1 shows the options for the financial condition of organizations with different values of this coefficient.

Table 1. The dependence of the financial condition on the coefficient of sustainability of economic growth

Coefficient value	Financial condition of a construction organization
QR ↑	Economic growth of the organization, increase of solvency, liquidity, expansion of production capacities, reduction of transaction costs
QR ↓	Economic downturn of the organization, a decrease in solvency, liquidity, a reduction in production capacity, an increase in transaction costs

Construction, including housing construction, is the "engine" of economic development.

Iliev et al. (2018) research the problematic issues of construction and make proposals for the development of construction until 2020 using the example of Bulgaria.

In an article by Bormpotsialou and Rovolis (2019), it was noted that in many economies, residential building activity is at the heart of economic growth, either boosting or weakening Gross Domestic Product (GDP), linking the housing market with economic structural components such as labor mobility, consumer expenditure, inflation rates as well as with other components with important effects on economy, such as the banking sector. Structural components may represent structural risks that accompany the development of housing construction in the following five Balkan countries: инфляция, Labor Mobility Inflation Rates, banking distress (Bormpotsialou and Rovolis, 2019, p.35).

A researcher from Belarus, Babina (2005) conducted research on the formation of a favorable climate to attract foreign investment. The author considers that one of the effective tools to guarantee the rights of foreign investors is to comply with the rules of immutability of legislation for investors on the date of their registration, which reduces institutional risks in business. According to the Organization for Economic Cooperation and Development (OECD), in some countries, overcoming administrative barriers at the stage of business organization increases costs. On average, globally to create a new company it is necessary to go through 10 procedures, which will take the entrepreneur 63 business days (if passing through these procedures is not delayed).

Below is an example of the implementation of the "deductions for social infrastructure" institutional norm in the construction complex of St. Petersburg. In 2013, in St. Petersburg, the Construction Committee introduced a formal norm of "deductions for social infrastructure". The Construction Committee has prepared a special account for donations from developers constructing residential buildings for up to 100 people. Developers were advised that making contributions would speed up the process of approving documentation for their projects. Developers were asked to deposit amounts calculated on the basis of the current standards for providing residents with social infrastructure facilities. Thus, the Committee calculated that on average one place in a school costs 1,044 thousand rubles, and a place in a kindergarten costs 1,200 thousand rubles.

Since 2018, a new initiative has appeared - housing developers will deduct 11 thousand rubles from each square meter for the development of social infrastructure. It is expected to receive a total amount of fees in the amount of 30 billion rubles. In this regard, it was planned to create a fund of social obligations – the initiative to create the fund belongs to the builders themselves. Developers calculated that if they would pay 3-4 thousand rubles to the fund from each square meter, then this amount will be enough for the construction of 40 kindergartens and the social problem will be solved in a few years, and the fund can be disbanded. It is about creating a norm that would help make contributions to the social structure transparent for all builders. But the opinions of builders are already emerging that the fund will not be disbanded. Covered by the fund, the authorities of St. Petersburg will solve other infrastructure problems of the city. Thus, the issue of the construction of social and engineering infrastructure is still not resolved at the legislative level. And in this case, the risks are borne by Smolny officials and city developers. And ultimately, additional financial investments are borne by consumers. The first time consumers pay taxes, and the second - when buying a new apartment. Opinions are expressed about the construction of social infrastructure at the expense of the city, and that the cost of square meters of housing for consumers of the residential accommodation is "out of reach", etc. Such a suspended state of developers is precisely the kind of institutional risk when there is no deduction rate for social infrastructure and it is not clear to developers how contributions will be made and whether they will be made in principle. As for construction, currently, builders do not have a unified approach for determining the costs of building social infrastructure facilities and how to transfer property to municipal and city authorities. Unification of rules and norms is required, which ultimately leads to stability in the construction market. Any government initiative should be considered in relation with long-term planning, and not be a search for a momentary solution to the problem on which it is necessary to report.

5. Conclusion

Previous considerations lead to the following conclusions.

The economic growth potential of the construction complex organizations is formed under the impact of risks:

- Institutional risks resulting in transaction costs associated with overcoming administrative barriers, most of which affect the cost of construction products.

- Production risks that affect the decrease in the efficiency of the use of available resources and the sustainability of economic growth of organizations.

Assessing direct losses from overcoming institutional risks is a complex and time-consuming process that is not performed at the micro or macro level. The inefficiency of state regulation through existing institutions in construction leads to institutional traps, the creation of new administrative barriers and the growth of institutional risks.

The stability of excessive administrative regulation leads to a decrease in the steady growth of the potential of construction organizations, which can be explained by the existence of officials who preserve and support such a mechanism in part for the purpose of enrichment. The abundance of regulations, ineffective norms and their weak impact on the economic and financial condition of entrepreneurial organizations in construction, leads to the logical question - why are so many institutional reforms (including those quite reasonable in design) that do not always lead to the expected result? And why do so many seemingly effective institutions not only not lead to economic growth of organizations, but also sometimes reduce it? From a methodological point of view, one of the reasons for this situation is the lack of a sense of institutions, which consists in the constant underestimation of the institutional systems complexity.

References

[1] Babina A 2005 Two-way highway for foreign investment *Bank messenger.*

[2] Bormpotsialou O and Rovolis A 2019 Housing construction as a leading economic indicator. *Studies in Business and Economics* **14(3)** 33-49.

[3] Chaparov B 2012 Opportunities for improving the economic sustainability of construction enterprises *Bulletin of the Bryansk State University* **3(2)** 76-80.

[4] Chepachenko N, Yudenko M and Palicki S 2018 Methodological aspects of research, measurement and evaluation of the development potential of construction organizations *Problems of social and economic development of Siberia. Publishing house of BrSU, Bratsk* **3(33)** 82-91.

[5] Fedoseev I et al. 2018 Research of the growth potential and development of construction organizations in the real estate market (*St. Petersburg: Publishing House of St. Petersburg State University of Economic).*

[6] Iliev P et al. 2018 The development of construction in Bulgaria in the context of the new program period for 2014-2020 (*Varna: Science and Economics)*

[7] Nandeeswaraiah K and Ramana A 2018 Role and performance of micro, small and medium enterprises. *International Journal of Management (JOM)* **6(5)** 10-17

[8] Oparin S, Chepachenko N and Yudenko M 2017 The role of social institutions in the activity of entrepreneurs in the construction sector *CBU International Conference on Innovations in Science and Education (Czech Republic: Central Bohemia University)*

[9] Yakovleva E et al. 2019 An empirical model of systematization of financial indicators by management functions as the basis for establishing the organization's innovative potential *Leadership and Management.* **6(2)** 73-79

Colors for optimal thermal comfort in university study hall - interdisciplinary theoretical and practical studies

I Marinova * and A Ivanova

Faculty of Architecture, University of Structural Engineering & Architecture (VSU) "L. Karavelov", 175 Suhodolska St., 1373 Sofia, Bulgaria

*E-mail: m_vania@hotmail.com

Abstract. Improving thermal comfort trough interior color design is a contemporary energy efficiency and healthy living environment tendency. The article considers scientific-applied task of choosing a color design for a specific university seminar hall. The concept integrates areas of architecture and design, color science, physics, chemistry, material science, psychology, ect., in the context of aesthetics, light-reflecting qualities and physical and emotional impact of colors upon inhabitants. Optimal color design considers general principles applied to a specific environment and requires integrated approach and interdisciplinary theoretical and practical studies in order to improve the process of creating such an optimal solution. The report examines contemporary interior color design scope. Characteristics of university hall and the requirements they raise are studied. An overview of colors' features as a means of improving thermal comfort is included. Stages of the design process and the involved specialists are listed. The advantages of nonlinear design process are applied. As result the succession and the degree of participation of different science fields specialists in the interdisciplinary theoretical and practical studies and in the choice of the final design are set.

1. Interior colour design scope

Contemporary interior color design is a complex task that meets requirements. In previous research authors differentiate colors in two groups according to the role they play in the living environment – traditional and contemporary [1].

1.1. Traditional role of colors in the living environment

Traditionally, colors carry information about the surrounding objects and denote separate functional zones in a single space. Most popular colors application is according to their aesthetic value. They also are given a symbolic meaning. At a later stage, the colors in the living environment are applied also according to the individual perceptions and preferences of the inhabitants.

1.2. Contemporary color application

Modern research reveals color properties which determine new application possibilities. The ability to influence the perception of objects and spaces leads to color application in order to visually complement and alternate the living environment characteristics. The application of colors according to their reflective properties assists the distribution of light and the accumulation of heat from solar electromagnetic energy by building elements. Most contemporary trend is the application of colors according to their physiological and psychological impact on the inhabitants.

2. Characteristics of a university study hall in reference to thermal comfort optimization

Permanent residents are students, 20 and more persons, average aged about 22 years. Temporary residents are professors, time of stay and age vary.

The variety of activities is also diverse - lecturing, conducting exercises, independent working, socializing, recreating.

The combination of various activities and occupants requires combined approach towards improved working and study environment [3] and meeting inhabitants' preferences [4]. Significant amount of objective and subjective factors are to be taken into account in order to create optimal thermal comfort.

Traditional color design of a university study hall meets the requirements for aesthetics, for emphasizing the functional zoning and in some cases - for light redistributing. Table 1 compares the traditional color scheme in a study room with modern requirements.

Table 1. Comparison of traditional study room color design with modern requirements

Modern requirements	Traditional color design
information about surrounding objects, functional zones marking	information about surrounding objects, functional zones marking
aesthetic value	aesthetic value
symbolic meaning	-
according to the perceptions and preferences of inhabitants	-
visual supplementation and adjustment of living environment characteristics	-
light distribution and heat storage	-
physiological and psychological impact on the inhabitants	-
new technologies in covering materials	-

The comparison illustrated in table 1 reveals that the application of colors according to those properties that contribute to increasing the thermal comfort of the occupants does not appear in the traditional color scheme of a study room.

3. Colors for optimal heat comfort in the living environment

Color, being a constant characteristic of interior, has rarely been considered for its thermal impact. Researches of color properties and of ways they influence the inhabitants reveal possible new functions in the architectural environment, which are directly related to improving energy efficiency.

Studies reveal that, when considering the psychological aspects of interior color design there are variations according to gender and age [4]. But they are little to none when physiological impact is concerned – colors have similar effects even if some of the inhabitants experience partial or complete color blindness [5].

When it comes to colors, three main characteristics to be considered – electromagnetically wave length, hue and saturation.

Physical property of colors is related mostly to the amount of solar radiation that is reflected or absorbed. Coloring structural elements in dark shades allows more solar energy to be accumulated by space surrounding elements and furniture. Disadvantage of dark surfaces is that they absorb and do not reflect light, so need for earlier switching on and greater intensity of artificial lighting appears.

Physiological effects include the stimulation of certain processes in human body, as well as some effects of synesthesia.

Psychological impact on colors is related to a subjective comfort criteria trough variety of associations color tones and shades induce.

Generally warm colors (red, orange, yellow) are associated with warmth and light. The physiological effects of red and some shades of orange, for example, are increased adrenaline levels, increased heart rate, muscle contraction and, as a result, a rise in body temperature. As a result, the inhabitants of an environment with predominant shades of orange and red have lower requirements for comfort temperature levels [6]. This also makes warm colors not suitable for interior of rooms that receive large direct and indirect heat revenues in the summer, as they can lead to increased requirements for cooling and ventilation [1].

The lighter shades of cold colors - green, blue, purple are associated with water, space, coolness, etc. In addition, green and blue tones have a calming effect on the nervous system and slow down the heart rate, resulting in a slight decrease in body temperature [6]. The occupants of rooms with predominant green, blue and violet shades have reduced requirements for air conditioning in summer, but experience discomfort during the cold seasons [1].

As additional measures to enhance the energy-saving properties of colors, dyes with improved reflectivity in the field of ultraviolet and infrared rays [2, 11, 12, 13], as well as completely new materials that respond to external factors [2, 11, 12, 13] are being developed. Such materials have a positive effect on the radiation temperature in the premises. The combination of materials, textures and colors can help to optimize the movement of air in the room, thus improving the effective temperature [7].

4. Stages of interior color design process

Finding an interior color solution can be considered as consisting of three main stages:

- Theoretical studies - justification of the need to update the color scheme and preliminary studies;
- Practical studies - research and development of the architectural potential of the room and choice of color design;
- Site management and monitoring.

An intermediate stage of negotiations to reach a consensus between the participants in the renewal appears between first and second stages [8].

In the first stage of theoretical research, the introductory and fundamental activity is the justification of the need and desire of the inhabitants for a more comfortable and more productive environment, combined with the desire to increase energy efficiency. The stage continues with a study of economic and technological opportunities for realization. This activity clarifies which services and products are essential for the specific solution. Similar cases are studied and compared with the specific situation.

The second is a stage of practical researches. First, the physical characteristics and condition of the selected room are studied. In connection with the optimization of thermal comfort, emphasis is placed on the volumetric-spatial solution, proportions, the functional solution and the number and location of communication openings, height and facade participation in connection with calculating factor shape, building materials and covering layers of enclosing structures in order to calculate their heat transfer coefficient. Next the priorities of color design are determined and variants are developed. To help the analysis in order to select most optimal option, an analysis with the respective software products of a virtual model of the room is applied, as well as research on a physical model with indicators and properties close to the original materials. A variant of a color design that best meets the defined priorities is chosen. Specific design plans are developed. Next is drawing up a plan and organizing the implementation and determining the participants in this process. The final activity at this stage is the realization of the color solution.

The third stage covers the administration of the implementation process and the management and monitoring of the project. It includes control of repair activities and development of the site to achieve its optimal capabilities and also the management and promotion of the site and its maintenance.

A study of possibilities is done at each stage of the project. Its main purpose is to reduce the risk and to form the basis of detailed design. It includes study of similar successfully held projects, assessment of the current situation and the most appropriate development strategy and selection of a specific interior color design. Following is a decision for appropriate investment, approval of an approach for renewal and drawing up a project development program.

Each of the aspects of improving thermal comfort trough interior color design project involves participation of specialists in various fields - optics, heat engineering, materials science, chemistry, mechanics, construction technology, medicine, biology, psychology, sociology, color science, finance, law, computer technology and more. Since it is a question of applying colors in a living environment, the intersection of all these areas is an architectural / design project. Mandatory permanent participation of users / occupants [9].

5. Participants in the process of optimizing the thermal comfort through an interior color design

The participants in the process are the residents, specialists-consultants and administrative authorities.

Studies reveal that occupants are the main driver of the process of finding an interior color solution. They are also the most active participant in the development, operation and further development of the specific site [9].

Sociological research shows that residents prefer to take the initiative to improve the environment they live in. They have a desire for change and a vision of what that change should be. They are the initiators of the organization of local organizations, which deal with the formation of the necessary teams of specialists, with the documentation process and the development of the sites. The improved living environment is a cause for pride, as people like to live in a place of which development they have actively contributed [9].

Therefore, public participation and establishing the interests of people are a key factor in projects to improve thermal comfort trough interior color design.

The role of administration authorities in the process of finding an interior color solution is mainly regulatory. Their main tasks and activities are:

- engagement in development programs;
- creation of normative and financial reliefs for stimulating the interest in improving the living environment;
- creation of organizations supporting and promoting these trends;
- creation of programs and funds for additional financing of projects for improvement of the living environment;
- creation of centers for information and communication with specialists

Specialists - consultants are the main participants in the processes of design, implementation and monitoring of thermal comfort trough interior colors. The team must include diverse specialists - architects, civil engineers, economists, lawyers, brokers. During the different stages of the process of improving thermal comfort trough interior color design some of them form the main and permanent core of the team and additional specialists are involved for temporary participation and / or one-time consultations. Practice shows that with such an organization, the team involved in the design process has a number of advantages, such as flexibility of solutions, the ability to work in a network and non-linear parallel development of the project stages [10].

Table 2 shows the groups of activities at each stage of the development of an interior solution and the necessary specialists-consultants for each stage.

IOP Conf. Series: Materials Science and Engineering **951** (2020) 012031 doi:10.1088/1757-899X/951/1/012031

Table 2. Specialists-consultants participation in activities and stages of interior color design development

Stage	Activities	Specialists-consultants
Preliminary study and justification (theoretical studies)	Initiative, connection with specialists	inhabitants
	Defining the basic requirements of the inhabitants	inhabitants, architect / designer, psychologist, sociologist
	Study of economic opportunities	financier, lawyer
	Study of technological possibilities	designer, technologist
	Exploring opportunities for realization	architect / designer, constructor, technologist, financier
	Study of similar projects	inhabitants, architect / designer, constructor,
Study of the potential of the room and realization of the color scheme (practical studies)	Physical characteristics and condition study	architect / designer, constructor
	Building materials study	architect / designer, constructor
	Setting priorities for a color scheme developing	inhabitants
	Options, virtual and real practical studies development	architect / designer, constructor, psychologist, computer technology, materials science, biology, medicine
	choice of a specific option	inhabitants, architect / designer, constructor,
	Design plans development	architect / designer, constructor, technologist, physicist, chemist, psychologist, financier
	Organizing the realization process	architect / designer, constructor, technologist, financier
	Color design completion	technologist
Implementation, management and monitoring of the project	Management of the implementation process	architect / designer, constructor, technologist, administration
	Project management and monitoring	inhabitants, architect / designer, constructor, financier, sociologist, administration

6. Conclusions

Main conclusion from the present study is that the task of optimizing interior thermal comfort of university study room through color design is complex and requires knowledge in a wide range of areas. The factors that need to be considered are a significant number. Some of them relate to the properties of materials and structures and to the organization of elements and spaces. Another concerns the occupants, as well as their interaction with the living environment specifically in terms of optimizing thermal comfort.

It follows from the above that the participation of various specialists is a mandatory element for complete project implementation. The proven advantages of interdisciplinary participation and the ability of simultaneous multi-task process are also to be considered.

According to their degree of participation, the specialists are to be divided into three groups - permanent participants, temporary participants and occasional consultants. Table 2 clearly reveals that the main core specialists in this case are architects/designers and constructors. The active participation of the occupants/users at each stage of the process is also mandatory.

In the preliminary theoretical studies, it is necessary to include as temporary participants in the process specialists in the fields of finance and technology. Psychologists, sociologists and lawyers are to be involved as consultants.

At the stage of practical research as temporary participants are technologists, computer specialists, psychologists, financiers. Specialists in physics, chemistry, biology and medicine should be involved as consultants.

Acknowledgments
The studies in this report are funded by science-research project №11, order № 256 / 27.05.2020. The authors express their gratitude to USEA (VSU) "Lyuben Karavelov", Bulgaria.

References

[1] Marinova I 2012 *Color properties as a means of improving living environment comfort* (Sofia: International Jubilee Scientific and Applied Conference 70 Years UACEG, report collection ISBN 978-954-724-049) pp 207 – 212
[2] Kumar A 2013 *Advances in the Building Materials for Thermal Comfort and Energy Saving* (Bentham Science Publishers: Recent Patents on Engineering)
[3] Hohmann T 2006 *New Aspects of Library Design* (Netherlands: LIBER Utrecht University Library Open Access Journals)
[4] Costa M, Frumento S, Nese M and Predieri I 2018 *Interior Color and Psychological Functioning in a University Residence Hall* (Switzerland: Frontiers in Psychology)
[5] Hettiarachchi and Emmanuel R 2017 *Colour as a psychological agent to manipulate perceived indoor thermal environment for low energy design ; cases implemented in Sri Lanka* (Scotland: Passive Low Energy Architecture (PLEA) - Design to Thrive Conference)
[6] Goreva GГорева Г. 2016 *Educational reference book on the subject of color science and color* (Minsk: Belarusian State University) p 36.
Amersfoort J, Balvert S, Hadlich S, Härter M and Liu F 2007 *Investigating the effects of altered colour perception on mood*
[7] Radivojević A, Đukanović B 2018 *Material Aspect of Energy Performance and Thermal Comfort in Buildings (Energy resources and building performance)* (Delft: TU Delft Open) pp 61-85
[8] 2019 *Project Development Guide* (Mosksw: Fund of the Unified Institute for Housing Development) pp 110-136
Manal O and Abowardah E 2016 Design Process & Strategic Thinking in Architecture (London: International Conference on Architecture & Civil Engineering)
[9] Jenkins P and Forsyth L 2009 *Architecture, Participation and Society* (Milton Park: Taylor & Francis Ltd)
[10] Novikova A 2015 *Network form of architectural design organization* (Nijnii Novgorod: Kazan State Architectural and Construction University)
[11] Koca G 2016 *Interior Finishing Materials (Developments in Science and Engineering* Chapter 43*)* ed R Efe, L Matchavariani, A Yaldir, L Levai (Sofia: St. Kliment Ohridski University Press) pp.588-601,
[12] Orf N 2009 *Thermeleon* (Greentech Media)
[13] Petrie T, Atchley J, Childs P and Desjarlais A 2007 *Energy Savings for Stucco Walls Coated with Cool Colors* (Oak Ridge: Oak Ridge National Laboratory)

AUTHOR INDEX